# Lecture Notes in Networks and Systems 984

### Series Editor

Janusz Kacprzyk ⓘ, *Systems Research Institute, Polish Academy of Sciences, Warsaw, Poland*

### Advisory Editors

Fernando Gomide, *Department of Computer Engineering and Automation—DCA, School of Electrical and Computer Engineering—FEEC, University of Campinas— UNICAMP, São Paulo, Brazil*

Okyay Kaynak, *Department of Electrical and Electronic Engineering, Bogazici University, Istanbul, Türkiye*

Derong Liu, *Department of Electrical and Computer Engineering, University of Illinois at Chicago, Chicago, USA*

    *Institute of Automation, Chinese Academy of Sciences, Beijing, USA*

Witold Pedrycz, *Department of Electrical and Computer Engineering, University of Alberta, Alberta, Canada*

    *Systems Research Institute, Polish Academy of Sciences, Warsaw, Canada*

Marios M. Polycarpou, *Department of Electrical and Computer Engineering, KIOS Research Center for Intelligent Systems and Networks, University of Cyprus, Nicosia, Cyprus*

Imre J. Rudas, *Óbuda University, Budapest, Hungary*

Jun Wang, *Department of Computer Science, City University of Hong Kong, Kowloon, Hong Kong*

The series "Lecture Notes in Networks and Systems" publishes the latest developments in Networks and Systems—quickly, informally and with high quality. Original research reported in proceedings and post-proceedings represents the core of LNNS.

Volumes published in LNNS embrace all aspects and subfields of, as well as new challenges in, Networks and Systems.

The series contains proceedings and edited volumes in systems and networks, spanning the areas of Cyber-Physical Systems, Autonomous Systems, Sensor Networks, Control Systems, Energy Systems, Automotive Systems, Biological Systems, Vehicular Networking and Connected Vehicles, Aerospace Systems, Automation, Manufacturing, Smart Grids, Nonlinear Systems, Power Systems, Robotics, Social Systems, Economic Systems and other. Of particular value to both the contributors and the readership are the short publication timeframe and the worldwide distribution and exposure which enable both a wide and rapid dissemination of research output.

The series covers the theory, applications, and perspectives on the state of the art and future developments relevant to systems and networks, decision making, control, complex processes and related areas, as embedded in the fields of interdisciplinary and applied sciences, engineering, computer science, physics, economics, social, and life sciences, as well as the paradigms and methodologies behind them.

Indexed by SCOPUS, INSPEC, WTI Frankfurt eG, zbMATH, SCImago.

All books published in the series are submitted for consideration in Web of Science.

For proposals from Asia please contact Aninda Bose (aninda.bose@springer.com).

Mustapha Hatti

Editor

# IoT-Enabled Energy Efficiency Assessment of Renewable Energy Systems and Micro-grids in Smart Cities

Harnessing the Power of IoT to Create
Sustainable and Efficient Urban Environments
Volume 1

 Springer

*Editor*
Mustapha Hatti ⓘ
UDES, Unité de Développement des Equipements
Solaires
Bou Ismail, Tipasa, Algeria

ISSN 2367-3370 ISSN 2367-3389 (electronic)
Lecture Notes in Networks and Systems
ISBN 978-3-031-60628-1 ISBN 978-3-031-60629-8 (eBook)
https://doi.org/10.1007/978-3-031-60629-8

This Springer imprint is published by the registered company Springer Nature Switzerland AG
The registered company address is: Gewerbestrasse 11, 6330 Cham, Switzerland

If disposing of this product, please recycle the paper.

# Contents

**Internet of Things and Sensors**

## MPPT and Optimization

## Robotics and Electrical Vehicle

## Materials in Renewable Energetic Systems

## Smart Systems and Communication

# Editors Biography

**Dr. Mustapha Hatti** was born in El-Asnam (Chlef), Algeria. He studied at El Khaldounia school, then at El Wancharissi high school, obtained his electronics engineering diplomat from USTHB Algiers, and his post-graduation studies at USTO–Oran (Master's degree and doctorate). Worked as research engineer, at CDSE, Ain oussera, Djelfa, CRD, Sonatrach, Hassi messaoud, CRNB, Birine, Djelfa, and senior research scientist at UDES/EPST-CDER, Bou Ismail, Tipasa, "Habilité à diriger des recherches" HDR from Saad Dahlab University of Blida, Algeria. Actually, he is Research Director in renewable energy. Since 2013, he is an IEEE senior member, he is the author of several scientific papers, and chapter books, and his areas of interest are smart sustainable energy systems, innovative system, electrical vehicle, fuel cell, photovoltaic, optimization, intelligent embedded systems. Chair of the Tipasa Smart City Association. An eBook editor at International Springer Publishers, guest editor and member of the editorial board of the journal Computers & Electrical Engineering, he has supervised, examined and reviewed several doctoral theses and supervised master's degrees. He organizes the conference entitled Artificial Intelligence in Renewable Energetic Systems. Enrolled in several European project in smart and sustainable cities.

# Internet of Things and Sensors

# An IoT-Based System to Control the Greenhouse's Microclimate

Seddiki Noureddine[(⊠)], Belghachi Mohammed, and MokedDem Kamal Abdelmadjid

Algeria: Department of Mathematics and Informatics, University of Tahri Mohamed Bechar, Bechar, Algeria
{Seddiki.nourddine,belghachi.mohamed,
mokeddem.kamal}@univ-bechar.dz

**Abstract.** In fact, data gathering and the creation of an active system with a range of actuators are the requirements for monitoring or control. Due to the intricacy of the greenhouse's microclimate, management measures can be used to increase plant production and quality while decreasing the use of energy. The microclimate is controlled by a microcontroller and sensors for multiple factors, including temperature, humidity, and soil moisture. The Internet of Things is an excellent infrastructure for collecting and processing this data. Interactive, observable elements are offered by web-based and mobile apps, which may have an impact on the greenhouse environment. The irrigation pump or the fan may switch on in real time based on the controller's optimum threshold values or user input. In this work, we propose a control system that can enhance crop productivity and generate a dataset for future research.

**Keywords:** greenhouse · control · microclimate · Internet of things

## 1 Introduction

The cutting-edge greenhouses are built to provide plants with protection from the elements outside and a climate conducive to intense food production. However, greenhouses are going through a profound change due to the quick transition to precision agriculture and the accompanying improvement in food security while safeguarding natural resources. This is because metering, communication, control, and monitoring systems have improved [1].

The management and automation of microclimate control within greenhouses have made agriculture in closed-field environments more sustainable. This is achieved through the reduction of resources such as water, fertilizer, and energy while increasing productivity and profitability [2]. The ideal environment for crop development is influenced by factors such as temperature, humidity, light intensity, and air circulation, collectively referred to as the microclimate. By accurately optimizing and controlling these parameters, an intelligent greenhouse can create an optimal environment for plant growth. This process of microclimate control involves monitoring and adjusting the environmental factors to provide the best possible conditions for the plants. This can be accomplished

M. Hatti (Ed.): IC-AIRES 2023, LNNS 984, pp. 3–13, 2024.
https://doi.org/10.1007/978-3-031-60629-8_1

manually or with the help of advanced technologies such as sensors, automated systems, and computer software.

Recent developments in sensor technology have enabled the achievement of the highest greenhouse crop yield and production levels to date. Automation experts can now take advantage of tailored solutions specifically designed for greenhouse applications, thanks to digital technologies such as the Internet of Things (IoT). By utilizing wireless sensors and IoT-enabled devices, greenhouse environments can be monitored and managed in real time from any mobile or fixed device with a secure internet connection [3].

In this paper, we showcase a greenhouse automation system that we developed in our lab for experimentation purposes. Additionally, we introduce a web-based backend infrastructure and mobile application that harnesses the collected sensor data on the greenhouse microclimate to enable control and automation. To achieve this, we have divided the paper into several sections. In Sect. 2, we discuss relevant works and innovations related to the greenhouse microclimate. In Sect. 3, we present the greenhouse parameters used to control the microclimate. Section 4 provides a detailed explanation of the various technologies used to construct the prototype and platform. We then discuss the outcomes in Sect. 5. Finally, in Sect. 6, we summarize the conclusions of this paper and highlight the untapped potential of our forward-looking ideas for this dataset.

## 2   Related Works

Recent research in precision agriculture emphasizes the significance of managing and monitoring the microclimate within and outside greenhouses. The techniques involved in controlling the environment include gathering, analyzing, and processing data, while also visually identifying climatic parameters to alert of potential issues. All these measures are aimed at increasing agricultural productivity and reducing input costs. The importance of contemporary technologies such as wireless sensor nodes, embedded devices, and IoT-based and cloud-based data collection platforms in achieving these goals cannot be overstated.

The use of novel technology such as IoT in the domain can be seen in [4]. In this paper, the authors use a model-based implementation in Matlab Simulink to analyze a greenhouse microclimate utilizing IoT, sensors, and Node MCU (ESP8266) as controller. The goal of this study is to better understand how to interpret microclimate data variances about plant needs at various growth stages. The model-based analysis offered a way to accept these uncertainties due to the nonlinear dynamics of the system (i.e., solar radiation and wind speed) and reduce the production risk despite even though the precise behavior of the greenhouse environment from raw data and its impact on the plants [8].

The authors of [5] designed and implement a remote monitoring and control system for the greenhouse based on IoT. The greenhouse model is divided into four sections for planting various plants. Different levels of soil moisture are maintained in each area. This is used as the basis for monitoring and analyzing plant development. The microclimatic conditions in the greenhouse model are controlled by an automated control greenhouse system, which is made up of various separate subsystems (watering, lighting, temperature control, air quality subsystem, etc.). The greenhouse IoT component uses the ThingSpeak cloud.

In [6], the objective was to gather information on air temperature, air humidity, and light levels using a group control device (ESP) in a greenhouse microclimate. To achieve this, an automated microclimate management system was developed using cloud architecture. This system comprises six fundamental components: a cloud server, a local object server, a mobile application, a group control device, a device module, and a sensor. The greenhouse complex's dependability is increased, and the possibility of data loss due to network failures is decreased, by replicating data from a local device to the cloud. Based on the test results, the user interface's reaction time can be deemed acceptable when it takes less than two seconds from the time that data packets are received until they are seen on the user's screen. Data availability and system dependability can be increased by duplicating functionality and replicating databases across on-premises object servers and cloud services.

The system designed here [7] is extremely complex, cutting-edge, and capable of measuring six microclimatic parameters in three separate greenhouse sections. A decision-making control system can be activated to adjust required parameter values by providing ventilation or by turning on and off processes of mist foggers in required areas of the greenhouse after comparing the collected data with expected values. In addition to this, the system continuously records to create databases. The system presented demonstrates its utility for measuring and revealing distinctions in the greenhouse microclimate in terms of the temperature, humidity, and soil moisture parameter monitored.

The primary contribution of this study [8] is the development of two-way data transmission based on the Internet of Things (IoT) utilizing Firebase Realtime Database as the platform for handling and tracking the environment of the smart greenhouse. Sensors and micro-controller are present as smart nodes, which may broadcast microclimate readings to a distant firebase to store the dataset. Users will be able to remotely operate the actuators in the smart greenhouse using the data that is provided on the Laravel-based website interface. By measuring the latency and packet loss data values, the architecture of a smart greenhouse-based two-way data communication system was analyzed and tested.

The main goal of the study [9] is to create an IoT-based system capable of producing temperature, humidity, and soil moisture sensors that are location-specific microclimatic characteristics. With the aid of the Cropwat program, which is based on macroclimatic parameters, the data is further examined and verified. By tracking various microclimatic indicators, farmers can assess their irrigation water requirements. This involves using sensors to measure soil moisture and temperature, which are then fed into the LoRA system. The system analyzes this data to estimate evapotranspiration. The Mcguinnes-Bordne formulation is used to evaluate the sensor data, and the results of this study effort open the way for the estimate of evapotranspiration in the microclimate setting.

Recent research in precision agriculture emphasizes the significance of managing and monitoring the microclimate within and outside greenhouses. The techniques involved in controlling the environment include gathering, analyzing, and processing data, while also visually identifying climatic parameters to alert of potential issues. All these measures are aimed at increasing agricultural productivity and reducing input costs. The importance of contemporary technologies such as wireless sensor nodes, embedded devices, and

IoT-based and cloud-based data collection platforms in achieving these goals cannot be overstated.

The use of novel technology such as IoT in the domain can be seen in [4]. In this paper, the authors use a model-based implementation in Matlab Simulink to analyze a greenhouse microclimate utilizing IoT, sensors, and Node MCU (ESP8266) as controller. The goal of this study is to better understand how to interpret microclimate data variances about plant needs at various growth stages. The model-based analysis offered a way to accept these uncertainties due to the nonlinear dynamics of the system (i.e., solar radiation and wind speed) and reduce the production risk despite even though the precise behavior of the greenhouse environment from raw data and its impact on the plants [8].

The authors of [5] designed and implement a remote monitoring and control system for the greenhouse based on IoT. The greenhouse model is divided into four sections for planting various plants. Different levels of soil moisture are maintained in each area. This is used as the basis for monitoring and analyzing plant development. The microclimatic conditions in the greenhouse model are controlled by an automated control greenhouse system, which is made up of various separate subsystems (watering, lighting, temperature control, air quality subsystem, etc.). The greenhouse IoT component uses the ThingSpeak cloud.

In [6], the objective was to gather information on air temperature, air humidity, and light levels using a group control device (ESP) in a greenhouse microclimate. To achieve this, an automated microclimate management system was developed using cloud architecture. This system comprises six fundamental components: a cloud server, a local object server, a mobile application, a group control device, a device module, and a sensor. The greenhouse complex's dependability is increased, and the possibility of data loss due to network failures is decreased, by replicating data from a local device to the cloud. Based on the test results, the user interface's reaction time can be deemed acceptable when it takes less than two seconds from the time that data packets are received until they are seen on the user's screen. Data availability and system dependability can be increased by duplicating functionality and replicating databases across on-premises object servers and cloud services.

The system designed here [7] is extremely complex, cutting-edge, and capable of measuring six microclimatic parameters in three separate greenhouse sections. A decision-making control system can be activated to adjust required parameter values by providing ventilation or by turning on and off processes of mist foggers in required areas of the greenhouse after comparing the collected data with expected values. In addition to this, the system continuously records to create databases. The system presented demonstrates its utility for measuring and revealing distinctions in the greenhouse microclimate in terms of the temperature, humidity, and soil moisture parameter monitored.

The primary contribution of this study [8] is the development of two-way data transmission based on the Internet of Things (IoT) utilizing Firebase Realtime Database as the platform for handling and tracking the environment of the smart greenhouse. Sensors and micro-controller are present as smart nodes, which may broadcast microclimate readings to a distant firebase to store the dataset. Users will be able to remotely operate the actuators in the smart greenhouse using the data that is provided on the Laravel-based website interface. By measuring the latency and packet loss data values, the architecture of a smart greenhouse-based two-way data communication system was analyzed and tested.

The main goal of the study [9] is to create an IoT-based system capable of producing temperature, humidity, and soil moisture sensors that are location-specific microclimatic characteristics. With the aid of the Cropwat program, which is based on macroclimatic parameters, the data is further examined and verified. By tracking various microclimatic indicators, farmers can assess their irrigation water requirements. This involves using sensors to measure soil moisture and temperature, which are then fed into the LoRA system. The system analyzes this data to estimate evapotranspiration. The Mcguinnes-Bordne formulation is used to evaluate the sensor data, and the results of this study effort open the way for the estimate of evapotranspiration in the microclimate setting.

## 3  Parameters of Microclimate in Greenhouse

To create the best climatic conditions for crop development while using the least amount of fuel and energy possible, a variety of factors are used both inside and outside the greenhouse. Solar radiation, air temperature, relative humidity, air flow rate, and carbon dioxide concentration are a few of these variables. Nevertheless, monitoring and climate control are made more difficult by the interaction of the variables that affect one another and by their dependency on the shifting ambient environment.

A. *Temperature*

Temperature's main function is to ensure that crop leaves develop at a young age. Nelson states in [10] that while greenhouse plants do best at temperatures between 15 and 30 degrees Celsius, most plants grow well at temps between 10 and 24 degrees. Nelson [10] found that the temperature difference needed for crop development between day and night should be around 8 to 10 °C under sunny situations. Depending on the type of plant, different temperatures are required for optimal plant development.

B. *Humidity*

The appropriate humidity levels for a plant are strongly dependent on water stressors, harsh weather conditions, the risk of pestilence/insects, ripeness, and the crop growth phase [11, 12]. As a consequence, optimal humidity levels are required for plant development, which is expressed in terms of the vapor pressure deficit, which is defined as the difference between the water vapor pressure at saturation and the actual water vapor pressure at the greenhouse temperature [11].

C. *Air Flowrate*

Poor air circulation inhibits plant activity and might cause humidity and disease control issues. The greenhouse's air circulation should be between 0.2 and 0.7 m per second. Plant development is hampered if carbon dioxide levels are not maintained. The exchange of air is what ventilation is all about. Moving the air breaks down this "stratified" air, reducing heating costs and keeping your plants comfy.

D. *Light Intensity*

This is the point at which photosynthesis is at its peak and plant growth is at its peak. Growth is slowed when the amount of light is lowered. Light saturation is the point at which an increase in light intensity no longer increases photosynthesis Maps, figures and tables.

E. *Carbon Dioxide Concentration*

The advantage of CO2 enrichment is mostly seen in increased crop yield due to improved photosynthesis. CO2 enrichment in the greenhouse is an important characteristic because it has a beneficial influence on crop development, provided that other growth parameters, such as water supply, are satisfied. CO2 levels should be provided to greenhouse crops to compensate for the significant drop in CO2 caused by photosynthesis, especially when adequate ventilation is not available.

F. *Solar Radiation*

The first and most important meteorological characteristic for determining a region's eligibility for protected cropping is solar radiation [13]. The primary source of heat gain is direct solar radiation intercepted in the greenhouse, and it contributes the most to the increase in the daytime temperature of the protected cropping environment. Furthermore, significant amounts of solar energy are collected and stored in the soil before being released at night [14].

# 4   Method and Materials

We utilized advanced technologies to gather crucial indicators for greenhouse growth, such as temperature and humidity levels inside and outside the greenhouse, soil moisture, light intensity, and airflow. These parameters were collected to develop comprehensive databases. Furthermore, we employed remote operation techniques in compliance with recommended procedures and effective technology. This allowed us to remotely operate various devices, including fans, lighting, and irrigation pumps. The primary objective of this study is to optimize the variables that create an ideal environment for plants to thrive in a greenhouse. We evaluated three control subsystems for these variables and provided recommendations for their potential application in an automated or autonomous sustainable greenhouse farming system.

Users can choose between an Android application or a Web application, both of which provide access to the database through a REST API in a Laravel project. The data is delivered and received in JSON (JavaScript Object Notation) format. Additionally, the app includes a user interface that allows for controlling the actuators.

The EPS32 controller, connected to the Wi-Fi network, sends the measured parameters to the local server every five seconds. More detailed design information regarding the built IoT-based data transmission system can be found in "Fig. 1".

The system utilizes an ESP-WROOM-32 controller, which is a dual-core 32-bit microcontroller with built-in Wi-Fi (802.11/b/g/n/e/i) and Bluetooth capabilities. The controller is programmed using the Arduino IDE through an ESP32 Dev Module add-on. It is configured as a client access point, allowing it to obtain an IP address from the gateway, connect to the Wi-Fi network, collect the sensed parameters, and transmit them to the server for storage.

The information sent by the gateway to the API address serves to specify the connection for the MySQL database's data storage. The information is then used on a mobile application interface built using Android Studio and a website-based interface created with Laravel as front-end and back-end. Using a web application framework (Laravel) with expressive and clean syntax, the API resources are specified and set up.

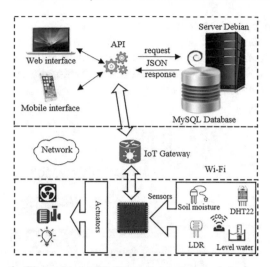

**Fig. 1.** The constructed IoT-based data transfer infrastructure.

Since the server is a basic device located in the lab with a public IP address. The server has a dual-core CPU, 2 Go of Memory, and 500 Go of hard drive space. We would rather install a Linux operating system, and we choose Debian 11, which is set up as a web server by installing apache2 and phpMyAdmin to build the database.

The "Fig. 2", illustrates the greenhouse model which is built of wood and wrapped in translucent plastic foil. The seedling starter in the center has sensors for measuring the soil moisture humidity. The irrigation system consists of a pump, a water level sensor, and two tanks. Moreover, there are three fans, where two create air and the other withdraws it. The light subsystem is made up of LEDs designed specifically for seedling growth and LDR sensors to gauge light intensity.

**Fig. 2.** The modal of the greenhouse constructed.

The automated control system is intended to regulate the greenhouse model's microclimate environment. As depicted previously, the model is made up of several autonomous subsystems, including temperature control, irrigation, lighting, remote monitoring and control, and IoT subsystem. The functional wiring diagram for all components of the greenhouse model is shown in "Fig. 3".

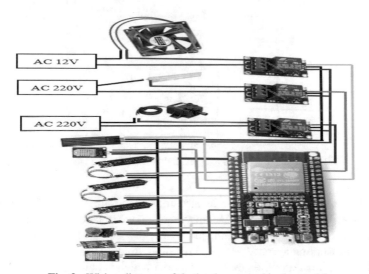

**Fig. 3.** Wiring diagram of the hardware used in the model.

The watering subsystem in "Fig. 4" is implemented with a water pump submerged in a large tank beneath the greenhouse to collect excess water droplets, which has a level sensor to detect when there is insufficient water and alert the user. A second tank is placed at a steeper angle to ensure that the automatic micro-irrigation (drip) will be filled by the needs of plants as determined by the tree soil moisture planted in a different pot. The slope will automatically give watering to the crops when the second tank is full or as needed. Experience has taught us that it takes a certain amount of time to fill the tank since the pump must be stopped so that there isn't an excess of water.

When conditions are darker than during the day, the lighting subsystem is used. That is made possible using a special LED meant to promote rapid development and larger yields (ultra-purple light due to the short wavelength), linked to the controller, and it will turn on when the intensity sensor notices that there is no daylight left. To give greenhouse plants the appropriate light during the day, the user can interact with the interface of the light subsystem at night if there is a need for light.

A group of fans was utilized to create another subsystem with a DHT22 and MQ2 gas sensor to regulate the indoor temperature and monitor the quality of the air. Cooling fans are activated to pull or generate air when the temperature, humidity, or air quality is below or upper the average required value.

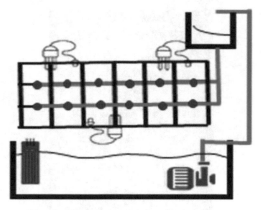

**Fig. 4.** The structural irrigation subsystem.

## 5 Results and Discussions

The model performs well with the relatively easy interactive user interface of the mobile and web-based applications shown in "Fig. 5", where the user can find a variety of make-up that aid them in keeping track of the greenhouse's internal microclimate and provide them with some information about the condition of the plants in the form of sweet notifications, as seen in "Fig. 5". If necessary, a system can automatically perform a single action or a set of actions, such as turning on a pump for irrigation when the soil moisture senses a need for it or turning on a fan to remove air.

For the dataset's development, it is quick and simple since every 5 s, a new record of the humidity (internal and external), temperature (internal and external), airflow, the amount of water in the tank, and the status of actuators (on/off) are recorded. All of these measurements are time and date stamped.

The use of APIs and JSON format (Requests/Responses) demonstrates our application's ability to co-use several programming languages and platforms for the collection and manipulation of data. Thus, our system is not limited by most of the software or hardware compatibility issues and can be supported by any device available to users. For example, we use the Java programming language in Android Studio to create mobile applications, and JavaScript on the web to access this JSON format (Fig. 6).

Our main objective is to help farmers achieve higher crop yields while conserving energy and resources, especially reducing water loss, by controlling the internal microclimate using this technology. In addition, we aim to collect and prepare the dataset for future research by integrating a deep-learning algorithm to predict microclimate parameters.

**Fig. 5.** control page

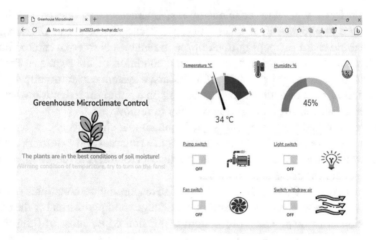

**Fig. 6.** The mobile and Web-based interface of the application

## 6   Conclusion

The use of digital devices and computer techniques is expected to rapidly expand the field of smart agriculture. Our system for tracking and managing various factors that affect the microclimate within a greenhouse has successfully achieved our goals of collecting data for characterizing these aspects. However, future work in this field must involve additional validation of the gathered data. Traditional methods for processing and organizing such vast amounts of data are impractical, so the adoption of modern tools and technologies such as cloud platforms, edge computing, and fog computing has become a standard practice to save costs and time.

# References

1. Ouammi, A.: Model predictive control for optimal energy management of connected cluster of microgrids with net zero energy multi-greenhouses. Energy **234**, 121–274 (2021)
2. Shamshiri, R.R., et al.: Advances in greenhouse automation and controlled environment agriculture: A transition to plant factories and urban agriculture (2018)
3. Rezvani, S.M., et al.: IoT-based sensor data fusion for determining optimality degrees of microclimate parameters in commercial greenhouse production of tomato. Sensors **20**(22), 64–74 (2020)
4. Shamshiri, R.R., et al.: Model-based evaluation of greenhouse microclimate using IoT-Sensor data fusion for energy efficient crop production. J. Cleaner Prod. **263**, 121–303 (2020)
5. Drakulić, U., Mujčić, E.: Remote monitoring and control system for greenhouse based on IoT. In: Advanced Technologies, Systems, and Applications IV-Proceedings of the International Symposium on Innovative and Interdisciplinary Applications of Advanced Technologies (IAT 2019), Springer, pp. 481–495 (2020)
6. Kachanova, O., Levonevskiy, D.: Cloud-Based Architecture and Algorithms for Monitoring and Control of an Automated Greenhouse Complex. In: Silhavy, R., Silhavy, P., Prokopova, Z. (eds.) CoMeSySo 2021. LNNS, vol. 231, pp. 910–921. Springer, Cham (2021). https://doi.org/10.1007/978-3-030-90321-3_76
7. Kolapkar, M.M., Sayyad, S.B.: Greenhouse Microclimate Study for Humidity, Temperature and Soil Moisture Using Agricultural Wireless Sensor Network System. In: Santosh, K.C., Gawali, B. (eds.) RTIP2R 2020. CCIS, vol. 1381, pp. 278–289. Springer, Singapore (2021). https://doi.org/10.1007/978-981-16-0493-5_25
8. Sofwan, A., Adhipratama, F.R., Budiraharo, K.: Data communication design based on internet of things architecture for smart greenhouse monitoring and controlling system. In: 2022 5th International Conference on Information and Communications Technology (ICOIACT), IEEE, pp. 205–209 (2022)
9. Mathew, T.E., Sabu, A., Sengan, S., Sathiamoorthy, J., Prasanth, A.: Microclimate monitoring system for irrigation water optimization using IoT. Measurement: Sens. **27**, 100727 (2023)
10. Nelson, P.V.: Greenhouse Operation and Management Prentice Hall: Hoboken (2003)
11. Amani, M., Foroushani, S., Sultan, M., Bahrami, M.: Comprehensive review on dehumidification strategies for agricultural greenhouse applications. Appl. Therm. Eng. **181**, 115–979 (2020)
12. Sultan, M., Miyazaki, T., Saha, B.B., Koyama, S.: Steady-state investigation of water vapor adsorption for thermally driven adsorption based greenhouse air-conditioning system. Renew. Energy **86**, 785–795 (2016)
13. Papasolomontos, A., Baudoin, W., Lutaladio, N.: 1. Regional working group on greenhouse crop production in the mediterranean region: history and development. Good Agric. Practices Greenhouse Vegetable Crops **46**(3), 656–785 (2013)
14. Buchholz, M.: The new generation of greenhouses. Unlocking the Potential of Protected Agriculture in the Countries of the Gulf Cooperation Council—Saving Water and Improving Nutrition, pp. 97–132 (2021)
15. Nam, T., Pardo, T.A.: Conceptualizing smart city with dimensions of technology, people, and institutions. In: Proceedings of the 12th Annual International Digital Government Research Conference: Digital Government Innovation in Challenging Times, pp. 282–291 (2011)

# A Review on Network Layer Attacks in Wireless Sensor Network and Defensive Mechanisms

Sana Touhami[1,3]([✉]), Mohamed Belghachi[1], and Achouak Touhami[2,4]

[1] Department of Mathematics and Computer Science, Tahri Mohammed University, Bechar, Algeria
touhamisana@gmail.com
[2] Department of Mathematics and Computer Science, Ali Kafi University Center, Tindouf, Algeria
[3] Information and Telecommunication Laboratory, Tahri Mohamed University, Bechar, Algeria
[4] Energetic in Arid Zones Laboratory, Tahri Mohamed University, Bechar, Algeria

**Abstract.** Wireless sensor networks (WSN) are considered as one of the most widely used networks for all-inclusive applications. They are organized into many sensor nodes. The deployment of nodes in these networks is not secure, and this may lead to security attacks. In this paper we are deliberating different types of attacks in network layer and defensive mechanisms against those attacks.

**Keywords:** Wireless Sensor Networks · Attacks · Defensive mechanisms · Network layer

## 1 Introduction

Wireless sensor networks represent novel decentralized wireless systems characterized by their low energy consumption, affordability, and compact dimensions [1, 2]. These networks find applications across a wide spectrum of domains, including healthcare, habitat monitoring, civilian and military use, traffic management, and environmental tracking [1, 3]. Within the environmental context, WSNs play a role in gauging parameters like temperature, pressure, noise, humidity, and others (Fig. 1).

The WSN (Wireless Sensor Network) [2, 4] functions through an extensive array of nodes, with each node being linked to one or more sensors. A typical sensor node comprises a radio transmitter, an interface circuit, a microcontroller, and a battery, with an antenna attached to the radio transmitter. These nodes operate under several constraints like restricted storage capacity, low power availability, minimal latency, narrow bandwidth, compact physical dimensions, and limited energy resources. These constraints on sensor nodes represent significant obstacles to ensuring security within the sensor network. Various forms of attacks have been documented during inter-node communication, whether they occur within the communication range or beyond it (i.e., insider attacks or outsider attacks). Consequently, security challenges arise in routing processes, such as data aggregation, route discovery, and data dissemination [1, 5].

M. Hatti (Ed.): IC-AIRES 2023, LNNS 984, pp. 14–23, 2024.
https://doi.org/10.1007/978-3-031-60629-8_2

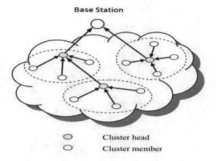

**Fig. 1.** Wireless Sensor Network.

Wireless Sensor Networks [1, 5] also fall under the category of ad hoc networks. The security objectives of such networks encompass both traditional network security concerns and the unique constraints posed by ad hoc networks [4, 6]. Security goals are categorized into primary and secondary objectives. Primary objectives, also known as standard security goals, encompass privacy, integrity, authentication, and availability. Secondary objectives include data freshness and self-organization [3]. Privacy pertains to the ability to shield messages from unauthorized listeners, while integrity ensures that incoming messages remain unaltered by potential attackers. Authentication verifies the identities of senders and receivers, preventing packet modification and injection of spurious packets. Availability ensures that resources are accessible for message transmission. Data freshness involves maintaining the recency of data, preventing the replay of previous messages.

## 2 Architecture of WSN

A Wireless Sensor Network (WSN) is constituted by an array of energy-efficient devices referred to as sensor nodes (SNs), strategically placed across a designated area for the purpose of monitoring atmospheric fluctuations. Interactions among these SNs establish a network, with certain SNs, known as sinks, serving as points of direct communication with end users. At the core of the WSN lies the sensor, responsible for gathering physical data related to environmental conditions, including but not limited to sound, moisture, intensity, and pressure, across diverse domains.

The sensor nodes possess several functions, encompassing data processing, communication, and managing network operations alongside multiple other SNs. This network's architectural depiction, as shown in Fig. 2, involves integral components such as the processing unit, sensing unit, power source, and communication module [7].

The sensing component comprises an arrangement of multiple sensors along with an analog-to-digital converter (ADC). By utilizing this combination, the sensors collect information and convey it in the form of sensed data.

The ADC's role involves conveying the data gathered by the sensor nodes (SNs) and recommending subsequent actions based on the sensed data. The communication module's purpose revolves around receiving requests or commands from the central processing unit (CPU). The CPU's primary function is to interpret these queries or

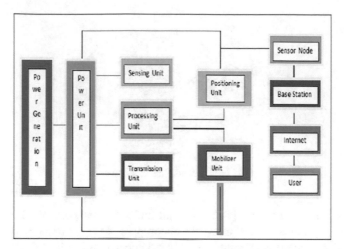

**Fig. 2.** Architectural diagram of WSN.

commands for the ADC, while also overseeing and controlling power distribution for the received data and calculating routes toward the sink.

The Power Distribution Unit's function is to supply power to all units within the WSN. Each SN unit comprises mechanisms for location identification (for locating purposes) and unit mobility (for sensor movement). The SNs carry out computations and transmit essential data across the network. In this scenario, SNs serve as routers, connecting with the battery-powered wireless network. WSNs are characterized by low energy consumption, scalability, fault tolerance, cost-effectiveness, and minimal maintenance requirements. These networks possess a limited bandwidth capacity and are programmable through software means.

## 3  Applications of WSN

Below are some WSN application fields [8]:

A. *Military Applications*

Sensor nodes play a role in conducting surveillance in military battlegrounds and are also involved in equipping smart missiles with guidance systems and identifying instances of weapons of mass destruction attacks.

B. *Medical Application*

The patient has the option to wear sensors, which proves highly beneficial for diagnosing and keeping track of the patient's condition. These sensors oversee the patient's physiological information, encompassing metrics like heart rate and body temperature.

C. *Environmental Applications*

This involves tasks such as identifying floods, enhancing agriculture precision, managing traffic, and detecting wildfires, among other things.

D. *Industrial applications*
   This encompasses the identification and analysis of industries, covering items like household devices, manufacturing plants, and supply networks, among others.
E. *Infrastructure Protection Application*
   This involves overseeing electricity grids, observing the distribution of water, and so on. The arrangement of sensor networks relies on protocols that are not interconnected, making it an inherent aspect (Fig. 3).

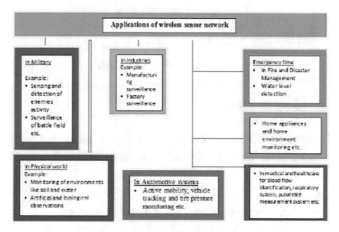

**Fig. 3.** Applications diagram of WSN.

# 4   The Internet of Things

The CERP-IoT, which stands for "Cluster of European research projects on the Internet of Things," provides a definition for the Internet of Things (IoT) as follows: It is a dynamic global network infrastructure that possesses self-configuration capabilities based on standardized and interoperable communication protocols. Within this network, both physical and virtual objects possess distinct identities, physical attributes, virtual personas, and intelligent interfaces. These elements are seamlessly integrated into the network [9].

This description illustrates the dual dimensions of IoT: time and space, enabling individuals to connect from anywhere at any time through interconnected devices like smartphones, tablets, sensors, CCTV cameras, and more [10] (Fig. 4). A connected object attains significance when it interacts with other objects and software components. For instance, a connected wristwatch gains value within a health and well-being focused ecosystem, transcending its mere timekeeping function.

The rise of the Internet of Things, which aims to blend the distinctions between computers and everyday items, can be attributed to two key factors: the widespread availability of computer resources and the acceptance of web services by users [11].

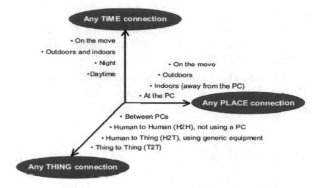

**Fig. 4.** A new dimension for the IoT [10].

## 5 Architecture of the IoT

From an architectural standpoint, the Internet of Things is structured into three primary tiers [12]: the data perception layer, the network layer, and ultimately the application layer.

A. *The Perception Layer*

The lowermost tier in the hierarchy, known as the perception layer, assumes the role of gathering data and recognizing it within its surroundings. This layer encompasses the physical components required for collecting contextual information from interconnected objects, which consist of sensors, RFID tags, cameras, GPS (Global Positioning System), and similar technologies.

B. *The Network Layer*

This layer ensures the reliable transmission of data produced within the perception layer and establishes connectivity between interconnected objects as well as between intelligent objects and other servers on the Internet. Conversely, the substantial growth in the number of Internet-connected objects anticipates a significant volume of data within the perception layer. Consequently, it has become essential to implement mechanisms and infrastructure for storing and processing this data efficiently and affordably on the Internet. This necessity is effectively fulfilled by cloud services [13], which offer flexible management of storage and processing resources through vast data centers situated on the Internet. These data centers are adept at handling the data load stemming from the Internet of Things.

It's important to highlight that the cloud employs a contemporary concept called Software Defined Networking (SDN). This concept aims to establish an abstract method of managing networks by separating decision-making and operational functionalities of network equipment. This approach facilitates the deployment of control tasks on platforms that are considerably more efficient than conventional switches, thereby reducing network latency and enabling the automation and self-configuration of extensive arrays of cloud servers.

C. *The Application Layer*

Concerning the application layer, it establishes intelligent service configurations and methods for managing various kinds of data originating from diverse sources, represented by different object types.

The architecture has the potential for expansion into a fourth layer, known as the middleware layer [14], positioned between the application layer and the two other tiers. This intermediary layer serves as the bridge connecting the hardware layer with the applications. It encompasses intricate functions for device management, along with responsibilities like data aggregation, analysis, filtration, and the regulation of service access. The middleware layer additionally conceals the intricacies of network operational mechanisms, simplifying the application development process for programmers.

# 6   Attacks on Network Layer

A. *Blackhole Attack*

The assault is referred to as a routing layer attack [6], wherein packets are transmitted through multiple nodes within the routing layer (Fig. 5).

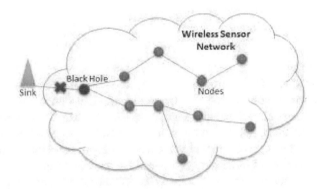

**Fig. 5.**   Blackhole attack.

This attack primarily focuses on causing harm. It presents a challenge to prevention or quick reduction [15]. It can cause temporary disruptions in networks. There are two categories: 1) Internal Black Hole Attack, where the attack originates within the network and involves a deceptive node creating a false route between sender and receiver nodes. 2) External Black Hole Attack, which emerges from outside the network and is akin to Denial of Service (DOS) attacks. Such attacks can lead to network congestion and inflict harm on the entire network.

B. *Sinkhole Attack*

A sinkhole attack [16] is more intricate compared to a black hole attack. Drawing from some understanding of the employed routing protocol, the attacker endeavors to divert traffic from a particular region through their own node. For instance, the attacker might publicize a misleadingly optimal path by promoting factors like power, bandwidth, or high-quality routes to a specific area. As a consequence, other nodes might perceive this attacker node as a better route than their current one and consequently reroute their traffic through it. Given that the affected nodes rely on the attacker for their communication, a sinkhole attack can amplify the impact of other attacks by positioning the attacker within the congested data flow. Numerous other attacks, including eavesdropping, selective forwarding, and black holes, can be exacerbated by sinkhole attacks (Fig. 6).

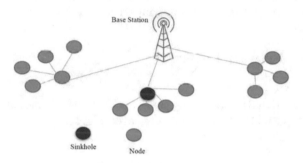

**Fig. 6.** Sinkhole attack.

C. *Wormhole Attack*

A wormhole attack [17] necessitates the involvement of two or more adversaries. These adversaries possess superior communication resources, such as increased power and bandwidth, compared to regular nodes. They can establish enhanced communication channels, referred to as "tunnels," between one another. Differing from various other attacks that occur within the network layer, these tunnels are physically established. As a result, other sensors are likely to inadvertently include these tunnels in their communication routes, effectively exiting the network under the surveillance of these adversaries (Fig. 7).

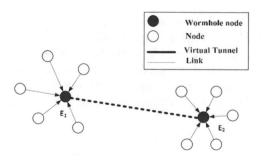

**Fig. 7.** Wormhole attack.

D. *Sybil Attack*

The attack described in [18] instills doubt in nodes and generates multiple identities, allowing the attacker to fabricate numerous indistinguishable nodes. These actions fundamentally undermine geographical routing protocols [19]. The primary objective of this attack is to disrupt multihop routing and distributed storage systems (Fig. 8).

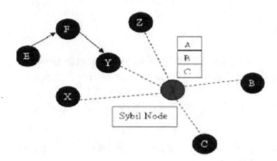

**Fig. 8.** Sybil attack.

E. *Hello Flood Attack*

Whenever a new node joins the network, a "Hello" message is dispatched. This message is usually transmitted with a heightened signal strength to guarantee its reception by network nodes and to make them aware of the new node's presence [20]. Hello flood attacks manifest when a malevolent node persistently disseminates Hello messages across the network, aiming to infiltrate the network by establishing connections through nearby nodes (Fig. 9).

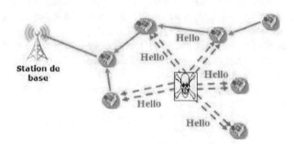

**Fig. 9.** Hello flood attack.

# 7   Defensive Mechnisms

External assaults on Wireless Sensor Networks (WSNs) can be thwarted by employing techniques like data encryption and authentication with either a pair-wise key or a common key. However, these methods prove inadequate when dealing with internal attacks perpetrated through compromised nodes within the WSN [21].

To counteract Sybil attacks, one approach is to validate the node's ID. Countering sinkhole attacks poses considerable challenges because the protective protocols employed against this attack establish routes based on energy measurements, which are intricate to validate. To combat this threat, geographic routing is adopted. This routing strategy builds a network topology using localized directives rather than instructions from a central base station [22, 23].

# 8  Conclusion

In this article, we discuss a brief survey of wireless sensor networks, internet of things, network layer attacks and defensive mechanisms of those attacks.

# References

1. Kaushal, K., Kaur, T.: A survey on attacks of WSN and their security mechanisms. Int. J. Comput. Appl. **118**(18) (2015)
2. Shah, P.K., Shukla, K.V.: Secure data aggregation issues in wireless sensor network: a survey. J. Inf. Commun. Technol. **2**(1) (2012). ISSN 2047-3168
3. Alam, S., De, D.: Analysis of security threats in wireless sensor network. Int. J. Wirel. Mob. Netw. (IJWMN), **6**(2) (2014)
4. Anand, G., Chandrakanth, H.G. Giriprasad, M.N.: Security threats & issues in wireless sensor networks. Int. J. Eng. Res. Appl. (IJERA) **2**(1) (2012). ISSN: 2248–9622 www.ijera.com
5. Alajmi, N.: Wireless Sensor Networks Attacks and Solutions. (IJCSIS) Int. J. Comput. Sci. Inf. Secu. **12**(7) (2014)
6. Modares, H., Salleh, R., Moravejosharieh, A.: Overview of security issues in wireless sensor networks computational intelligence, Modelling and Simulation (CIMSiM). In: Third International Conference on 20–22 Sept. 2011 (2011)
7. Yogeesh, A.C., Shantakumar, B.P., Premajyothi, P.: A survey on energy efficient, secure routing protocols for wireless sensor networks. In: International Journal of Engineering and Computer Science, vol. 5, Issue 8, p. 1770217709 (2016)
8. Azeem, A., khan, K., Pramod, A.V.: Security architecture framework and secure routing protocols in wireless sensor networks – survey. In: International Journal of Computer Science & Engineering Survey, vol. 2, No. 4, pp. 189–204 (2011)
9. Cluster of European Research Projects on the Internet of Things. Vision and Challenges for Realising the Internet of Things (2010)
10. Challal, Y.: Sécurité de l'Internet des Objets: vers une approche cognitive et systémique. HDR (2012). UTC
11. Thebault, P.: La conception à l'ère de l'Internet des Objets: modèles et principes pour le design de produits aux fonctions augmentées par des applications. Thèse soutenue le 31 mai 2013, ParisTech
12. Atzori, L., Lera, A., Morabioto, G.: The Internet of Things : a survey. Comput. Netw. **54**(15), 2787–2805 (2010)
13. Aceto, G., Botta, A., Donato, W., Pescapè, A.: Cloud monitoring: a survey. Comput. Netw. **57**(9), 2093–2115 (2013)
14. Granjal, J., Monteiro, E., Sá Silva, J.: Security in the integration of low-power wireless sensor networks with the internet: a survey. Ad Hoc Netw. **24**, 264–287 (2015)
15. Singh1, R., Singh, D.K., Kumar, L.: A review on security issues in wireless sensor network. **1**(1), 01–07 (2010). JISCE ISSN: 09768742 & EISSN: 0976–8750

16. Karlof, C., Wagner, D.: Secure routing in wireless sensor networks: attacks and countermeasures. Elsevier's AdHoc Netw. J. **1**(2–3), 293–315 (2003). Special Issue on Sensor Network Applications and Protocols
17. Hu, Y., Perrig, A., Johnson, D.: Wormhole detection in wireless ad hoc networks (2002). http://citeseer.ist.psu.edu/hu02wormhole.html
18. Shi, E., Perrig, A.: Designing secure sensor networks. Wirel. Commun. Mag. **11**(6), 3843 (2004)
19. Gruteser, M., Schelle, G., Jain, A., Han, R., Grunwald, D.: Privacy-aware location sensor networks. In: Proceedings of the 9th USENIX Workshop on Hot Topics in Operating Systems, (HotOS IX) (2003)
20. Wallgren, L., Raza, S., Voigt, T.: Routing attacks and countermeasures in the RPL-based Internet of Things. Int. J. Distrib. Sens. Netw. **9**(8), 794326 (2013)
21. Kim, H., Chitti, R.B., Song, J.: Novel defense mechanism against data flooding attacks in wireless ad hoc networks. IEEE Trans. Consum. Electron. **56**(2), 579–582 (2010)
22. Sharma, K., Ghose, M.K.: Wireless Sensor Networks: An Overview on its Security Threats. IJCA Special Issue On Mobile Ad-Hoc Networks. Manets, pp. 42–45 (2010)
23. Singh, V.P., Jain, S., Singhai, J.: Hello flood attack and its countermeasures in wireless sensor networks. In: IJCSI International Journal of Computer Science Issues, vol. 7, Issue 3, No 11, pp. 23–27 (2010)

# IoT-Driven Automation in Vertical Farming: A Survey

Amina Bourouis[1](✉), Achouak Touhami[2], Tariq Benahmed[1], Sana Touhami[1], Kamal Abdelmajid Mokeddem[1], and Khelifa Benahmed[1]

[1] Department of Maths and Computer Science, Tahri Mohammed University-Béchar, Béchar, Algeria
amina.bourouis@univ-bechar.dz
[2] Department of Mathematics and Computer Science, Ali Kafi University Center, Tindouf, Algeria

**Abstract.** By 2050, the world's population is estimated to hit the 9 billion mark. This will necessitate, at minimum, the doubling of food production world-wide to ensure continuous human survival. Agriculture is without doubt the most prolific food source, but with climate change and lack of arable lands, the traditional approach will not be sufficient to avoid world hunger. Vertical farming is a precision agriculture domain seeking to enhance yield per space while shortening supply chains by bringing the tech-enhanced farm to urban areas and large cities. Being a new and still in development area, though, Vertical Farming requires study and research, especially when it comes to automatization using IoT, AI and robotics. In this work we conduct a survey on the exiting works of this domain, seeking to detect the best practices and the open issues as a foundation for proposing a new optimized automatized Vertical Farming system in arid and semi-arid regions.

**Keywords:** Vertical Farming · IoT · Precision Agriculture · AI · Agriculture 4.0

## 1 Introduction

Food security is one of the most urgent and important issues world-wide. Agriculture is responsible for most of the food production in almost all countries. With climate change causing severe shifts in agricultural environments though, it becomes necessary to reconsider open-field agriculture and move towards more controlled climates such as in greenhouse, hydroponic and vertical farming schemes.

Though several definitions exist, we can define Vertical Farming (VF) as large-scale crop cultivation systems housed within enclosed structures. These systems meticulously regulate the plant environment using electric lighting, climate control mechanisms, and hydroponic techniques. VF systems, as their name suggests, often prioritize space utilization by cultivating plants vertically. This is typically achieved through horizontal stacking of layers, although vertical walls are also utilized for this purpose [1]. Vertical farming presents a promising solution to several issues in traditional agriculture by maximizing land use efficiency and enabling year-round crop production independent

M. Hatti (Ed.): IC-AIRES 2023, LNNS 984, pp. 24–33, 2024.
https://doi.org/10.1007/978-3-031-60629-8_3

of external climate factors. In addition to that, with this rigorous climatic control, VF systems can provide more nutrient and flavorful products with better visuals and less use of chemicals and other pollutants [1, 2].

Given its possible benefits, VF is emerging as an adjunct to traditional agriculture, aiming to enhance sustainable food production to meet the needs of a rapidly expanding global population amidst escalating climate challenges. Initially, the progress of vertical farming systems primarily concentrated on technological advancements, encompassing design innovation, the automation of hydroponic cultivation, and the utilization of advanced LED lighting systems. However, recent studies have shifted their focus towards the resilience and circularity aspects of vertical farming, exploring ways to enhance its adaptability and promote a closed-loop system (Van Gerrewey et al. 2021). In fact, these farms can now cultivate a diverse range of crops within urban areas, including but not limited to cities in several countries such as China, Holland, South Korea, Japan, Canada, Italy, the United States, Singapore, the United Arab Emirates, and England [2].

With increased development across the world, VF systems emerged into four main types according to size and purpose [3]:

1. A plant factory with artificial lighting (PFAL) refers to an expansive vertical farm situated within a dedicated structure, specifically designed for large-scale cultivation.
2. A container farm denotes a modular vertical farm that is housed within a shipping container, providing a portable and adaptable cultivation environment.
3. An in-store farm represents a vertical farm strategically positioned at the point of sale or consumption, such as retail stores or restaurants, enabling fresh produce to be grown on-site.
4. An appliance farm refers to a vertical farming system integrated into a home or office setting, functioning as a self-contained unit for personal or small-scale cultivation purposes.

Figure 1 illustrates the four different types of VF systems.

With this range of use and scale, the technologies backing VF have also diversified and improved across the years, including the use of the most recent in IoT, Cloud and Edge computing.

IoT, short for the Internet of Things, refers to a network of interconnected physical devices embedded with sensors, software, and internet connectivity. These devices, known as *smart* or *connected* devices, can collect and exchange data, interact with each other and humans, and enhance automation and convenience. Common examples of IoT devices include smartphones, home appliances, security cameras, and agricultural equipment. They gather data through sensors, transmit it to the cloud or other devices, and can be remotely controlled or monitored. The data generated by IoT devices enables insights, process automation, better decision-making, and operational optimization across various fields such as agriculture, healthcare, transportation, and smart homes and cities. IoT technology has the potential to revolutionize industries by enabling greater connectivity, data-driven insights, and intelligent automation [6].

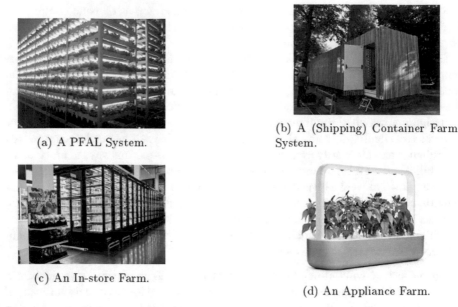

(a) A PFAL System.

(b) A (Shipping) Container Farm System.

(c) An In-store Farm.

(d) An Appliance Farm.

**Fig. 1.** Types of VF Systems [4, 5].

With such a wide range of application potential, it is no wonder that the use of IoT and related technologies like Big Data, Cloud and Fog computing, and Blockchain bore a new, ongoing, revolution termed industry 4.0 and agriculture 4.0 as a natural evolution across time as seen in Fig. 2 [7].

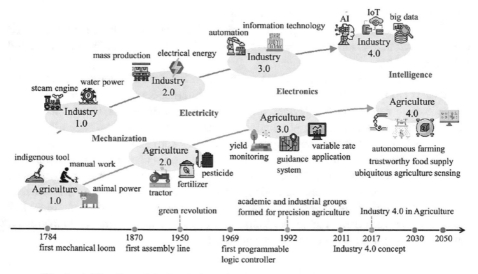

**Fig. 2.** A Timeline of the Evolution of Industry and Agriculture (LIU et al. 2020).

Thus, VF automatization using these technologies becomes a natural progression in both research and development. In this paper, we are interested not only in the potential of VF in building sustainable urban food supply-chains, but also their full automatization using IoT, robotics and AI. Our study focuses on surveying the most recent cutting-edge research in the domain, seeking to understand the necessary components for its implementation and the best findings to grantee its success. This work is therefore, a building foundation on which we hope to propose our own fully-automated sustainable closed-loop VF farming in arid and semi-arid regions especially African ones.

The remainder of this work is organized as follows:

1. Section 2 introduces some basic notions and preliminaries.
2. Section 3 details our survey of exiting works ending with a synthesis.
3. In Sect. 4, we summarize our findings and list some paths for future work.

## 2 Preliminaries

In this section we give some background notions necessary to understand the subject matter of this paper.

### A. *Precision Agriculture and Vertical Farming*

Utilizing advanced information, communication, and data analysis techniques in decision-making processes, precision agriculture (PA) is a management strategy that aims to improve crop production while minimizing water and nutrient losses and negative environmental impacts. PA encompasses practices such as site-specific crop management, target farming, variable rate technology, and grid farming. These terms are often used interchangeably with PA. Moreover, PA has extended beyond crop production and found applications in viticulture, horticulture, pasture management, and livestock production [8].

Vertical Farming can be considered as a type of PA, where we seek to maximize food production on a minimal footprint by growing crops vertically, either above or below one another instead of horizontally, i.e., next to each other. This approach can be implemented outdoors using natural sunlight or indoors with the aid of artificial lighting. Each solution offers its own advantages and disadvantages. Indoor vertical farming benefits from precise environmental control, allowing farmers to finely regulate various factors and ultimately increase crop yield. Additionally, indoor systems eliminate the need for herbicides and pesticides since only the intended crops are grown in the controlled growing medium. Space utilization is also highly efficient, as crops can be cultivated in underground tunnels or repurposed underground structures, enabling food production near urban areas. Furthermore, indoor vertical farming is less vulnerable to the effects of climate change compared to outdoor methods [9].

In addition to their vertical layout, many VF systems replace soil with water or fog that are filled with nutrients. This led to the appearance of three main types of soil-less VF [10, 11]:

1. Hydroponics: In this type of VF, the plants' roots are suspended in a nutritional solution.

2. Aeroponics: In this type of VF, the plants are suspended in air receiving a spray/mist (fog) of the nutritional solution directly to the roots. This has the benefit of reducing water usage by up to 90%.

3. Aquaponics: Like Hydroponics, in this VF the plants' roots are also suspended in water. The difference lays in the merger between aquaculture and agriculture. This is to say that the planting water is also used to grow fish. Organic waste from the fish provides the necessary nutrients necessary for plant growth.

Using these soil-less approaches allows for a VF system to become a closed-loop system. In fact, A closed-loop hydroponic system is the central circular component of a VF. The term closed-loop hydroponic system refers to the circulation and reuse of the nutritional solution via capture, disinfection, enhancement and then reproduction to the roots. In simpler terms, in a closed-loop VF system, the plants receive the necessary nutrients via water (deep water, ebb and flow, mist/spray…etc.). The excess water and precipitation are then recollected and reused. To avoid the spread of disease, disinfection of the solution is necessary (example using hydrogen peroxide, filters, heat, ozone, or UV radiation). Furthermore, the recollected water may need rebalancing after changes in levels of nutrients, pH and or phytotoxic organic acids exuded by the roots [3].

## B. *IoT and Sensor Technologies*

The fundamental concept of the Internet of Things (IoT) is to connect objects that are currently not part of a computer network, enabling communication and interaction between them and people. It involves making objects smarter and connecting them through an intelligent network. By enabling remote sensing and control of objects and machines across a network, IoT allows for a tighter integration between the physical world and computers, resulting in improvements in efficiency, accuracy, automation, and advanced applications. The world of IoT is vast and complex, encompassing various components, protocols, and technologies. Rather than viewing IoT as a single technology domain, it is helpful to see it as an umbrella term that includes different concepts and technologies, often specific to particular industries. While IoT brings numerous benefits in terms of productivity and automation, it also presents challenges such as scaling the large number of devices and processing vast amounts of data. These challenges also require the use of other technologies such as AI, Cloud, Fog and Edge computing, energy-efficient networking, robotics, Big Data and Data analytics [6].

To simplify the idea, in an IoT-enabled environment, sensors collect data about their surroundings including temperature, humidity and light intensity in agriculture, for example, and relay this data to gateways then to servers (the sensors and their controllers are considered the Edge, the gateways the Fog and the servers hold the Cloud backend). This data, generated by one or several systems over time, can then be processed and used in different ways, from data analysis to predictive modelling. With the help of AI including Machine Learning and Deep Learning, better decision making is possible. Once the decisions are made, the distant control of the system is possible by relaying the decision of the actuators (another Edge component) where they are carried out: opening valves, turning on lights…etc. Based on these simple principles, a system can be automated, requiring little to no human intervention. In agriculture, this could not only improve crop yield, but also lower the need for farming hands and other support personnel in a farm [12] (Fig. 3).

**Fig. 3.** IoT Concepts [12].

## 3 Related Works

The vertical farming concept goes back to the 600BC in the form of one of the seven wonders of the world: The Hanging Gardens of Babylon. Thus, VFs were in the beginning a human dream, which can be best captured in the fact the closest to possibility came about in a Belgian comic strip under the name *"Suske en Wiske"*. Today's VF theory, however, started in the early 2000 with Dickson Despommier and Toyoki Kozai et al.. These works finally opened the world to serious VF considerations, that remained in the labs at first, to spread now to real businesses specializing in either creating VF systems or producing using VF systems are This turn of events going back as far as 2010, was due to advancements in many axes such as LED lightning technologies, sensors, embedded systems...etc. [3, 11].

As it is, VF boasts of several advantages including [11]:

- Reduced water usage (up to 95%).
- Reduced land cost since less space is needed for cultivating the same amount of produce.
- Water filtration with recycling of city wastewater.
- Increased yield per unit area, and year-round cultivation of even seasonal plants.
- Ability to move farming indoors and into urban areas, where large spaces is not necessary, due to stacked up layout.
- Resilience against natural disasters especially weather variations, hurricanes, flooding, droughts..etc.
- Uses less chemicals including herbicides and pesticides.
- Shorter supply chain, improved food security and increased job opportunities.
- Bringing plants into the city has great positive impact on society, including mental and physical health and air quality.

But even with all these advantages in both economic, environmental and social aspects, and though promising from a research perspective and seemingly necessary given the urbanization of the modern world, VF faces many challenges that must be addressed before its large-scale implementation [11, 13]:

- Huge initial investment and scale-up, as well as difficulty in return on investment.
- More suited to leafy greens and similar plants rather than more filling ones.
- Controlled environments are suited for one type (or of the same conditions) of plants and are hard to adapt to cultivating several types of crops at the same time.
- Energy and maintenance high costs.
- Natural phenomena such as cross pollination must manualized.
- Dependency on artificial lightning.
- Know-how and training necessary in initial stages.
- Public perception that sees plants growing in VF as unnatural, especially those of soil-less cultivation and within urban pollutants, despite the evidence to the contrary (In the USA, VF-grown plants are considered organically grown for example).

Solving these issues and several more, and empowered by new IoT technologies, several studies have been conducted seeking the improvement of traditional VF systems.

In [14], the authors developed a cyber-physical system for an SOA (Service Oriented Architecture)-based platform of a an indoor VF system. Their proposal was a PFAL, plug and play, system that uses Cloud, Fog and sensor/actuator technologies. The implementation used a Raspberry Pi Model 3, air temperature, pressure, humidity, CO2, soil temperature and humidity and eTVOC sensors. LEDs and two valves (supply and drainage of water) were used as actuators. Their software included two local clouds: a controller and a management one, with an SOA-based Web implementation and UI. Though modular and using well established technologies, the drawback of this system consists in the use of scheduling for light and water rather than intelligent control as well as lack of energy efficiency and automated growth tracking.

In [9], the authors also based on SOA platforms for the creation of a modular, indoor or outdoor VF system. The main advantages of this work were the possibility of using LoRaWAN or Wi-Fi, the development of decision models, being cost-effective, plug-and-play approach where sensors can be added. Furthermore, the authors carried out extensive benchmarking for performance evaluation for the Raspberry Pis they used as microcontrollers to track their performance under different data loads. With a Web UI, several types of sensors including GrovePi and LoRaWAN plug and sense ones, the main drawback of this work was the manual decision models rather than the more intelligent AI and ML based ones.

The work in [15] focused on the development of an IoT-based system sensing temperature, humidity and pH. The collected data is then uploaded and stored to be later used in Big Data Analytics. The VF system contained an Arduino Uno, pH, temperature and humidity sensors, a Node MCU for connectivity, TDS, Salinity sensor (EC), water level sensor and LDR sensor for light intensity. Things speak was used as backend. The authors also used a Linear Regression algorithm for predictions with the goal of predicting sensor data errors. They also developed an accompanying mobile application for distant control of motors and LEDs, which are the system's actuators. The system showed great results in adjusting sensor data, but there was no exploration for better algorithms, and or the actual use of sufficiently Big Data Analysis due to the small scale of the experiment.

Another work is that in [16] where an Arduino, temperature and ultrasonic sensors, with pH and EC (detects salinity) sensors and valves system is introduced. This system boasted a water recycling and rainwater harvesting module as well as an android mobile app. It could detect water levels and nutrients levels in a hydroponic cultivation environment. Sensed data is stored for analytics with the objective is growth tracking and automated control for small farms. Though innovative in its approach, no actual robotics system was described, and the automation is dependent on thresholds (pH...etc.) rather than more intelligent approaches.

In [17], the researchers conducted a comparative study between their developed IoT-based VF system and a traditional VF system, where plants growth, health and visuals were compared. The system made use of an Arduino Uno, UV lamps, water pump, soil moisture and temperature sensors as well as air temperature ones. It had a Web UI, with automatic control of actuators (lamps and pump). In their conclusion, their IoT system proved better after the first weeks than the traditional one. This works main drawback was the lack of soil or water quality sensors, and not using any AI approaches which could have further improved the findings.

A most promising positioning work is that of [18]. In this paper, the authors propose a theoretical system that is based on modular growth boxes with Reinforcement Learning algorithms for the purpose of positioning their in-progress VF project "AgrarSense" as a fully automated, intelligent system. This work is the most recent one and boasts the possibility of leveraging the more recent works in precision agriculture for VF especially regarding the use of AI approaches. We will take inspiration from this work the most in designing our own system.

As a synthesis, we can say that though VF has a long history, it is only now beginning to benefit from IoT and AI advancements. This is probably due the many implementation challenges which must addressed by researchers. This however, does not derail from the presence of very successful VF-based companies such as [3, 13]:

- AeroFarms (PFAL; Newark, NJ, USA).
- Bowery Farming (PFAL; New York, NY, USA).
- Jones Food Company (PFAL; North Lincolnshire, UK).
- Spread (PFAL; Kyoto, Japan).
- Agricool (container farm; Paris, France).

And several more in many parts of the world.

## 4 Conclusion

In this work, we conducted a background study on the most recent works in VF systems which use IoT or similar technologies. Our review includes existing work exploration, detection of the advantages and disadvantages of VF systems and the shortcomings of existing IoT-enabled VF systems as a synthesis to be used for our own system design. As a result, we can say that VF has a lot of potential in improving food security, but only if more intelligent and cost-effective implementations are made, something that the literature seems to still behind on, if compared to other precision agriculture domains.

In our future works, we take inspiration from this survey to propose our own Tiny-DL (Deep Learning implemented at the edge) fully automated IoT-based system that is focused on growing plants in arid and semi-arid regions especially in northern Africa.

# References

1. Ji, Y., Kusuma, P., Marcelis, L.F.M.: Vertical farming. Curr. Biol. **33**(11), R471–R473 (2023). https://doi.org/10.1016/j.cub.2023.02.010
2. Kalantari, F., Tahir, O.M., Joni, R.A., Fatemi, E.: Opportunities and challenges in sustainability of vertical farming: a review. J. Landsc. Ecol. **11**(1), 35–60 (2018)
3. Van Gerrewey, T., Boon, N., Geelen, D.: Vertical farming: the only way is up? Agronomy **12**(1), 2 (2021)
4. Butturini, M., Marcelis, L.F.M.: Vertical farming in Europe: present status and outlook, p. 77 (2019). https://doi.org/10.1016/B978-0-12-816691-8.00004-2
5. Kozai, T., Niu, G., Takagaki, M., Eds.: Chapter 3 - PFAL Business and R& D in the World: Current Status and Perspectives. In: Plant Factory, San Diego: Academic Press, pp. 35–68 (2016). https://doi.org/10.1016/B978-0-12-801775-3.00003-2
6. Hanes, D., Salgueiro.: IoT Fundamentals: Networking Technologies, Protocols, and Use Cases for the Internet of Things. Pearson India (2017)
7. Liu, Y., Ma, X., Shu, L., Hancke, G.P., Abu-Mahfouz, A.M.: From Industry 4.0 to Agriculture 4.0: current status, enabling technologies, and research challenges. IEEE Trans. Ind. Inform. **17**(6), 4322–4334 (2020)
8. Sishodia, R.P., Ray, R.L., Singh, S.K.: Applications of remote sensing in precision agriculture: a review. Remote Sens. **12**(19), 3136 (2020)
9. Jandl, A.: IoT and edge computing technologies for vertical farming from seed to harvesting. PhD Thesis, Wien (2020)
10. Mir, M.S., et al.: Vertical farming: the future of agriculture: a review. Pharma Innov. J. **11**(2), 1175–1195 (2022)
11. Sharma, S., Dhanda, N., Verma, R.: Urban vertical farming: a review. In: 2023 13th International Conference on Cloud Computing, Data Science & Engineering (Confluence), pp. 432–437. IEEE (2023)
12. Mokeddem, K.A., Noureddine, S., Amina, B., Khelifa, B., Tariq, B., Boubakeur, L.: IoT and WSNs technology for control in the greenhouse agriculture – review. In: 2022 3rd International Conference on Embedded & Distributed Systems (EDiS), pp. 136–141 (2022). https://doi.org/10.1109/EDiS57230.2022.9996500
13. Gupta, M.K., Ganapuram, S.: Vertical farming using information and communication technologie. Infosys (2019)
14. Haris, I., Fasching, A., Punzenberger, L., Grosu, R.: CPS/IoT ecosystem: indoor vertical farming system. In: 2019 IEEE 23rd International Symposium on Consumer Technologies (ISCT), pp. 47–52. IEEE (2019)
15. Chand, J.G., Susmitha, K., Gowthami, A., Chowdary, K.M., Ahmed, S.K.: IOT-enabled vertical farming monitoring system using big data analytics. In: 2022 Second International Conference on Advances in Electrical, Computing, Communication and Sustainable Technologies (ICAECT), pp. 1–6. IEEE (2022)
16. Shrivastava, A., Nayak, C.K., Dilip, R., Samal, S.R., Rout, S., Ashfaque, S.M.: Automatic robotic system design and development for vertical hydroponic farming using IoT and big data analysis. Mater. Today Proc. **80**, 3546–3553 (2023)

17. Gustilo, R.C., Guillermo, D., Dim, F.: Automated Iot-Enabled Vertical Farming: Planting Atypical Crops in an Urban Environment. J. Southwest Jiaotong Univ. **58**(1) (2023)
18. Hegedűs, C., Frankó, A., Varga, P., Gindl, S., Tauber, M.: Enabling scalable smart vertical farming with IoT and machine learning technologies. In: NOMS 2023–2023 IEEE/IFIP Network Operations and Management Symposium, pp. 1–4 (2023). https://doi.org/10.1109/NOMS56928.2023.10154269

# Attacks and Countermeasures in Wireless Sensor Networks

Ahmed Saidi[✉]

Department of Mathematics and Computer Science, Faculty of Exact Sciences,
University of Ghardaia, Ghardaia, Algeria
ahmedduc@gmail.com

**Abstract.** Today wireless sensor networks (WSN) play an important role in most recent applications such as the internet of things, smart grid, VANET, and underwater sensor networks. Security is a fundamental aspect in WSN. However, WSN suffers from various constraints such as unsecured communications, fault tolerance, and limited resources, which make security a challenging task. In this paper, we present a survey on different attacks in WSNs and their corresponding countermeasures.

**Keywords:** Wireless Sensor Networks · attacks · vulnerabilities ·
countermeasures · trust management system

## 1 Introduction

WSNs are interesting areas that provide a strong base for many emerging technologies to be realized in the near future [1].WSN is a special type of ad hoc network that consists of several nodes, perhaps even thousands of nodes spread randomly in a monitored area. Each sensor performs two principal objectives. It senses the event occurred, then it informs the base station (BS or sink) once the event was detected. The advantages of WSNs such as ease of deployment, low cost and simplicity make this type of network ideal for various applications such as environment monitoring, military, critical infrastructure monitoring, Etc. However, WSN suffer from different vulnerabilities and attacks. Security is a fundamental aspect that is not totally covered. For instance, in health monitoring domain, a high level of security is required in order to protect the data of the patients. In military applications, the confidentiality and integrity of information is essential to protect data from being captured by the enemy. This paper aims to survey the most and recent WSNs attacks and their countermeasures proposed by the research community. It is organized as follows; section two presents the characteristics and the vulnerabilities of WSN, section three provides attacks and their descriptions including our main taxonomy of attacks, section four offers some countermeasures, synthesis and evaluation in section five, finally the conclusion is presented in section six.

# 2   Characteristics and Vulnerabilities of WSN

## A. *WSN Characteristics*

WSN share some characteristics of ad hoc network, but it has its own specifications as illustrated in Table 1 [2]:

**Table 1.** WSNs characteristics

|  | WSN CHARACTERISTICS |
|---|---|
| Scalability | Unlike an ad hoc network, WSN consists of thousands of sensors each one trying to communicate with the base station. In this case, the communication protocols play an important role and must be efficient in order to secure communication in the network |
| Auto configuration | In WSN the network topology changes frequently because some sensors add to the network or some of them are dead. The auto configuration is the sensor's abilities to adapt to the dynamic change of the network |
| Environment | The sensors are often deployed en mass in dangerous places such as battlefields, in a biologically or chemically contaminated area. Therefore, they must operate unattended in a remote geographic area |

## B. *WSN Vulnerabilities*

WSN suffer from different vulnerabilities and threats, in order to develop useful security approaches it is necessary to know and understand these vulnerabilities [2, 3] (Table 2):

**Table 2.** WSNs vulnerabilities

|  | WSN VULNERABILITIES |
|---|---|
| Wireless connectivity | The first vulnerability comes from the open wireless communication; any attacker can sense the message exchanged between the sensors or can send a strong signal in the same frequency used by sensors in order to produce noise and interference |
| Open deployment | The sensors are generally deployed in a distant and unsecured area, therefore, physical access to the nodes is possible; The attacker can reprogram the nodes or inject malicious data into them, or damage the nodes completely |
| Resource limitation | Most security solutions are often faced with limited resources of sensors such as energy and small memory with minimum processor performance which makes it difficult to design robust and secure security schemes |

## 3  Our Main Taxonomy of Attacks

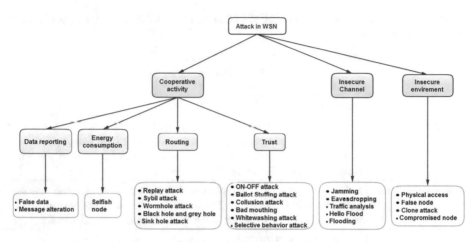

**Fig. 1.** Attacks taxonomy

In order to design solid defenses against attacks, it is necessary to know the classes and types of attacks [4]. In the literature, we found several attack classifications [5, 6]. Some researchers classify them based on the capacity of the attacker, for example when the attacker uses powerful devices such as a laptop or has the same capabilities as a sensor node. Other researchers classify the attacks based on their nature, (passive or active). In passive attacks, the goal of the attacker is to obtain access to network resources without modifying the data. In active attacks, the attacker starts modifying the data or disturbs the network operations. In [7] they divide the ensemble of attacks into three categories; routing attacks, data attacks, and trust attacks. Routing attacks affect the communication and the link stability between the nodes, for instance, the malicious node drops packets or not forwards them to their destination. The second class is the data attacks that target the integrity and the confidentiality of messages exchanged in the network. The attacks of the trust model consist of providing false trust value, the attacker tries to increase its reputation or decrease other node's reputation by sending false recommendation. In [8] they classify the attacks in four categories:(*Interruption*: The attackers block the communication between the nodes, *Interception*: The attacker gains unauthorized access to the data exchanged between the nodes, *Modification*: The attacker modifies the data exchanged between the nodes, and *Manufacturing*: The attacker injects false data into the network). In [9–11] they represent the taxonomy of attacks as a list of items, each item describes some kind of attacks. The authors in [12] use a two dimensional matrix to represent the attacks, their taxonomy associates the attack with its corresponding potential damage. It gives a better understanding of attacks compared to the representation of the list, but it didn't cover all types of attacks. In this paper, we have designed our main taxonomy, after analyzing the characteristics and vulnerabilities of WSN. We represent the taxonomy of attacks as a tree because it illustrates more effectively where the attack is coming from and gives a better understanding of the vulnerability and functionality exploited by the attackers. Figure 1 shows our taxonomy composed of three categories.

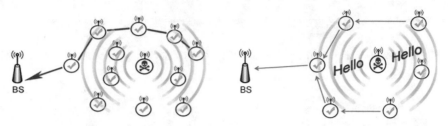

Fig. 2. Jamming attack                    Fig. 3. Hello flood

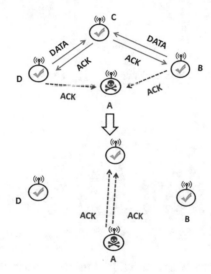

Fig. 4. Replay attacks

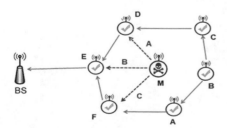

Fig. 5. Sybil attacks

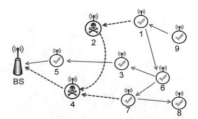

Fig. 6. Wormhole attack

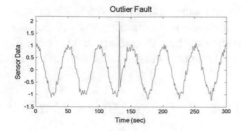

**Fig. 7.** Sensor signal with the outlier fault [3]

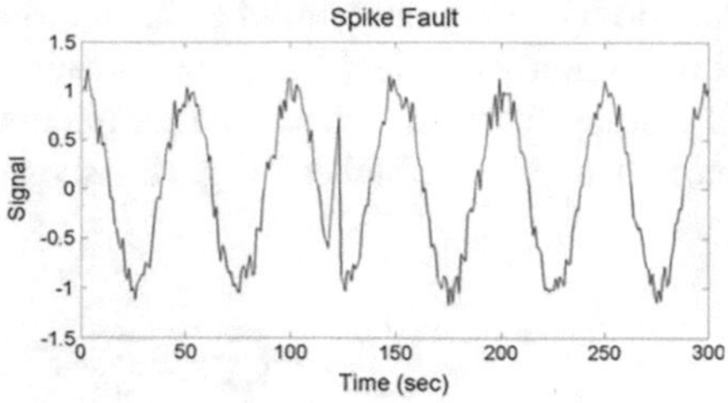

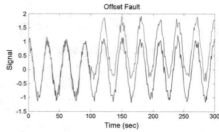

**Fig. 8.** Sensor signal with the spike fault [3]

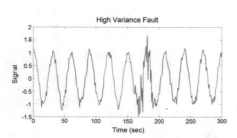

**Fig. 9.** Sensor signal with the high variance fault [3]

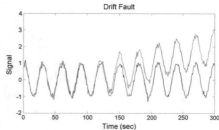

**Fig. 10.** Sensor signal with the offset [3]

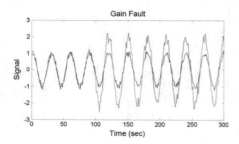

**Fig. 11.** Sensor signal with the gain fault [3]

**Fig. 12.** Sensor signal with the drift fault [3]

## A. *Attacks Related to the Insecure Communication Channel*

Include all types of attacks related to the vulnerability of wireless channel as described in Table 3:

**Table 3.** WSNs connectivity attacks

|  | WSN connectivity attacks |
|---|---|
| Jamming | The attacker uses a powerful device to generate strong radio waves. It sends these waves at the same frequency used by the sensors in order to disturb the communication between them (Fig. 2) [3] |
| Eavesdropping | The attacker analyzes the network traffic by capturing the packets exchanged between the nodes in order to extract sensitive information [3] |
| Traffic analysis | The attacker is not interested in the data. It analyzes the traffic to find were to locate the BS and attack it, which makes the network completely out of service [2] |
| Flooding | The attacker communicates with target nodes and each time this node tries to close communication, the attacker repeatedly initiates a new connection request until the exhaustion of the target node's resources [3] |
| Hello flood | Sensor nodes use hello packets in the initial network phase in order to recognize their neighbors. When the attacker has a wide radio range, the distant nodes that receive the hello packet, route the data to the attacker thinking that it's part of their neighbors (Fig. 3) [3] |

## B. *Attacks Related to the Insecure Environment*

Include all types of attacks related to the operating conditions of sensor nodes as they are deployed in unprotected area (Table 4):

**Table 4.** Node capturing attacks

|  | Node capturing attacks |
|---|---|
| Physical access | The nodes are prone to physical damage due to their functionality in an unsecured area. The attacker can compromise the nodes, export the data from their memories or destroy them completely [2] |
| False node | The attacker adds one or more malicious nodes in the network in order to inject malicious data, which could be extended to all nodes in the network [2] |
| Clone attack | The adversary captures some of the nodes and extracts their data from them, such as their identities and their keys. Then, it makes a copy of this node (clone) in order to inject the cloned nodes in the network. Because the cloned node has a valid identification, they are able to participate in network operations, hence it launches attacks and obtains sensitive information [13] |

## C. *Cooperative Activity Attacks*

The cooperative work of sensor nodes is very vulnerable since there is no guarantee that all nodes are cooperative in the network. In routing activity, due to the limited communication range and the constraint of resources, the sensor nodes use multi hop communication to deliver the information to the sink. The cooperative activity is often exploited by the attackers, for example a malicious node drops the packet or forward faulty data to the sink. In the trust management system, the sensor nodes exchange recommendations between each other. Thus, malicious nodes collude to increase their reputation or send false recommendation about the trusted nodes in the network. Table 5 demonstrates routing attacks in WSNs.

**Table 5.** Cooperative activity attacks in WSNs

|  | Cooperative activity attacks |
|---|---|
|  | Routing attacks in WSNs |
| Node outage | The node outage is the situation when a node such as a cluster head ceases to function. In this case, the routing protocols must be sufficiently robust to this failure by providing alternative routes [2] |
| Replay attacks | When two nodes communicated between each other, the control packets exchanged between them can be heard by other nodes in the network. The attacker uses these control packets to create the illusion of presented nodes in the network which actually didn't exist [3]. The attack is illustrated in Fig. 4, the attacker node A uses the ACK packet of node B to communicates with C when B and D close the connection with C |
| Sybil attacks | This is one of the most dangerous attacks in WSN. The malicious node presents multiple identities to several nodes in the network [14]. In Fig. 5, the malicious node M presents multiple identities of A, B, C to D, E, and F respectively. When D wants to communicate with A, it sends the message to M. In clustering architecture, the malicious node tries to usurp the cluster head (CH) identity in order to get the data from the cluster members (CM) or simply sends wrong data to the BS |
| Wormhole | The malicious node locates itself close to the sink and establishes the wormhole path with another malicious node in the network. The attacker's objective is to present the illusion of a short path to the sink and the node required to forward the data uses this path. The data passing through this path is captured by the malicious nodes. The routing algorithms based on the number of hops are very vulnerable to this kind of attack [3]. Figure 6 describes the attack technique; nodes 2 and 4 are malicious and nodes 9, 1, 3, 6, 7, and 8 use the wormhole path to send their data to the sink |

*(continued)*

**Table 5.** (*continued*)

| | Cooperative activity attacks |
| --- | --- |
| | Routing attacks in WSNs |
| Black hole and grey hole | In the black hole attack, the malicious node receives the data sent from the nodes and does not forward it to its destination and does not notify the sender of this failure. The grey hole is a special case of black hole attack. The attacker hides by forwarding just part of the data which makes it difficult to be detected. In order to do that, the malicious node appears more attractive to others nodes by showing that it has high energy and a high degree of connectivity [15] |
| Sink hole | This attack is similar to the black hole attack. But in this case, the attacker compromises one or several nodes and manipulates their routing information, so all their traffic will be redirected toward the attacker [16] |
| | *Abnormal energy consumption attacks* |
| Selfish node | This is a compromised node which does not participate in routing operations in order to conserve its energy [3] |
| | *Data reporting attacks* |
| Message alteration | The malicious node alters messages exchanged between the nodes by adding wrong data or corrupts some parts from it [4] |
| | *Trust attacks in WSNs* |
| On-off attack | A malicious node alternates its behavior from bad to good and from good to bad in order to remain undetected from trust models [7] |
| Ballot stuffing attack | A malicious node sends a false recommendation about trusted nodes in the network [17] |
| Collusion attack | The malicious node behaves like a normal node in order to be trusted by other nodes in the network [18] |
| Bad mouthing | The compromised node sends a good recommendation about a malicious node in the network in order to increase its reputation (e.g. selected as cluster head) [19, 20] |
| Whitewashing attack | A special type of bad mouthing attack; the malicious node gives zero recommendation about a good node in the network [21] |
| Selective behavior attack | The malicious node behaves well for the majority of the nodes but it behaves badly for some specific target nodes. In this way, the malicious node reputation remains positive but it still damages the nodes in the network [22] |

A trust methodology is widely applied in WSN. From a social science perspective, it is can be defined as the degree of confidence between humans based on their previous interactions, on which, an expectation of their future behaviors is founded. The same mechanism is applied in WSN to protect the network against internal attacks and the compromised nodes. Using cryptography and authentication is difficult in most cases considering the minimum computational power of the nodes which make trust useful. However, trust technique suffers from various attacks as described in Table4. Data sensing is very vulnerable to misbehavior nodes. This category of nodes sends faulty data readings, they appear in the case of false data injection attacks, tampering attacks, and compromised nodes. According to [3] and [23], the data faults can be classified on the base of data-centric and system-centric faults. Data-centric faults correspond to the sensor readings itself, they change data semantic produced by the sensor. System centric faults correspond to the degree of correlation and similarity between the sensor node and its related neighbors. System centric faults are identified using a healthy data model produced by healthy sensors to detect faulty sensors. This technique requires that the nodes in WSN measure the same event (phenomena). Table 6 illustrates the data-centric faults and system-centric fault.

**Table 6.** Data-centric faults and system-centric in WSNs

| Data-centric faults | Outlier fault | The sensor returns a data reading whose value is significantly outside of the range defined by the previous data readings [23]. The acceptable range of data readings is defined by the system designer (Fig. 7) |
|---|---|---|
| | Spike fault | It is a set of data values that falls or rises more rapidly during the healthy sensor readings [23]. Like outlier fault, spike fault is not continuous, it happens from time to time and sensor back to its correct readings after the spike (Fig. 8) |
| | Variance fault | Often called "noise fault". In this type of fault, the senor returns a data value that is different from the mean value of its readings by significantly higher or lower variation (Fig. 9) [23] |
| System-centric faults | Offset fault | It occurs when the sensor returns a different data reading (offset) from the phenomenon being observed (Fig. 10) [23] |
| | Gain fault | The sensor returns a constantly different data reading (dissimilar) from the healthy sensor readings by a fixed value (Fig. 11) [4] |
| | Drift fault | It occurs when sensor readings drift away from the real values by a value that increases with time (Fig. 12) [4] |

## 4  Wsn Countermeasures

### A. *Prevention Techniques*

WSNs countermeasures can be classified either on prevention techniques or detection techniques. Prevention techniques focus on eliminating attack occurrences regardless of where the attack is coming from. It is out of the scope of preventing techniques to detect attackers. These techniques rely on cryptography and authentication to prevent external nodes to access the network resources. Some of the prevention techniques are described as follows:

- **Cryptography** It is a set of techniques used to protect information and guarantee its confidentiality. Cryptography techniques are classified into symmetric and asymmetric techniques [24]. In symmetric techniques, the same key is used for encrypting and decrypting the message [24]. These techniques assign a unique key for each user which makes exchanging keys among users a challenging operation. In contrast, asymmetric techniques use two keys one is public and the other is private. The public key is used for encrypting messages whereas the private key is used for decryption [24].
- **The hash function** is a function used in cryptography to ensure the integrity of transmitted messages. The hash function is applied to the transmitted message and generates a small version of it called "hash values" [36], then the sander joins hash values with the transmitted message and sends it to the receiver. Once receiving this message, the receiver applies the hash function to the received message and compares the produced result "calculated hash values" with the hash values sent by the sender. If they match, then the message integrity is ensured.
- **MAC authentication** The hash function ensures message integrity by comparing hash values and the original message. However, hash function alone does not guarantee the authentication of the sender. For example, if an adversary sends a faulty message with faulty hash values attached to it, therefore, it is necessary to guarantee message authentication. Mac algorithm is a symmetric cryptography technique which provides message authentication by exchanging symmetric key between the sender and the receiver [24].
- **The digital signature** among useful features of asymmetric cryptography is signing the data using the private key [24]. As we said earlier, asymmetric cryptography uses two keys one is public which is known by all and the other is private which is known only by its owner.
- **The digital certificate** digital signature is used to check from the trustworthiness of the data sent. However, a digital certificate is used to verify the trustworthiness of the sender itself. Also, it used to protect the public key from spoofing [24].
- **Detection Techniques**

As we mentioned earlier, prevention techniques focus on preventing attack occurrences regardless of where the attack is coming from and who is the attacker. These techniques are not interested in detecting the attacker. On the contrary, detection techniques can detect the attacker using an IDS (shortening of Intrusion Detection System). The IDS monitors the network and sensors activities and emits alerts (alarms) when detecting malicious activity. Based on the detection techniques, IDSs can be classified into signature-based IDS and anomaly-based IDS [25].

- **Signature-based IDS** signature-based IDS compares node activity with the attack signature to decide if the node is malicious or not [25]. This identification technique is efficient in detecting already known attacks. However, it is not efficient in detecting unknown attacks. As a result, signature-based IDS must constantly update its signature database by adding new signatures.

- **Anomaly-based IDS** anomaly-based IDS observes the node behavior (activity) and compares it with the system's normal behavior [26]. If their behaviors are different (dissimilar), the node is detected as malicious. This identification technique detects unknown attacks or new intrusions. However, it suffers from high false-positive rates [25]. As a result, most current IDS applies both detection techniques to increase detection accuracy.

- **Collaborative based IDS** In general, IDS suffers from two problems: First, identifying new attacks by constantly updates the signature database. Second, finding the optimal sensitivity level to avoid false alerts (false positive rates). However, if these IDSs form a network, cooperate and share their knowledge, they achieve more accurate detection. For example, when an IDS detects new attacks, it alerts its collaborators to block similar attacks when they occur. The collaboration between IDSs creates a CIDN (shortening of Collaborative Intrusion Detection Network [25]) that allows a strong intrusion detection capability. However, collaboration among IDS introduces several challenges such as preventing dishonest IDSs peers from participating in collaboration or detecting selfish IDSs that are not participating at all. To mitigate these issues, a trust based IDS can be used to identify malicious and dishonest nodes.

## C. Trust Based IDS

As previously outlined, the trust provides a high-security level for network operations such as routing, CH election, and key exchange. However, trust comes with several challenges, This is generally related to the characteristics of WSNs that are substantially different from other wired and wireless networks. Depending on the type of evidence, purpose, and scope trust models can be classified into two categories [27]:

- **Certification-based trust models** Based on the performance of PKI (shortening of Public Key Infrastructure) and PGP (shortening of Pretty Good Privacy) crypto-graphic approaches, certification-based trust models aim to create trust relationships between the nodes using the certificate as evidence. More precisely, the node is considered trustworthy to other nodes as much its certificate is valid. We mean by a valid certificate, the node that is certified by a CA (shortening of Certificate Authority which is the most trustworthy node in the trust system) or any nodes that are trustworthy in the trust system. After the node is identified and gains the trust of the CA, the CA gives this node a certificate signed by the CA private key. This certificate is used as evidence to prove that the node is already authenticated to the system and we can trust it. Since the CA public key is known by all nodes in the network, each node can check from the validity of this certificate. The mechanism of the digital certificate is already mentioned in the prevention techniques. To increase the system security, the certificate validity is restricted by time. If the certificate is expired it can be revoked by the CA. To do this, the CA broadcasts periodically a certification revocation list. Furthermore, the certificate can be revoked for other reasons for example: if it contains errors or the private key of its issuer is compromised, then this certificate must be revoked immediately. Depending on the number of CAs and their organization, certification-based trust models are classified into Authoritarian models and Anarchic models [28].

- **Behavior-based trust models.** In behavior-based trust models, the trust level of the node is evaluated by its behavior. In WSN, the subject node continuously monitors the behavior of the target node. Since there is more than one trust type communication, data, and energy, the subject node gathers different evidences depending on the trust type that the subject wants to evaluate. For example, evidence related to the communication trust includes forwarding successfully packets to other nodes in the network. Data trust evidence includes data consistency and integrity. Energy trust includes consuming the energy normally compared to other nodes in the network. As the energy of malicious node may drop abnormally after an attack.

D. Limitation of Trust Models

   Certification-based trust models build the trust decision based on the certificate. As much the node owns a valid certificate, it considered trustworthy in the system. These trust models are useful for the operation of access control or key exchange. However, the behaviors of sensor nodes are not included in the trust decision nor the trust evaluation, it is generally out of the scope of certification-based trust models to evaluate the behavior of sensor nodes [27]. Unlike certification-based trust models, behavior-based trust models aim to build trust relationships between the nodes based on their behaviors. Each node establishes trust relationships with its neighbors by continuously monitors their behaviors. This type of trust model assumes that the node is already authenticated to become a member of the network. Therefore, it is out of the scope of behavior-based trust models to authenticate the nodes or check from their certificates unless there is a heterogeneous approach that combines between the two trust models [27].

## 5 Conclusion

As the area of WSN applications increase the need for new security solutions also increases. The WSN characteristics and their vulnerabilities make traditional security mechanisms such as cryptography and authentication difficult to apply due to the limited resources of the nodes and minimum computational power. Trust management is better suited to WSN and represents a security solution if it's applied in an efficient and a reliable way. The challenge of trust management techniques is how to exchange the accurate trust value between the nodes without it being altered by the malicious nodes in the network. In this paper we aim to survey the different attacks in WSN and their corresponding countermeasures. We believe that there is no security solution that covers all types of attacks. We conclude that WSN security systems should be designed with respect to the characteristics of the sensors. Security system design also depends on where it is applied because each application has its specific requirements and therefore its corresponding degree of security needs.

## References

1. Rehmani, M.H., Pathan, A.S.K.: Emerging Communication Technologies Based on Wireless Sensor Networks: Current Research and Future Applications. CRC Press, Boca Raton (2016)

2. Benahmed, K.: Surveillance Distribuée pour la Sécurité d'un Réseau de Capteurs Sans Fil. University of Ahmed Ben Bella, Oran, Alegria (2011)
3. Selmic, R.R., Phoha, V.V., Serwadda, A.: Wireless Sensor Networks: Security, Coverage, and Localization. Springer International Publishing (2016). https://doi.org/10.1007/978-3-319-46769-6
4. Islam, M.S., Rahman, S.A.: Anomaly intrusion detection system in wireless sensor networks: security threats and existing approaches. Int. J. Adv. Sci. Technol. **36**, 1–8 (2011)
5. Bidan, C.: Sécurité des systèmes distribués: apport des architectures logicielles. University of Rennes 1, Rennes, Brittany, France (1998)
6. La, O.A.: Fiabilité de Dissémination dans les Réseaux de Capteurs Sans Fil. University of Science and Technology Houari Boumediene, Oran, Alegria (2008)
7. Ye, Z., Wen, T., Liu, Z., Song, X., Fu, C.: An efficient dynamic trust evaluation model for wireless sensor networks. J. Sens. **2017**, 1–16 (2017)
8. Stallings, W.: Network and Internetwork Security: Principles and Practice. Prentice Hall, Englewood Cliffs (1995)
9. Cohen, F.: Information system attacks: a preliminary classification scheme. Comput. Secur. **16**, 29–46 (1997)
10. Icove, D.S., Karl, V.W.: Computer Crime: A Crime Fighter's Handbook. O'Reilly & Associates California, Sebastopol (1995)
11. Jayaram, M.: Network security a taxonomic view. In: European Conference on Security and Detection. London, UK. New York, NY, USA, pp. 124–127, IEEE (1997)
12. Perry, T.S., Paul, W.: Can computer crime be stopped? The proliferation of microcomputers in today's information society has brought with it new problems in protecting both computer systems and their resident intelligence. IEEE Spectr. **21**, 34–45 (1984)
13. Jaballah, W.B., Conti, M., Filè, G., Mosbah, M., Zemmari, A.: Whac-A-Mole: smart node positioning in clone attack in wireless sensor networks. Comput. Commun. **119**, 66–82 (2018)
14. Newsome, J., Shi E., Song, D., Perrig, A.: The Sybil attack in sensor networks: analysis & defenses. In: Proceedings of the 3rd international symposium on Information processing in sensor networks. Berkeley, California. New York, NY, USA: ACM, pp. 1–10 (2004)
15. Pathan, A.-S.: Security of self-organizing networks: MANET, WSN, WMN. CRC Press, VANET (2016)
16. Wang, Y., Attebury, G., Ramamurthy, B.: A survey of security issues in wireless sensor networks. IEEE Commun. Surv. Tutorials **8**, 2–23 (2006)
17. Wang, Y., Chen, R., Cho, J., Tsai, J.: Trust-based task assignment with multi objective optimization in service-oriented ad hoc networks. IEEE Trans. Netw. Serv. Manage. **14**, 217–232 (2017)
18. Fang, W., Zhang, C., Shi, Z., Zhao, Q., Shan, L.: Beta-based trust and reputation evaluation system for wireless sensor networks. J. Netw. Comput. Appl. **59**, 88–94 (2016)
19. Sun, Y., Han, Z., Liu, K.: Defense of trust management vulnerabilities in distributed networks. IEEE Commun. Mag. **46**, 112–119 (2008)
20. Vijaya, K., Selvam, M.: Improving resilience and revocation by mitigating bad mouthing attacks in wireless sensor networks. Int. J. Sci. Eng. Res. **4**, 1–5 (2013)
21. Raj, M., Kumar, G., Kusampudi, K.: A survey on detecting selfish nodes in wireless sensor networks using different trust methodologies. Int. J. Eng. Adv. Technol. (IJEAT) **2**, 197–200 (2013)
22. Lopez, J., Roman, R., Agudo, I., Fernandez, C.: Trust management systems for wireless sensor networks: Best practices. Comput. Commun. **33**, 1086–1093 (2010)
23. Lee, K.-S., Lee, S.-R., Kim, Y., Lee, C.-G.: Deep learning based real time query processing for wireless sensor network. Int. J. Distrib. Sens. Netw. **13**, 1–10 (2017)
24. Raja, K.N., Beno, M.M.: Secure data aggregation in wireless sensor network-Fujisaki Okamoto (FO) authentication scheme against Sybil attack. J. Med. Syst. **41**, 107 (2017)

25. Fung, C., Boutaba, R.: Intrusion Detection Networks: A Key to Collaborative Security. CRC Press, Boca Raton (2013)
26. Pathan, A.S.K.: The State of the Art in Intrusion Prevention and Detection. CRC Press, Boca Raton (2014)
27. Aivaloglou, E., Gritzalis, S., Skianis, C.: Trust establishment in sensor networks: behaviour-based, certificate-based and a combinational approach. Int. J. Syst. Syst. Eng. 1(1–2), 128–148 (2008)
28. Omar, M., Challal, Y., Bouabdallah, A.: Certification-based trust models in mobile ad hoc networks: a survey and taxonomy. J. Netw. Comput. Appl. 35(1), 268–286 (2012)

# A System for Monitoring Hydroponic Plants in Greenhouse Using WSN, IoT and RnE

Achouak Touhami[1(✉)], Amina Bourouis[2], Nawal Touhami[3], Sana Touhami[2], Tariq Benahmed[4], and Khelifa Benahmed[2]

[1] Department of Mathematics and Computer Science, Ali Kafi University Center, Tindouf, Algeria
touhamiachouak66@gmail.com
[2] Department of Mathematics and Computer Science, Tahri Mohammed University, Bechar, Algeria
[3] Department of Electrical Engineering, Ahmed Draia University, Adrar, Algeria
[4] Department of Mathematics and Computer Science, Tamaghasset University, Tamaghasset, Algeria

**Abstract.** Hydroponic cultivation offers a promising solution to address environmental and food-related challenges by minimizing the use of fertilizers and water. Our research presents a greenhouse-based design for monitoring hydroponic plants, using wireless sensor networks (WSN), hydroponic irrigation system, Internet of Things (IoT) technologies and renewable energy system (RnE). The WSN consists of diverse sensors, including those for water level, water temperature, and pH levels. The sensors are connected to a cluster head, which is linked to a microcontroller. Through the Internet, the microcontroller transmits the collected data to the farmer. The farmer evaluates the data by comparing it to predetermined thresholds and, using a laptop or smartphone, sends commands to the associated actuator. The cloud serves as the storage system for the collected data. The system features automated control capabilities, allowing for adjustments in temperature, pH levels, and water levels. Furthermore, the proposed design can be easily customized to accommodate different hydroponic conditions and systems.

**Keywords:** Greenhouse · Hydroponic plants · WSN · Hydroponic irrigation system · IoT · RnE

## 1 Introduction

Agriculture is universally acknowledged as a vital profession that plays a significant role in producing food, textiles, fuel, and natural resources. With the world's population projected to surpass 8 billion by 2025 and approach nearly 10 billion by 2050 [1], the importance of agriculture becomes even more pronounced. The population growth will cause a high demand on fruits and vegetables [2, 3]. Moreover, achieving ecologically sustainable agricultural output heavily relies on technological advancements and innovative research. Modern technologies like the Internet of Things (IoT), sensors, cloud

M. Hatti (Ed.): IC-AIRES 2023, LNNS 984, pp. 48–58, 2024.
https://doi.org/10.1007/978-3-031-60629-8_5

computing, big data, and artificial intelligence (AI) are expected to play a pivotal role as policy solutions for agricultural expansion [4, 5]. These technologies have the potential to enhance crop scale and significantly contribute to boosting agricultural output in an environmentally sustainable manner [6].

Due to recent technological advancements, greenhouses have become increasingly popular as a sustainable solution to overcome agricultural challenges and enhance crop yields [3]. A greenhouse is a specially constructed environment made of transparent materials where plants can be cultivated rapidly and thrive in optimal conditions, anyplace and before the developing season [3, 7].

Kevin Ashton coined the term "Internet of Things" (IoT) to refer to the network of interconnected physical devices, known as "things," which encompass a wide range of objects such as appliances, smartphones, infrastructure components, industrial machinery, agricultural equipment, and more [8]. The IoT encompasses a diverse set of paradigms that find application in various domains, ranging from smart homes to precision agriculture. Precision agriculture, specifically, aims to develop smart farms that incorporate innovative technologies and interconnected devices into the farming environment, including hydroponic greenhouses [9]. All these efforts are aimed at addressing the issues and challenges faced by this crucial sector [10].

Hydroponic farming is one of the farming methods in a greenhouse building [11]. Hydroponics is a method that involves growing plants without soil, utilizing water, nutrients, and oxygen instead [11–14]. The advantages of hydroponic systems include improved land utilization, higher crop yields, increased quantity and quality of production, more efficient use of fertilizers and water, and easier control of pests and diseases [13]. Fluctuations in pH levels, electrical conductivity of nutrients, water temperature, air temperature, and light intensity have a significant impact on the growth of hydroponic plants, particularly vegetable crops. Without regular and periodic monitoring, the growth of plants can be compromised, leading to suboptimal quality of the vegetable crops. To ensure the production of high-quality vegetables, farmers must consistently monitor and address variations in pH levels, electrical conductivity of nutrients, water temperature, air temperature, and light intensity in the hydroponic growing medium [11, 13, 15]. By integrating IoT technology, cultivators can remotely monitor and control hydroponic systems from any location [16, 17].

In this study, we present a greenhouse-based design for hydroponic plants. The design incorporates a WSN system, hydroponic irrigation system, RnE, and IoT technology. Furthermore, the system includes algorithms that monitor water temperature, pH levels, and detect water levels in both the main tank and hydroponic tank.

The rest of this paper is structured as follows, in Sect. 2, we discuss the existing research conducted in this particular area of study. In Sect. 3, we formulate a proposal design for monitoring a hydroponic plant of greenhouse. We, also, propose the algorithms that monitor water temperature, pH levels, and detect water levels in both the main tank and hydroponic tank. Finally, in Sect. 4, we conclude the paper and outline our future research directions.

## 2 Related Works

The hydroponic greenhouse monitoring system was initiated a few years ago and has been the subject of discussion in multiple works. In light of recent advancements in electronics and computing, researchers have shown growing interest in employing sensor technologies and leveraging the Internet of Things (IoT) for remote monitoring of hydroponic greenhouses.

In their work [13], the researchers put forward a monitoring system that incorporates a pH sensor, electro conductivity sensor, water temperature sensor, air temperature sensor, light sensor, GSM/GPRS technology, Open Garden Shield, Open Garden Hydroponic, and Arduino Uno as the main board or microcontroller. During the experiment, the researchers recorded the results of number of leaves and plant height for each type of plant as follows: lettuce had 6 leaves and a height of 3.6 cm, red spinach plants had 6 leaves and a height of 3.8 cm, and mustard plant bok choy had approximately 6 leaves with a height of 4.2 cm. The sensor and the system demonstrated reliable performance during the two-week testing period on lettuce, red spinach, and mustard bok choy. The authors in [18] proposed a hydroponic system that requires several parameters such as the water temperature, water level, acidity (pH), and the concentration of the nutrient (EC/PPM). The initial step involves monitoring and collecting information from Nutrient Film Technique (NFT) Hydroponic farmers, followed by a systematic evaluation and analysis of the collected data. They constructed a hydroponic monitoring and automation system that utilizes sensors connected to the Arduino Uno microcontroller, Wi-Fi module ESP8266, and Raspberry Pi 2 Model B microcomputers as the webserver. The system is designed with the Internet of Things (IoT) concept, enabling each hydroponic farming block to communicate with the webserver (broker). The web interface serves as the system's platform, enabling users to monitor and control the NFT hydroponic farming operations. The management systems for the NFT hydroponic web interface utilize a responsive web framework, such as Bootstrap, along with JQuery and JavaScript libraries for the front-end development. The results demonstrate that this system effectively enhances the efficiency and effectiveness of monitoring and controlling NFT Hydroponic Farms for farmers. In [11], on the other hand, the authors developed a prototype of a smart greenhouse specifically designed for hydroponic plants. The hardware components of the smart greenhouse were assembled using the Arduino microcontroller, DHT11 sensor, pH sensor, TDS sensor, DS18b20 temperature sensor, ultrasonic sensor, and ESP8266 WiFi module. The monitoring system includes water quality information from hydroponic plants and enables the recording of farming activities, starting from planting preparation all the way to web-based harvesting. The test results of the smart greenhouse monitoring system demonstrate its capability to display the conditions of hydroponic plants and effectively control the humidification process. In a not far off approach, the authors of [15] proposed a monitoring and controlling system for hydroponics precision agriculture that leverages the Internet of Things (IoT) concept and incorporates fuzzy logic. IoT is employed to facilitate continuous monitoring of plants' nutritional and water requirements, while fuzzy logic is utilized to regulate the supply of nutrition and water to the plants. The experimental results indicate that the proposed system yields improved growth for lettuce and bok choy plants, particularly in terms of leaf size. The authors of [12] implemented a system capable of recording sensor values and adjusting nutrient values as needed. This project focuses on implementing drip control techniques in hydroponic farming by creating an interface that enables continuous monitoring of pH and various sensors. The interface also incorporates plant positioning through camera capture and facilitates monitoring through a mobile application. Using big data, the supply of nutrient values is monitored and recorded. The recorded data becomes valuable for the further automation of the irrigation system. In their study [17], the authors utilized IoT technology to develop control devices and remotely monitor hydroponic plants. This approach aimed to simplify the process for cultivators in controlling and monitoring parameters such as plant color, temperature, nutrient levels, and pH value of the hydroponic plant water. Control and monitoring can be performed conveniently using a smartphone application. During the testing of hydroponic plant conditions, the collected data indicated an average error of 1.8% for air temperature, 4.8% for water pH, 6.6% for plant color, and 7% for water nutrients. Hydroponic plants equipped with the TCS3200 sensor have a monitoring capability of 53.3%. The testing of the tool control, aimed at nutritional improvement, was conducted using the fuzzy Mamdani method. The results showed an 88.75% increase in the probability of adding nutritional value and a 0% probability of decreasing nutritional value. The tool control for enhancing the pH value of hydroponic plant water has been effectively executed. The research in [19] focused on the advancement of a smart greenhouse system designed for hydroponic gardens. This system incorporated the integration of Internet of Things (IoT) technologies and enabled monitoring through mobile devices. In order to accomplish the objectives of the study, the researchers constructed a functional hydroponic greenhouse system featuring fully grown plants. The system incorporated automation for the analysis and monitoring of water pH level, light, water supply, greenhouse temperature, and humidity. These parameters were connected and linked to ThingSpeak for data integration. Alpha testing is conducted by the respondents to verify the readiness of the developed system for both beta testing and the subsequent evaluation process. The smart greenhouse monitoring system that was developed underwent testing and evaluation to ensure its reliability, functionality, and usability in accordance with the ISO 9126 evaluation criteria. The prototype, along with the accompanying mobile application, was tested and evaluated by a group of respondents that included casual plant owners as well as experts in hydroponic gardening. The evaluation covered monitoring various parameters, resulting in scores of 7.77 for pH level, 83 for light, 27.94 °C for water temperature, 27 °C for greenhouse temperature, and 75% for humidity. The descriptive assessment for both software and hardware were deemed "Very Good," with a mean average of 4.06. These results indicate that the developed technology is considered useful and highly recommended. In [14] the authors focus on a case study involving the application of fuzzy control methods to hydroponic strawberry crops (Fragaria vesca). The results demonstrate that, in this approach, the hydroponic strawberry crops exhibit increased foliage and larger fruit size compared to natural cultivation systems that rely on default irrigation and fertilization practices without considering variations in the mentioned variables. The conclusion drawn is that the combination of modern agricultural techniques, such as hydroponics, along with diffuse control, enables the enhancement of crop quality and optimization of required resources. The authors in [16] presented an approach based on the Internet of Things (IoT) for observing and managing a hydroponic system. The system incorporates a variety of sensors to evaluate crucial factors such as temperature, humidity, nutrient levels, pH, and water levels. The sensors are connected to a microcontroller or single-board computer, which collects and transmits the data to a cloud-based platform for storage and analysis purposes. A web or mobile application enables users to remotely access the data and efficiently manage the system. The system is equipped with automated control functionality, which adjusts humidity, temperature, and nutrient levels based on sensor data and predefined criteria. The proposed system offers a comprehensive and efficient approach for regulating and monitoring hydroponic systems, and it can be readily adapted to different hydroponic conditions and systems. Through experiments and obtained results, they demonstrate the practicality and effectiveness of their proposed system.

## 3 Our Suggested Approach

### 3.1 The System Design

This work is the extension of our work [20]. The composition of our system design (as shown in Fig. 1) consists of:

- Wireless sensor network system.
- Hydroponic irrigation system.
- Renewable energy power system.
- Internet system.

The wireless sensor network system comprises the following components:

1. *Water level sensors:* Measure the level of the water. We have, in our design, two water level sensors: one in the water tank and the other in the hydroponic tank.
2. *Ph water sensors:* Measure the Ph in the water.
3. *Temperature water sensor:* Measure the temperature of the water.
4. *Cluster head:* Gathers all the data collected by sensors and sends it together.
5. *Microcontroller:* Sends the data to the farmer via Internet.
6. Actuators in:
   a. **Cooling system:** Cools the water of the hydroponic tank.
   b. **Heating system:** Heats up the water of the hydroponic tank.
   c. **Drones:** Add various liquid solutions to regulate the Ph of the water in the hydroponic tank.
   d. **Water pump:** Transports water from the well to the tank, if the tank is empty of water.
   e. **Electro-valve:** Moves water from the tank to the hydroponic tank via the sprinklers.

The hydroponic irrigation system functions as a circuit that is closed. It contains:

1. Well.
2. Water pump.
3. *Water tank:* Stores water coming from the well and the hydroponic tank.
4. Electro-valve.
5. *Sand filter:* Filters the sand from the water before distributing it from the tank.

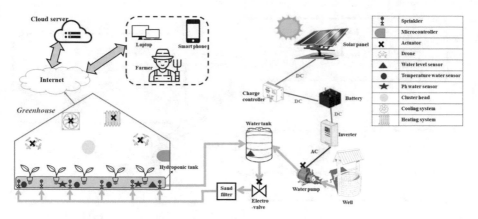

**Fig. 1.** Our proposed system design.

6. *Hydroponic tank:* Contains the hydroponic plants.
7. *Sprinklers:* Distribute the water in the hydroponic tank.

The renewable energy power system contains as show in [20].
And, the Internet system consists of:

1. *Internet:* Transforms the collected data from the greenhouse to the farmer and cloud server.
2. *Farmer:* Evaluates the data by comparing it to the predetermined threshold and sends a command to the associated actuator using a laptop or a smart phone.
3. *Cloud server:* To save the captured data in a database.

We can implement our proposed approach in any region. But, to implement it, we need the four systems with materials mentioned above.

### 3.2 Control System

Before to start controlling the system, you need to know:

1. The type of the greenhouse and the region.
2. The sort of plant developed in a greenhouse and the season to define the threshold.

Initially, the system verifies the condition of our solar power (SP) and contrasts it against a predefined target value. In the event that our SP falls below the designated threshold, we activate an alternative energy source, such as wind energy.

After our system is activated (Fig. 2), the deployed water level sensors, temperature water sensors (TW) and Ph water sensors (Ph) gather the data, save it in a database (DB) located in a cloud server and transmit it to the cluster-head (CH). Then, the CH gathers all the data collected by sensors and sends it together to the microcontroller. The microcontroller sends the data to the farmer via Internet. The farmer evaluates the data by comparing it to the predetermined threshold and sends a command to the associated actuator using a laptop or a smart phone. The new data will be saved in a DB.

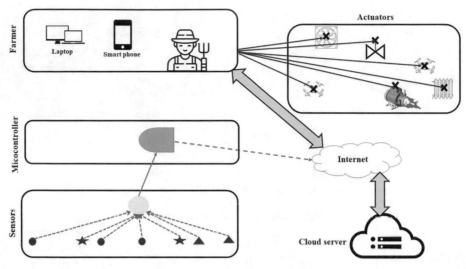

**Fig. 2.** The network architecture.

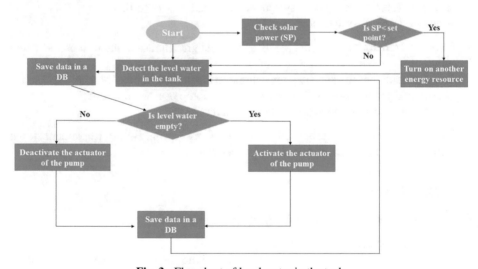

**Fig. 3.** Flowchart of level water in the tank.

If the level of the water in the tank (Fig. 3) is empty, the farmer activates the actuator of the water pump to replenish the tank from the well. Else, the farmer deactivates the actuator of the water pump.

If the level of the water in the hydroponic tank (Fig. 4) is empty, the farmer activates the actuator of the electro-valve to refill the tank from the water tank. Before refilling the hydroponic tank, the water will be filtering from the sand using a filter. If the level of the water in the hydroponic tank is full, the farmer deactivates the actuator of the electro-valve.

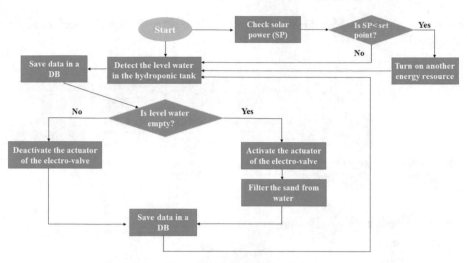

**Fig. 4.** Flowchart of level water in the hydroponic tank.

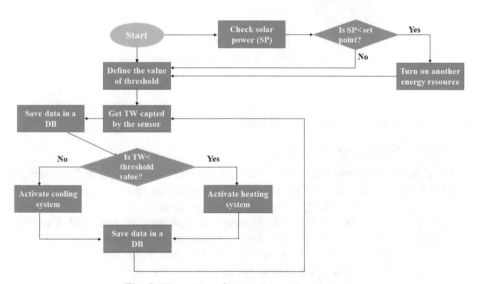

**Fig. 5.** Flowchart of temperature water control.

If the TW (Fig. 5) is below the threshold (=25 °C), the farmer activates the actuator of the heating system for keeping the TW above the threshold. Else, the farmer activates the actuator of the cooling system for cooling the water.

A liquid solution with a pH below 7 is categorized as acidic, whereas solutions with a pH above 7 are identified as basic or alkaline solutions. If the Ph (Fig. 6) is below the threshold (=7.0), the farmer activates the actuator of drone 1 where drone 1 contents a product based on NaOH (Sodium hydroxide) or KOH (Potassium hydroxide). If the Ph is above the threshold, the farmer activates the actuator of drone 2 where drone 2 contents a product based on $H_2SO_4$ (Sulfuric acid), $H_3PO_4$ (Phosphoric acid) or $HNO_3$ (Nitric acid).

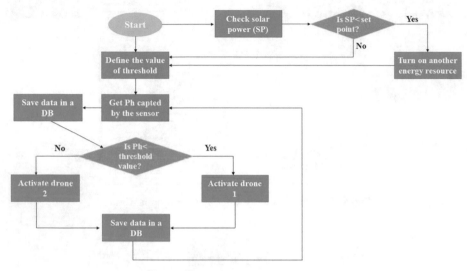

**Fig. 6.** Flowchart of Ph water control.

### 3.3   Advantages of the Proposed System

The benefits of this system as follows:

1. *Clear and easy system to use:* The system is designed to be user-friendly, making it easy for farmers to understand and operate. Its interface and controls are intuitive, allowing for efficient management of hydroponic plants.
2. *High quality of crops:* By optimizing water use and providing controlled weather conditions, the system ensures that crops receive the appropriate amount of water at the right time. This helps in promoting healthier and higher-quality crop growth, leading to better yields and improved agricultural productivity.
3. *Optimization of water use:* The system is equipped with advanced technology that allows for precise monitoring and control of water usage. It takes into account factors such as weather conditions, and crop requirements, ensuring that water is efficiently utilized and minimizing wastage.
4. *Automation of the irrigation process:* The system automates the irrigation process, reducing the need for manual intervention. This saves time and labor for farmers, allowing them to focus on other important tasks while ensuring that plants receive the necessary water for optimal growth.
5. *Sustainable and renewable energy:* The system is based on sustainable and renewable energy sources, such as solar power. Bu utilizing clean energy, it reduces reliance on non-renewable sources and minimizes environmental impact, making it an eco-friendly solution for irrigation of hydroponic plants.

Overall, this system offers a range of advantages that enhance agricultural practices, promote efficient water management, and support sustainable farming methods.

## 4 Conclusion and Perspectives

This paper introduces a proposed design for monitoring hydroponic plants in a greenhouse, inspired by exiting studies. Our proposed design incorporates wireless sensor networks (WSN), hydroponic irrigation system, renewable energy system (RnE), and Internet of Things (IoT) technologies. Additionally, we present algorithms responsible for controlling water levels in the tank and hydroponic tank, as well as monitoring water temperature and pH levels.

Our future plans involve implementing this design by utilizing affordable and available components to establish a hydroponic plant in a greenhouse. Furthermore, we aim to develop the required software for remote control and surveillance, disease/pest detection, prediction, and any other intelligent algorithms essential for achieving optimal monitoring.

## References

1. F. a. A. Organization: The Future of Food and Agriculture—Trends and Challenges. Food and Agriculture Organization of the United Nations, Rome, Italy (2017)
2. Kirci, P., Ozturk, E., Celik, Y.: A novel approach for monitoring of smart greenhouse and flowerpot parameters and detection of plant growth with sensors. Agriculture 12(10), 1705 (2022). https://doi.org/10.3390/agriculture12101705
3. Touhami, A., Benahmed, K., Parra, L., Bounaama, F., Lloret, J.: An intelligent monitoringof greenhouse using wireless sensor networks. Smart Struct. Syst. 26(1), 117–134 (2020). https://doi.org/10.12989/sss.2020.26.1.117
4. Achouak, T., Khelifa, B., García, L., Parra, L., Lloret, J., Fateh, B.: Sensor network proposal for greenhouse automation placed at the South of Algeria. Netw. Protoc. Algorithms. 10(4), 53–69 (2019). https://doi.org/10.5296/npa.v10i4.14155
5. Maraveas, C., Bartzanas, T.: Application of internet of things (IoT) for optimized greenhouse environments. AgriEngineering. 3(4), 954–970 (2021). https://doi.org/10.3390/agriengineering3040060
6. Ahmad, B., et al.: Evaluation of smart greenhouse monitoring system using raspberry-pi microcontroller for the production of tomato crop. J. Appl. Res. Plant Sci. 4(1), 452–458 (2023). https://doi.org/10.38211/joarps.2023.04.01.54
7. Touhami, A., Benahmed, K., Bounaama, F.: Monitoring of greenhouse based on internet of things and wireless sensor network. In: Bouhlel, M.S., Rovetta, S. (eds.) SETIT 2018. SIST, vol. 147, pp. 281–289. Springer, Cham (2018). https://doi.org/10.1007/978-3-030-21009-0_27
8. Tamboli, A.: Build Your Own IoT Platform: Develop a Flexible and Scalable Internet of Things Platform. Apress, Berkeley, CA (2022). https://doi.org/10.1007/978-1-4842-8073-7
9. Bourouis, A., Benahmed, T., Mokeddem, K., Benahmed, K., Lairedj, A.: IoT for smart apiculture: issues and solutions. In: Proceedings of the 3rd International Conference EDiS (2022)
10. Krishnan, S., Rose, J.B.R., Rajalakshmi, N., Prasanth, N.: Cloud IoT Systems for Smart Agricultural Engineering. CRC Press (2022)

11. Arif, S., Fathurrahmani, F.: The prototype of the greenhouse smart control and monitoring system in hydroponic plants. Jurnal Teknologi Informasi dan Komunikasi. **10**(2), 131–143 (2019). https://doi.org/10.31849/digitalzone.v10i2.3265
12. Akhil, A., Prakash, G.: Automated hydroponic drip irrigation using big data. In: Proceedings of the 2$^{nd}$ International Conference on Inventive Research in Computing Applications, pp. 370–375 (2020)
13. Baihaqi, S., Syahril, E., Heru, P., Roy, G., Ulfi, A., Fahmi, F.: Remote monitoring system for hydroponic planting media. In: The International Conference on ICT for Smart Society (2017)
14. Olalla, E.M., Flores, A.L., Zambrano, M., Limaico, M.D., Iza, H.D., Ayala, C.V.: Fuzzy control application to an irrigation system of hydroponic crops under greenhouse: case cultivation of strawberries (fragaria vesca). Sensors **23**(8), 4088 (2023). https://doi.org/10.3390/s23084088
15. Surantha, N.: Intelligent monitoring and controlling system for hydroponics precision agriculture. In: 7t$^{h}$ International Conference on Information and Communication Technology (2019)
16. Thakur, P., Malhotra, M., Bhagat, R.M.: IoT-based monitoring and control system for hydroponic cultivation: a comprehensive study. Research Square (2023). https://doi.org/10.21203/rs.3.rs-2821030/v1
17. Untoro, M., Hidayah, F.: IoT-based hydroponic plant monitoring and control system to maintain plant fertility. INTEK Jurnal Penelitian. **9**(1), 33–41 (2022). https://doi.org/10.31963/intek
18. Nyoman Crisnapati, P., Nyoman Kusuma, W., Komang Agus Ady, A., Hermawan, A. Hommons: hydroponic management and monitoring system for an IOT based NFT farm using web technology system. In: 2017 5th International Conference on Cyber and IT Service Management (CITSM) (2017)
19. Bernardino, J., Cajoles Doctor, A.: Development of IoT smart greenhouse system for hydroponic gardens. Int. J. Comput. Sci. Res. **7**, 2111–2136 (2023). https://doi.org/10.25147/ijcsr.2017.001.1.149
20. Touhami, A., Touhami, S., Touhami, N., Benahmed, K., Bounaama, F.: Design of smart irrigation system in the greenhouse using WSN and renewable energies. In: Hatti, M. (ed.) Advanced Computational Techniques for Renewable Energy Systems, pp. 126–131. Springer International Publishing, Cham (2023). https://doi.org/10.1007/978-3-031-21216-1_14

# Implementation of Smart Parking Using an IoT Based System

Dalila Cherifi$^{(\boxtimes)}$, Assil Belhacini, Mohamed Reda Boulezaz, and Menouar Barkat

Institute of Electrical and Electronic Engineering, University of Boumerdes, Boumerdes, Algeria
da.cherifi@univ-boumerdes.dz

**ABSTRACT.** Due to the proliferation in the number of vehicles to control on the road, traffic problems are inevitable. This is due to the fact that the current transportation infrastructure and car park facilities are unable to cope with the influx of vehicles on the road. *In this work, we are heading to automate the Parking process using an IOT based system. It consists of three main parts: The self-parking car, the mobile application, and the smart parking. The late gathers information about the current state of the available spots by checking the measurements received from sensors, and then it sends the data to the user through database that could be accessed via mobile application so the user can visualize his destination and choose the appropriate spot, the final step of the process will be done when the application gives instructions to the car which will follow them to park automatically without the need of manual assistance.*

**Keywords:** Smart Parking · IoT · Automate · Sensors

## 1 Introduction

Throughout the years, the number of cars is continuously increasing; recently a remarkable growth in the new registered cars has been shown. According to a research paper published in 2016 by the "US auto-industry giant WardsAuto" there are 1.42 billion cars worldwide, which makes it roughly a car for each 5 people [1]. However, the numbers are in fact a lot bigger in big urban cities where the population is much concentrated, which creates problems in many aspects such as traffic jams and crowded parking areas. Various measures have been taken in the attempt to overcome the traffic problems. Although the problem can be addressed via many methods.

One of the critical concerns is the issue of parking those vehicles. The increasing personal vehicles than public transport make parking slots may not be enough to support all those vehicles. Even when there are enough parking slots, it still takes a long time to find an available slot because the people do not know the exact location of the available parking slot. Consequences of this will increase fuel consumption thus not eco-friendly. It is very difficult and frustrating as well to find parking space in most metropolitan areas, especially during the rush hour such as mall, hospital and so on. Therefore, how can we organize the process of parking in a way such that we achieve to save time and

M. Hatti (Ed.): IC-AIRES 2023, LNNS 984, pp. 59–65, 2024.
https://doi.org/10.1007/978-3-031-60629-8_6

fuel? The objective of this work is to develop and implement the car park management system introduced, which is the smart parking system.

This article consists of four sections. The first section is a review and a description about smart parking and the impact of IOT systems on this field. The second section explains the parking implementation process and describes the hardware used to achieve it, and it emphasizes the development of a self-parking car. The third section includes the software design which is composed of: parking control, database, and user interface. The final fourth section is a discussion of the results obtained from our experimental implementation, then the article ends with a general conclusion.

## 2   Review about Smart Parking

Parking problems are widely common, especially for big cities. By 2023, market spending for smart parking products and services is expected to grow at 14% and surpass $3.8b according to an IoT analytics report [2] the growth of market spending is good news because it will force people to try to find a solution to these traffic problems instead of taking no action. The smart parking systems were initially implemented in Europe, USA and Japan, but later on as the other countries started developing, these smart parking systems are being installed in these countries as well. With further advancement in the smart parking systems the problem of finding vacant spaces and all the hassle is going to be deprived. IOT is the network of physical devices embedded with electronics, software, sensors, actuators and network connectivity which have the ability to identify, collect and exchange the data. Each thing is uniquely identifiable through its embedded computing system and able to interoperate within the existing internet infrastructure [3].

## 3   Smart Parking Implementation

We have proposed an IOT aware automatic smart parking system. The system is implemented by using sensors technologies for occupancy checking of the parking slots and Microcontroller for storing data and perform the connectivity with the IoT gateway. The internet of things (IoT) is a widespread technology across all industries providing features like monitoring, analysis, prediction and control. It is an immersive, ambient networked computing environment built through the relentless proliferation of smart sensors, cameras, software, databases, and huge data centers. An IoT architecture can be categorized into sensors or actuators, internet gateway, edge IT and data center, and cloud. Microcontrollers are designed to enable connectivity and control in all the things which could be connected to the internet, and they can be integrated into almost anything ranging from industrial equipment such as automobile engine control systems, implantable medical devices, remote controls, office machines, and other embedded systems [4]. To realize this system, we made a parking system prototype, it consists of 10 spots with different sensors to detect presence of vehicles on each spot (Fig. 1).

- **Infrared obstacle detector HW-201:** This type of sensor uses reflective properties of IR that can be used due to its various features which are the less power consumption, easy to interface and compatibility with different microcontrollers also because it requires only three pins to connect it: + 5v, GND, and control pin.

**Fig. 1.** Data flow diagram.

- **Ultrasonic HC-SR04 sensor:** This sensor can both transmit and receive the signals. The received signals are interpreted, measuring the time interval between sending and receiving of the signal, and frequency of the received signal. This is then used to determine the presence of a vehicle [5]. we have used this sensor due to its outstanding performance and accuracy and fast response because of its high-end specifications.
- **Pressure sensor:** A Force Sensor is defined as a transducer that converts an input mechanical load, weight, tension, compression or pressure into an electrical output signal (load cell definition) [6].
- **Image processing detection**: An image processing algorithm is used to detect empty parking areas from aerial images of the parking space. The algorithm processes the image, extracts occupancy information concerning spots, and their positions. The system also reports if individual parking spots are occupied or not. The image segmentation technique is applied to the acquired image, and subsequently, the image is enhanced, and the parking slot is detected. It is essential to mention that employing the same procedure for the detection of vacant slots. The pre-processing of an image includes the gray scale conversion and filtering of the input images that aims at selectively removing the redundancy present in captured images without affecting the details of the original information of an input image. Image pre-processing that may have huge positive effects on the quality of the image segmentation and the feature extraction [7].
- **Microcontroller**: These sensors are all directly connected to a microcontroller to provide it with data. After it updates the data accordingly, then it will send it to the database. The microcontroller used in our prototype is Raspberry pi 4 model b [8]. Raspbian OS is official Operating System available for free to use. This OS is efficiently optimized to use with Raspberry Pi. Raspbian have GUI which includes tools for Browsing, Python programming, office, games, etc. Due to these characteristics, we deduce that the raspberry pi 4 model b is the more suitable for implementing the parking system, because the build in Wireless networking and high performance of the CPU, and the compatibility with all the sensors that can provide us information about the spots.
- **Autonomous CAR**: A self-driving car, also known as an autonomous vehicle (AV or auto), or driverless car, it is a vehicle capable of sensing its environment and operating without human involvement. A human passenger is not required to take control of the vehicle at any time, nor is a human passenger required to be present in the vehicle at all. An autonomous car can go anywhere a traditional car goes and do

everything that an experienced human driver does [9]. The self-parking car is classified in the level three of driving automation scale, because this segment includes cars
that automate only the parking process as a task with few other detection capabilities.
The general block diagram of the complete hardware implementation is represented
in the following (Fig. 2):

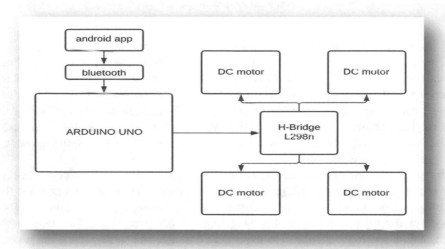

**Fig. 2.** Car schematic diagram.

In order to control and move our car, we used the open-source Arduino Software
(**IDE**), which makes it easy to write code (a set of C/C + + instructions) and upload it to
the board, by using built-in functions and libraries, as well as the possibility to include
more libraries available on the net [10]. It runs on Windows, Mac OS X, and Linux.
After wiring, we are ready to make the Arduino code for enabling the communication
between the Arduino board and the smartphone. As follow we can read the incoming
data from the phone using the built-in function **Serial.read()**, and store it in a variable.
After the car receives the empty position via Bluetooth it executes the instruction which
is a custom function consisting of a movement set that indicates the appropriate path
which leads to the desired destination.

## 4   Database and User Interface

This section discusses the software design of the automated parking going from getting
information of the spots to storing them in database and the final step is to deliver this
information to the end user through UI (user interface) which is in this case an android
application.

- **Data Processing:** The data collected by the sensor devices and organized by the raspberry pi must be stored in a proper form so that it can be transformed into the information we are looking for and give access to store, retrieve, and change this data. Database is the right place to manage this information because of its ideal structural architecture.
- **Storing Data:** For secure and fast dataflow, we used POSTGRESQL DB, which is a powerful and open-source object-oriented relational database management system that is able to securely support the most complex data workloads, the tool to handle data conception is DBMS (Database management system).
- **User Interface: Mobile application:** The android app was built from scratch with android studio using Java language it has three main functionalities. The launcher activity consists of the list of Bluetooth paired devises, so that we can connect with our HC-05 Bluetooth module under the name SPP-CA and default password 1234. If there is no Bluetooth connection and the aim is just to check the parking status and drive manually to an empty spot, the user can skip this step by clicking the upper button. The user can see graphically the state of each spot in the parking. Moreover, he has the ability to choose the desired parking zone, and the button get spots is to refresh the data. The interface is presented in Fig. 3.

**Fig. 3.** Parking screen layout.

## 5   Experimental Results and Discussion

In this part, we will go through our implementation results, and we will discuss the efficiency of our prototype and how accurate it can be, in order to test the possibility of its realization in the real world. In the beginning, we can notice in Fig. 4 the initial state of the parking spots is vacant; we could visualize it from the user interface of the application.

After clicking on the desired spot, the car immediately received the instruction via Bluetooth connection, thus the car started moving towards that position as shown in Fig. 5. The average speed measured of the car was near 3.6 km/h. that is a reasonable

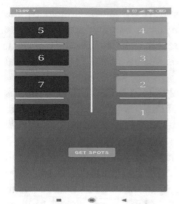
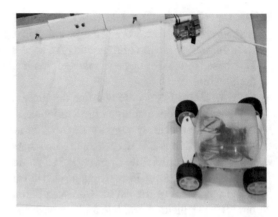

**Fig. 4.** Initial Position of the car.

**Fig. 5.** Parking in progress.

speed due to the car size and the power of the motors used in our car; however, this speed can be adjusted according to the car size by changing the power supplied using code.

When the car finally arrived at the desired position, it activated the parking function and automatically stopped, then, to check whether the parking has updated the spots status, we opened the application and we noticed that a change has occurred with an occupied spot at position 2, which is where our car has parked. The update time of the database was near 4 s and that was due to 2 s of delay between each checking cycle sent by the microcontroller to the database and 2 s of information transferring by HTTP requests. From this experiment, we conclude that our implementation was successful at the software side (Fig. 6).

 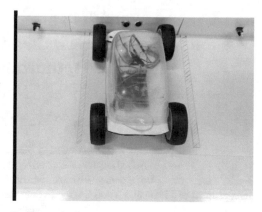

**Fig. 6.** Car parked.

## 6    General Conclusion

In this work, we have presented an IoT-based smart parking system to address urban parking challenges. Our system includes a self-parking car, a mobile app, and smart parking infrastructure. It uses various sensors to monitor parking spot occupancy, and the mobile app provides real-time parking spot information. The self-parking car autonomously parks in the chosen spot.

Our experiments showed successful navigation and real-time updates. Our system has the potential to alleviate urban parking problems, reduce fuel consumption, and enhance transportation efficiency. It holds promise for future smart city applications and urban planning.

## References

1. Cutcher-Gershenfeld, J., Brooks, D., Mulloy, M.: The Decline and Resurgence of the U.S. Auto Industry. Economic Policy Institute, Bulgaria (2015). https://policycommons.net/artifa cts/1414280/the-decline-and-resurgence-of-the-us/2028542/. Accessed 03 May 2024. CID: 20.500.12592/p60242
2. Ashhwath, C., Rohitram, V., Sumathi, G.: Smart parking system using MQTT communication protocol and IBM cloud. J. Phys. Conf. Ser. **2115**(1), 012013 (2021)
3. Gillis, A.S.: What is IoT (internet of things) and how does it work? IoT Agenda, TechTarget 11 (2020)
4. Lutkevich, B.: Microcontroller (MCU). IoT Agenda (2019)
5. Components101: Ultrasonc Sensor working. https://components101.com/sensors/ultrasonic-sensor-working-pinout-datasheet. Accessed 03 May 2024
6. What is a force transducer, what are the different types of force transducers and how do they work? https://www.futek.com/force-transducer. Accessed 03 May 2024
7. Krig, S.: Computer Vision Metrics: Survey, Taxonomy, and Analysis, pp 39–83. Springer Nature (2014). https://doi.org/10.1007/978-1-4302-5930-5
8. Raspberry, P.I.: http://www.raspberrypi.org. Accessed 03 May 2024
9. SYNOPSYS: What is an Autonomous Car. https://www.synopsys.com/automotive/what-is-autonomous-car.html. Accessed 03 May 2024
10. Arduino.cc: The Arduino software IDE. https://docs.arduino.cc/learn/starting-guide/the-ard uino-software-ide. Accessed 03 May 2024

# Dynamic Load Balancing in Cloud-IoT Based on Neural Network Classification and GGA Optimization

S. Benabbes[1]([✉]) and S. M. Hemam[2]

[1] ICOSI Laboratory, Department of Mathematics and Computer Science,
Abbes Laghrour University, Khenchela, Algeria
benabbes.sofiane@univ-khenchela.dz
[2] Department of Mathematics and Computer Science, Abbes Laghrour University, Khenchela,
Algeria
hemam.sofiane@univ-khenchela.dz

**Abstract.** The Cloud IoT paradigm, designed to combine Cloud Computing (CC) and the Internet of Things (IoT) benefits, is increasingly utilized for extensive services and addressing users' connectivity, data processing, and storage needs. However, achieving load balancing in Cloud-IoT setups remains challenging due to dynamic and diverse connected objects and Cloud resources. This paper presents a novel approach that employs neural networks for task classification from connected objects, coupled with a task scheduler merging genetic and grasshopper optimization algorithms. This aims to dynamically balance loads in the Cloud-IoT environment. The proposed method is compared with recent alternatives, assessing metrics like makespan, throughput, resource use, energy consumption, and cost. Experimental results compellingly demonstrate the superiority of our approach across these factors.

**Keywords:** IoT · Cloud Computing · Cloud-IoT · Dynamic load balancing · SGGA · Neural networks

## 1 Introduction

The Internet of Things (IoT) and Cloud Computing are two emerging paradigms in distributed computing. The IoT enables the transformation of everyday objects into intelligent entities (Benabbes, S. and Hemam, S. M., 2020), while Cloud Computing provides a scalable and reliable infrastructure for data processing and storage. The fusion of these two technologies has led to the emergence of CloudIoT, which offers a number of advantages over traditional IoT solutions (Botta, A. and Donato, W. and Persico, V. and Pescapé, A., 2016). One of the challenges of CloudIoT is load imbalance, which can occur when IoT devices generate workloads that are unevenly distributed across the available resources. This can lead to resource underutilization and prolonged response times (Shradha, J. and Jayshree, J. and Chandraprbha, K., 2019). To address this challenge, we propose a novel approach based on task classification and task scheduling. Our approach first classifies tasks based on their characteristics and priorities. This classification helps to group similar tasks and gain a better understanding of their resource needs. We then use a genetic algorithm and grasshopper optimization algorithm to schedule tasks at both the Cloud and IoT levels (Benabbes, S. and Hemam, S. M., 2023). This scheduling takes into account the current resource loads and specific constraints such as cost and energy consumption. Our approach offers several advantages. First, task classification reduces the search space and better manages workload variability. By grouping similar tasks, we can predict their resource requirements more accurately and distribute them evenly. Second, the use of the genetic algorithm and grasshopper optimization algorithm enables efficient task scheduling while considering system constraints. The outcomes acquired reveal that, in terms of achieving the optimal solution timeframe, makespan, average resource utilization ratio, energy consumption rate and cost, the suggested approach demonstrates greater efficiency when contrasted with the latest methodologies (BGA, HGOW-ABC, GWO, and ACO) in Cloud environment and (GTO, TGA, BGA and Jellyfish) in IoT environment.

The organization of our paper unfolds as follows: commencing with the introduction, the subsequent section (Sect. 2) will delve into relevant prior research. The mechanics of the proposed approach and its distinct components will be outlined in Sect. 3. Functionality of the proposed approach will take the spotlight in Sect. 4, followed by the presentation and discussion of experimental results in Sect. 5. Finally, our paper will draw to a close with a conclusive section (Sect. 6).

## 2 Related Work

Related works constitute a pivotal segment of any research study or project. They enable us to position our work within a broader context and showcase how it aligns with current trends and recent developments in our field of interest. Reviewing related works is also crucial for pinpointing gaps in existing research and identifying opportunities for new investigations. In this paper, we will meticulously examine relevant related works pertinent to our study. The central focus of our work revolves around two main aspects: task classification sent by IoT devices and dynamic load balancing in Cloud Computing. We commence by scrutinizing prior research that has centered on the same dimensions as our study, delving into the methodologies, approaches, and outcomes of these endeavors. Lastly, we present our own approach and determine the final trajectory for our research.

### 2.1 Classifications in the IoT Environment

To classify tasks or -in general- data within the IoT environment, several studies and approaches have employed various techniques.

Jairam, N., (2020) introduced a system called Classification and Scheduling of Information-Centric IoT Applications (CS_IcIoTA). It identifies application needs and categorizes them into diverse classes. The scheduler assigns tasks from these categories either to local fog nodes or to remote leased cloud nodes for execution. This is based on the tasks' current resource requirements and parameter vectors. Simulation results indicate that CS_IcIoTA minimizes the average time interval and service cost by up to

11.45% and 10.60%, respectively. It also maximizes the average fog node utilization to 77.83%. Authors Nwogbaga, N. E. and Latip, R. and Affendey, L. and Suriani, R. and Amir, R.A., (2021) put forth a two-pronged approach. The first strategy involves classification using a convolutional neural network to ascertain the impact of canonical polyadic decomposition on data. The second strategy is a rank accuracy estimation model. The outcome of the proposed method indicated that the suggested methods outperformed deep learning layers in terms of data compression. Nascita, A. and Cerasuolo, F. and Monda, D.D. and Garcia, J. T. A. and Montieri, A. and Pescape, A., (2022) employed state-of-the-art deep learning-based traffic classifiers, evaluating their efficacy in IoT attack classification. This aimed to distinguish different attack classes from benign network traffic. Experimental results underscored the need for advanced deep learning architectures tailored with specific input data for IoT attack classification. Cunha, A. and Borges, J. and Loureiro, A., (2022) proposed a Convolutional Neural Network (CNN) model that classifies attack types following accurate botnet detection evaluated with the N-BaIoT and Bot-IoT datasets. The model achieved F1-scores and precision rates of 98% for N-BaIoT and 100% for Bot-IoT. Solatidehkordi, Z. and Ramesh, J. and Al-Ali, A.R. and Osman, A. and Shaaban, M., (2023) presented a classification of smart home devices using the Long Short-Term Memory (LSTM) deep learning architecture trained on the latest version of the Plug-Load Appliance Identification Database (PLAID). The results and information are accessible to end-users or utility providers through a mobile application connected to the same database.

## 2.2  Load Balancing in the Cloud Computing Environment

In this section, we delve into the most recent works that address the issue of load imbalance within the Cloud environment.

Several swarm intelligence algorithms and behavior-based algorithms have been proposed and applied for task scheduling in the Cloud environment. There are two types of these algorithms: (1) those based on exploiting the best solution among previous results, known as local search, and (2) those based on exploring new regions of the solution space or suddenly prospecting a new solution search space. The most interesting works in this context are reviewed below. We first present the category of behavior-based algorithms, followed by the swarm intelligence algorithms. Authors Gulbaz, R. and Siddiqui, A.B. and Anjum, N. and Alotaibi, A.A. and Althobaiti, T. and Ramzan, N., (2021) introduced the Balancer Genetic Algorithm (BGA) to enhance makespan and load balancing. BGA incorporates a load balancing mechanism that considers the actual load assigned to virtual machines. The need for multi-objective optimization for improved load balancing and makespan is also emphasized. Simulation demonstrated significant enhancement in makespan, throughput, and load balancing. Benabbes, S. and Hemam, S. M., (2023) proposed a scheduler based on the pairing of the genetic algorithm and the grasshopper optimization algorithm for dynamic load balancing in CloudIoT, where the AG's mutation component was replaced by a grasshopper optimizer. This approach improved makespan, quality of service, throughput, and resource utilization. In the category of swarm intelligence algorithms, numerous works exist. In this chapter, we cite the most recent and significant ones. Ouhame, S. and Hadi, Y. and Arifullah.A., (2020) integrated the Grey Wolf Optimizer (GWO) with Artificial Bee Colony (GWO-ABC) to enhance resource allocation in the Cloud. This technique improved load balancing parameters in Cloud Computing by 1.25%. Shafahi, Z. and Yari, A., (2021) employed the Ant Colony Optimization (ACO) algorithm for dynamic task scheduling. The scheduler behaves like an ant searching for food. The results of experimental simulations for both approaches exhibited superior performance compared to others. They reduced task execution times, improved system resource utilization, and maintained system balance. This section has provided a comprehensive review of previous research and recent developments in our field of interest. We have examined studies that focused on similar axes, as well as approaches that addressed related aspects of our subject. Our hybrid approach for dynamic load balancing has been proposed, comprising three levels: A Recurrent Neural Network (RNN) to classify tasks sent by IoT devices, and two task schedulers—one at the IoT level and the other at the Cloud level. The scheduler at the IoT level is designed to pair with the Genetic Algorithm (GA) and the Grasshopper Optimization Algorithm (GOA), creating the Scheduler GA-GOA Algorithm (SGGA). This load balancer aims to enhance five parameters: makespan, throughput, energy consumption, cost, and resource utilization (QoS).

## 3   Proposed Architecture

In this section, we present our approach that enables dynamic load balancing in CloudIoT. The proposed approach consists of three main phases. In the first phase: Tasks captured by Internet of Things (IoT) devices are initially sent to a classifier. This classifier is responsible for creating three classes of tasks using a supervised classification technique. The first class contains notification tasks (tasks that require no resources), the second class contains tasks to be executed on other IoT devices, and the third class pertains to tasks that will be executed on Cloud Computing. In the second phase: We focus on tasks sent to IoT devices, which are received by a scheduler. This scheduler must ensure dynamic load balancing while minimizing global execution time and energy consumption on one hand, and maximizing resource utilization (Quality of Service - QoS) on the other hand. In the third phase: We address tasks sent by the scheduler to the Cloud. The role of the scheduler is to ensure dynamic load balancing by minimizing global execution time (makespan), throughput, energy consumption, cost, and maximizing resource utilization. This architecture is illustrated in Figure (e.g., Fig. 1). It comprises IoT devices, virtual machines, a classifier, and a load balancer used at two levels. In the following sections, we describe the components of the proposed architecture.

### 3.1   Connected Object (IoT)

IoT devices send their tasks for processing to the classifier. The tasks are characterized by their sets of instructions, identifiers, size (KB), worst-case execution time (MIPS), priority (predefined order from 0 to 5), and the requested resource for task processing (predefined list of 0 to 150 resources).

### 3.2   Virtual Machine (VM)

A virtual machine consists of a set of software and hardware geographically distant. It processes the tasks emitted by the load balancer.

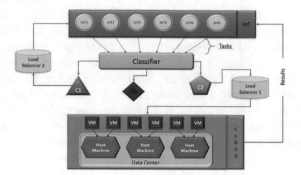

**Fig. 1.** Architecture of the proposed approach.

### 3.3 Classifier

The classifier classifies tasks into three classes based on size, WCET, priority, and the required resource using the proposed neural network.

### 3.4 Load Balancer 1

This component acts as an intermediary between the tasks to be processed and the IoT devices. IoT devices send their tasks for processing to the scheduler (Load Balancer 1). The scheduler selects the most suitable resource, in this case, an IoT device, for each task. To achieve this goal, this component uses the algorithm we proposed, which is a combination of genetic algorithm and grasshopper optimization algorithm.

### 3.5 Load Balancer 2

The role of this component is to dynamically balance the task load at the Cloud level. IoT devices send their tasks to the scheduler (Load Balancer 2), which maps them to different virtual machines. This component works in the same way as Load Balancer 1, but the difference lies in the parameters to be optimized.

## 4 Functionality of the Proposed Model

Our approach operates across three levels. The first level involves classifying tasks into three categories based on size, worst-case execution time, priority, and resource requirement. Class 0 includes tasks that require no processing (notifications), Class 1 contains tasks to be handled by IoT devices, and Class 2 comprises tasks for Cloud Computing. This classification is achieved through the application of the proposed neural network, consisting of three layers, with a hidden layer of 24 neurons, as depicted in Figure (e.g., Fig. 2).

The second level takes place at the IoT devices. Once tasks are classified, Class 1 contains tasks sent to the SGGA2 load balancer. This load balancer orchestrates the scheduling of these tasks, minimizing overall execution time and energy consumption.

The third level addresses load balancing at the Cloud level. In this context, SGGA1 is invoked to achieve load balancing by mapping tasks to different virtual machines. This minimizes overall execution time (makespan), improves throughput, reduces energy consumption, cuts costs, and maximizes resource utilization (QoS).

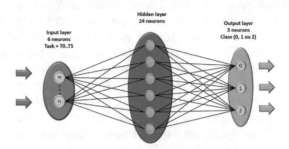

**Fig. 2.** Proposed neural network model.

We have developed a task scheduler based on both the grasshopper optimization algorithm and the genetic algorithm. While the genetic algorithm yields highly satisfactory results due to its flexibility and robustness, it encounters limitations in terms of its mutation operator (Benabbes, S. and Hemam, S. M., 2023), particularly as task numbers increase. To address this limitation, we propose substituting the mutation operator with the grasshopper optimization algorithm. This algorithm is categorized as a novel metaheuristic algorithm, drawing inspiration from the behavior of grasshoppers (Benabbes, S. and Hemam, S. M., 2023). The detailed architecture is summarized in Figure (e.g., Fig. 3).

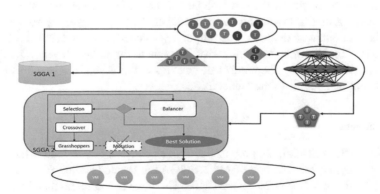

**Fig. 3.** Detailed architecture of the proposed approach.

In the following subsections, we elucidate the functioning of each component within the proposed architecture.

## 4.1 The Classifier

The proposed classifier is a neural network with three layers. To train the proposed neural network, we synthesized a dataset that facilitates model training and ensures the credibility of its classes. The Task vector constituting the dataset is defined as follows:
   Task (task_id, task_size, task_WCET, task_priority, task_resource, task_class).
   Where: task_id is the task identifier, task_size is the task size (KB), task_WCET is the worst-case execution time of the task (s), task_priority is the task priority ranging from 0 to 5, with 0 being the highest priority, task_resource is the resource required by the task, ranging from 0 to 150 based on a predefined list of resources, and task_class represents the task class, initialized to 0. To complete the dataset synthesis, we created three classes: Class 0 is created with the condition task_resource = 0, while the other two classes are generated randomly. To validate our dataset, we applied an intuitive supervised classification approach commonly used in machine learning, known as the k-nearest neighbors algorithm (KNN) with k = 3.
   We defined a scoring function (Score) based on the four parameters of a task (task, WCET, priority, and resource). Each parameter is multiplied by a predefined weight, allowing differentiation in terms of importance between parameters. In our case, the importance follows the order: size, priority, WCET, resource. Formula (1) represents the function that calculates the score.

$$Score = (task\_size * 0.5) + (task\_WCET * 0.3) + (task\_priority * 0.4) + (task\_resource * 0.01) \tag{1}$$

After synthesizing the dataset, we proceed to introduce the classifier, which is a neural network. This neural network features a simple architecture and interesting advantages including simplicity, flexibility, speed, the capability to handle large amounts of data and generalize them effectively, and high performance. Our model comprises 3 layers: an input layer, a hidden layer, and an output layer. The input layer represents the task vector, thus containing 6 neurons. The output layer consists of 3 neurons that represent the 3 classes. To determine the number of neurons in the hidden layer, we conducted a series of tests by varying the number of neurons in this layer. The best accuracy was achieved with 24 neurons in the hidden layer.

## 4.2 Balancers

In addition to the classifier, the proposed approach consists of two load balancers (task schedulers), namely SGGA1 and SGGA2. SGGA1 is situated at the IoT level and interacts with them, while SGGA2 is placed at the Cloud level and interacts with the virtual machines. Each load balancer has its own optimization function.
   Both load balancers (SGGA1 and SGGA2) receive tasks sent by the classifier. Tasks from class 1 are mapped to IoTs by SGGA1, while tasks from class 2 are mapped to various virtual machines in the Cloud, as depicted in Figure (e.g., Fig. 4). To achieve robust and reliable load balancing, we've developed a task scheduler that combines both the grasshopper optimization algorithm and the genetic algorithm. The genetic algorithm, due to its flexibility and robustness, yields highly satisfactory results. However,

it suffers from the heaviness of its mutation operator (Benabbes, S. and Hemam, S. M., 2023), especially with an increased task count. To address this limitation, we propose replacing the mutation operator with the grasshopper optimization algorithm (see Figure. e.g., Fig. 4), classified among new metaheuristic algorithms, drawing inspiration from grasshopper behavior (Benabbes, S. and Hemam, S. M., 2023).

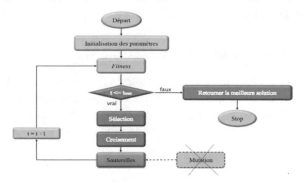

**Fig. 4.** GGA scheduler flow diagram.

The load balancer initializes the position matrix (solutions) randomly. Rows and columns correspond to solutions and tasks, respectively. To illustrate the general operation of the load balancers, let's consider a case where the number of solutions is 100. Each cell in the matrix holds the CPU speed of a VM or IoT. After initialization, SGGA calculates the fitness function of each solution (row) at each iteration. Subsequently, solutions are sorted in ascending order based on their calculated fitness values. The best solution resides at the top of the population (100 solutions). To enhance the quality of the matrix's population, the top 7% of solutions are selected. The genetic algorithm's selection and crossover operators are then applied to produce 50% of new populations. At the end of the iteration, the grasshopper optimization algorithm is employed to generate the second half of the population (the lower 50%) based on values from the first half.

The components SGGA1 and SGGA2 each rely on a fitness function. For the IoT environment, SGGA1 employs the fitness function specified by formula (2) (Benabbes, S. and Hemam, S. M., 2023):

$$f(P_i) = makespan + AvgLoad \qquad (2)$$

And for the Cloud environment, SGGA2 utilizes the fitness function indicated by formula (3) (Jia, L. and Li, K. and Shi, X., 2021).

$$F(P_i) = w1 * TExec\ (P_i) + w2 * IoTLoad\ (P_i) + w3 * CoutExec\ (P_i) \qquad (3)$$

## 5   Experimental Results and Discussion

We implemented our approach in the CloudSim environment and developed the classifier using Python. Subsequently, we proceeded with the validation of our proposal and compared it with recent approaches within the same context.

In this section, we will present various comparison graphs between our approach and methods such as (BGA, HGWO-ABC, GWO and ACO) for load balancing at the Cloud level. Additionally, we will compare our approach with Group (GTO, TGA, BGA and Jellyfish) for load balancing at the IoT level.

To provide the experimental context, we set the number of iterations to 1000 (tmax = 1000), utilized 30 host machines, and employed 50 virtual machines. The total memory of the host machine was set at 16,384 MB.

## 5.1 Experience 1: Evaluation of the Waiting Time for the Optimal Solution

In this experiment, we vary the number of iterations from 100 to 1000 in increments of 100, while keeping the number of tasks fixed at 500. Through Figure (e.g., Fig. 5), it becomes evident that our approach yields superior results in terms of waiting time to reach the optimal solution and the number of iterations. This translates to a significant economic gain.

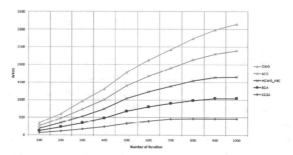

**Fig. 5.** The results of the time needed to reach the optimal cloud solution.

## 5.2 Experience 2: Analysis of Makespan Performance Across Varying Task Counts

The objective of this experimentation is to measure the makespan of our approach and compare it with that of other methods. We varied the number of tasks from 100 to 1000 in increments of 100, while maintaining the number of iterations at 1000. Subsequently, we recorded the maximum time required to execute a set of tasks. Figure (e.g., Fig. 6) reveals that for the first two sets, there is no notable difference between the approaches in terms of execution time. However, starting from the 7th set, our approach stands out significantly due to its superior performance.

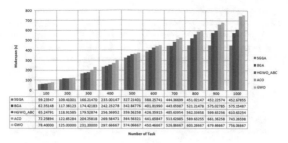

**Fig. 6.** Results of makespan in relation to the number of tasks at the Cloud level.

### 5.3 Experience 3: Evaluation of Average Resource Utilization Ratio

We evaluate the average resource utilization ratio (AVG_UR), a metric ranging from 0 to 1. We analyze this metric based on the number of tasks and compare the results with other approaches. In this experiment, we vary the number of tasks from 100 to 1000 in increments of 100, while fixing the number of iterations at 1000. We then observe the average resource utilization rate based on the number of tasks. Figure (e.g., Fig. 7) shows that our approach outperforms BGA, HGWO-ABC, ACO, and GWO in terms of resource utilization by margins of 3.27%, 0.61%, 6.68%, and 7.26%, respectively. This indicates that SGGA offers superior performance in terms of average resource utilization compared to other approaches.

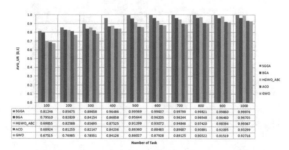

**Fig. 7.** Results of resource utilization rates based on number of tasks at the cloud level.

### 5.4 Experience 4: Comparative Analysis of Energy Consumption in IoT Environment

Energy consumption within the IoT environment poses another challenge. Hence, we evaluate this metric in this experiment and compare it with the same four approaches. In this experimentation, we set the number of iterations and tasks to 1000. At the conclusion of the experiment, we calculate the energy consumption rate. Figure (e.g., Fig. 8) demonstrates that our approach achieves a gain of 5.12%, 17.93%, 27.04%, and 37.09% compared to the Jellyfish, BGA, GTO, and TGA approaches, respectively, confirming our theoretical assumptions.

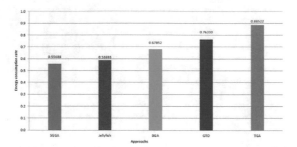

**Fig. 8.** IoT energy consumption rate comparison.

## 5.5   Experience 5: Cost Analysis of Distributed Task Completion in IoT

In the final experiment, we measured the overall cost. This pertains to the cost required to complete all tasks distributed to various connected objects. Figure (e.g., Fig. 9) shows that the Jellyfish approach exhibits a better cost performance, surpassing our approach by a slight margin of 1.47%, and exceeding the other approaches GTO, BGA, and TGA by 4.20%, 12.70%, and 21.14% respectively. This underscores the need for further refinement of this metric in future work.

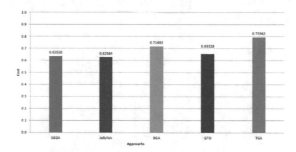

**Fig. 9.** Overall cost comparison in the IoT environment.

## 6   Conclusion

In this paper, we propose a novel approach for dynamic load balancing in CloudIoT, based on a neural network classifier. This classifier distributes tasks sent by connected devices into three classes: the first-class groups tasks processed by the IoT devices themselves, the second-class groups tasks handled by the Cloud, and the third class includes notification tasks. A scheduler based on the combination of Genetic Algorithm (GA) and Grasshopper Optimization Algorithm (GOA) is employed at both the Cloud and IoT levels to map tasks to appropriate devices.

We evaluated the performance of our approach using various metrics, comparing it to the most recent and relevant works in the field. The results demonstrate that our approach achieves efficient dynamic load balancing in CloudIoT, reducing the average task execution time, increasing resource utilization rate, and lowering overall cost and energy consumption rate.

# References

1. Benabbes, S., Hemam, S.M.: An approach based on genetic and grasshopper optimization algorithms for dynamic load balancing in CloudIoT. Comput. Inform. **42**(2), 387–396 (2023)
2. Benabbes, S., Hemam, S.M.: An approach based on (Tasks-VMs) classification and MCDA for dynamic load balancing in the CloudIoT. In: Smart Energy Empowerment in Smart and Resilient Cities: Springer International Publishing, vol. 102, pp. 387–396 (2020). https://doi.org/10.1007/978-3-030-37207-1_41
3. Botta, A., Donato, W., Persico, V., Pescapé, A.: Integration of cloud computing and Internet of Things: a survey. Future Gener. Comput. Syst. **56**, 684–700 (2016)
4. Cunha, A., Borges, J., Loureiro, A.: Classification of botnet attacks in IoT using a convolutional neural network. In: Proceeding of the 18th ACM International Symposium on QoS and Security for Wireless and Mobile Networks on 18th ACM International Symposium on QoS and Security for Wireless and Mobile Networks, pp. 63–70 (2022)
5. Gulbaz, R., Siddiqui, A.B., Anjum, N., Alotaibi, A.A., Althobaiti, T., Ramzan, N.: Balancer genetic algorithm - a novel task scheduling optimization approach in cloud computing. Appl. Sci. **11**(14), 6244 (2021)
6. Jairam, N.: Classification and scheduling of information-centric IoT applications in cloud-fog computing architecture (CS_IcIoTA). In: Proceeding of the 14th International Conference on Innovations in Information Technology (IIT), pp. 82–87 (2020)
7. Jia, L., Li, K., Shi, X.: Cloud computing task scheduling model based on improved whale optimization algorithm. Wireless Commun. Mobile Comput., 1–13 (2021)
8. Nascita, A., Cerasuolo, F., Monda, D.D., Garcia, J.T.A., Montieri, A., Pescape, A.: Machine and deep learning approaches for IoT attack classification. In: Proceeding of the Conference on Computer Communications Workshops (INFOCOM WKSHPS), pp. 1–6 (2022)
9. Nwogbaga, N.E., Latip, R., Affendey, L., Suriani, R., Amir, R.A.: Investigation into the effect of data reduction in offloadable task for distributed IoT-fog-cloud computing. J. Cloud Comput. **10**(1), 1–12 (2021)
10. Ouhame, S., Hadi, Y., Arifullah, A.: A hybrid grey wolf optimizer and artificial bee colony algorithm used for improvement in resource allocation system for cloud technology. Int. J. Online Biomed. Eng. (IJOE) **16**(14), 4–17 (2020)
11. Shafahi, Z., Yari, A.: An efficient task scheduling in cloud computing based on ACO algorithm. In: Proceeding of the 12th International Conference on Information and Knowledge Technology (IKT) (pp. 72–77) (2021)
12. Solatidehkordi, Z., Ramesh, J., Al-Ali, A.R., Osman, A., Shaaban, M.: An IoT deep learning-based home appliances management and classification system. Energy Rep. **9**, 503–509 (2023)

# Blockchain-Based Identity Management

A. Berbar$^{(\boxtimes)}$ and A. Belkhir

Laboratory of Computer System, USTHB, Algiers, Algeria
Ahmedberbar@hotmail.Com

**Abstract.** E-government or the use of ICT means for the delivery of government services, is an effective way to facilitate administrative procedures for both citizens and businesses. The sine qua non condition remains their identification. Frequently plagued by misuse, data theft, and lack of regulation, traditional solutions for the management of digital identity, often centralized, show their limits. This is where blockchain technology can be exploited to increase the level of security of elaborate systems such as government platforms. This work therefore proposes to replace the National Identity Register of such a platform with its alternative blockchain; resulting in not only more secure identity data but also improved performance.

**Keywords:** Digital Identity · service provider · blockchain · trusted third party

## 1 Introduction

Today, the development of technology and the increasing availability of the Internet facilitate communication and interaction of entities in various fields of activity. This interaction between the fields of activities led to creating and organizing a cross-domain management to facilitate the tasks and answer services. Among these aspects, digital identity meets this requirement. Several governments have used information technology and communication; most countries have launched e-government programs in the context of the country's transformation into an information society. Governments want to ensure that citizens, businesses, academia, and public entities have more easy, faster, and cost-effective access to public services. To improve citizen satisfaction with government services, we are deploying a secure e-government solution that uses smart cards for identification. Smart card use tends to spread in different areas to facilitate access to services, and management services that require identification. Our objective is the deployment of an e-government solution based on identification by smart card. The goal is to design and implement a secure identity services platform-based smart card.

ICTs (Information and Communication Technologies), allow the governments to have new and powerful tools to facilitate better and faster communication with the citizens. Recently, numerous governments supplied important efforts [1–7], to set up their e-government services. The concept of identity, in the digital world, also lives a radical evolution. Context-aware management domains allow having a flexible and dynamic policy management model for context-aware service platforms [8]. Since e-government

M. Hatti (Ed.): IC-AIRES 2023, LNNS 984, pp. 78–85, 2024.
https://doi.org/10.1007/978-3-031-60629-8_8

services are distributed over several domain activities, Context-Aware Management Domains seem to be an appropriate solution for e-government services.

Digital identity is essential for unifying and standardizing government services offered to citizens and facilitating their high integration. Strong digital identity management facilitates the distribution of social assistance to needy citizens, secure online payments, border control, effective management of government personnel, and prevention of tax evasion. With their characteristics of decentralization, immutability and auditability among others, blockchains seem to be a particularly suitable solution for digital identity management [9, 10]. Indeed, their high level of security goes particularly well with the sensitive nature of identity data.

Section 2 presents the identity management and the blockchain usage. Section 3 defines the e-government platform for managing services-based identity. Section 4 presents similar works; we shall end with a conclusion and perspectives.

## 2   Identity Management

### 2.1   Definition of Identity

The National Registry ensures recording, storing, and communicating information relating to the identification of citizens. It is a database that contains all the basic information of citizens and an archive of deceased citizens. The National Regis-try ensures the sharing of identifying information of individuals between different public services. The information stored in the ID card includes the name and surname, Address, date and place of birth, Sex, Nationality, and National ID Number (NIN). A NIN is a unique number assigned to each citizen [11]. It is used for the identification and strong authentication. The national identification number includes eighteen (18) digits. The citizen's profile is modeled by a tree structure [8]. Every branch is bound to a domain of life. Besides the information of basic identification (for example, the identifier or elements of civil status), the identifier number is information that allows to distinguish without ambiguity an identity of a citizen in a given context. Two entities cannot share the same identifier. The citizen's profile can include more data according to the needs. These data are both dynamic and static. The static data are tied to the person and never change in time for example (the date of birth). The dynamic characteristics are data, which vary in time, for example (civil status or health status). The complete profile of the digital identity consists of a set of profiles distributed through domains. The adopted e-government model is based on four main actors [8,14]:

- National Identity Registry (NIR) and Certification Authority:
- Trusted third party (TTP)
- Service providers
- Users

We propose to quadruple each NIR, and TTP to balance the total load on four deployed instances taking into account the population density which differs completely from the south to the north of the country. For the north of the country, which hosts 85% of the total population in a strip not exceeding 250 km from the coast, three instances of each TTP have been planned, located in high-density population centers, namely:

Algiers, Constantine, and Oran. The south of the country will host a central body at Ain Salah.

The introduction of a blockchain would make it possible to obtain a global shared NIR replicated at the TTP level. Each of these instances would host its own copy of the ledger kept consistent with the other copies thanks to the concerted process of adding blocks that characterize the blockchain. Unlike the existing system where each TTP only sees the data at its level without any other reference point, the slightest discrepancy between the copies of the NIR thus shared would be quickly detected, corrected compared to the majority version, and its author identified.

### 2.2 What is stored on the blockchain?

The identity data of Algerian citizens for each citizen are as follow:

- National Identity Number (NIN)
- Full name
- Gender (Female/Male)
- Date and place of birth
- NINs of the father and the mother
- Family status (Single/Married/Divorced/Widowed) and NIN(s) of spouse(s) if applicable
- Date of death if applicable

The identity data of Algerian legal entities for each company are as follow:

- National identity number NIN.
- Denomination.
- Legal form or status [68]:

  - EURL-Single-Person Limited Liability Company
  - SARL-Limited Liability Company
  - SPA-Société par Action
  - SNC-Société en Nom Collectif

- Registration in the commercial register
- Date of constitution
- Place of residence
- NIN of the owner(s) if applicable
- Date of dissolution if applicable

## 3 E-Government Platform

Instances of the NIR register are processes running at the level of the P2P network that constitute the TTP nodes. These processes work to replicate the ledger between them, to propose new transactions, to reach consensus [12] between the validating nodes concerning these transactions, and to satisfy requests on the data. Each node is identified by a public key (Fig. 1).

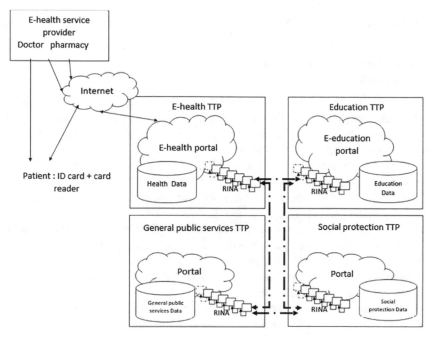

**Fig. 1.** Blockchain-based e-government architecture.

The most basic data for a blockchain is therefore the transaction, which traditionally represents an interaction between two elements. In the broad sense, the transaction represents a change of state of the ledger. In our case, for natural persons, four changes of state can occur, corresponding to four types of possible transactions:

- At the birth of an individual, the National Identity Number NIN, the name, the first name, the sex (Female/Male), the date and the place of birth, the NINs of the father and the mother and the condition family (single) of the new citizen must be filled in. This is done on the condition that the NIN does not exist.
- At marriage, the family status of the two individuals changes to 'Married' by specifying the NIN of the spouse. This is done on the condition that the two individuals exist, are alive, and of legal age that the family condition is not already at the value 'Married' for the woman. For the man, the family condition must initially be 'Single', 'Divorced', 'Widowed' or 'Married' but with less than four wives. Additional information such as the wedding date is not stored on the blockchain but on the database of the general public services TTP.
- On divorce, the family condition changes from 'Married' to 'Divorced' for the woman. This is only the case for the man if he was previously married to only one woman; otherwise, his family condition remains at 'Married'. The spouse's NIN is not deleted to keep the history. This is done if both individuals are alive and married.
- On death, the date of death is recorded. If the individual is married and female, the family condition changes from 'Married' to 'Widowed'. This is only the case for the man if he was previously married to only one woman; otherwise, his family condition remains at 'Married'. The same treatment is applied to the spouse. His NIN is not

deleted in order to keep the history. This is done on the condition that the individual is alive.

For legal persons, six changes of status can occur, corresponding to six types of possible transactions:

- When the company is created, the NIN, the name, the legal status, the registration in the commercial register, the date of incorporation, the place of residence and the NIN of the owner(s) if applicable are instructions. This is done if the NIN does not already exist and the eventual owners are alive and of legal age.
- At the change of denomination, the new denomination is entered. This is done if the company exists and is still in operation (not dissolved)
- When the legal status changes, the new legal status is recorded. This is done if the company exists and is still in operation (not dissolved).
- When the registered office is transferred, the new place of domiciliation is entered. This is done if the company exists and is still in operation (not dissolved).
- At the change of owner(s), the new owner(s) is (are) recorded. This is done on the condition that the business exists and is still in operation (not dissolved) and that the new owner(s) are alive and of legal age.
- When the company is dissolved, the date of dissolution is entered. This is done if the company exists and is still in operation (not dissolved).

The consensus algorithm is necessary for the creation of a block within the blockchain. The chosen algorithm is pBFT [13] for the following reasons:

- The block proposal comes from the client (or leader), and the order changes by round robin.
- The validation of the blocks/transactions is done by verification of the signature, which corresponds to our case where the nodes are defined by their public key.
- There is no reward mechanism, which corresponds well to our case because it is not a tokenized platform.

The segregation of duties provided by the platform is achieved by using Transaction Processors - TP - and client applications. These two types of entities are part of the overall architecture of the framework.

To create a blockchain in Hyperledger Sawtooth, we must implement what is called one or more transaction families that encapsulate business functionality for an application.

A transaction family consists of three main entities:

- The TP, with each validator node executing at least one TP after its registration.
- The customer
- The data model.

Several families of transactions can be applicable on the same blockchain, knowing that it can be configured in order to refuse the execution of transactions belonging to a specific family.

In order to emulate our government blockchain network, we will duplicate the previous node into nine: Four representing the General Public Services TTPs, four others the Economy TTPs.

These eight nodes represent all of our validators. The ninth node represents any other TTP, or the national node, that can access read-only.

These nine nodes were emulated under docker on a computer with the following configuration: 7th generation i5 processor with an Ubuntu Focal OS with 60GB of memory (Fig. 2).

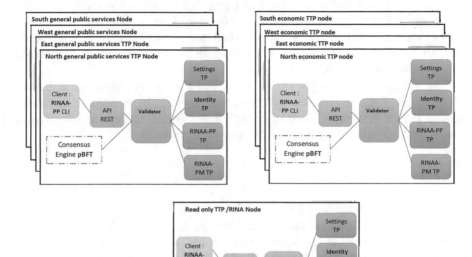

**Fig. 2.** Emulated NIR network.

By default, the sawtooth blockchain is public, we must make it private by setting the write access of the different clients via the implementation of two access policies:

- The first Allows only Individual Family customers of the four generals public TTPs North, South, East and West to offer Individual Family transactions
- The second Allows only customers of the legal person family of the four economic TTPs North, South, East and West to offer transactions of the legal person family

Also, we use the public keys to identify all the nodes accepted in the Sawtooth network.

For a local node, a latency of 1 to 2 s has been observed. Whereas for nine nodes, a valid transaction requires a duration of 4 s. This duration remains below the standards and recommendations adopted for the ergonomics of human-machine interactions on the web (10 s) [19].

In order to test our blockchain, we replaced the physical person TP of the second validator node with a malicious TP, the attempt to execute a malicious transaction was unsuccessful. By replacing more TPs in the same way, the results exceeded the predictions since the Byzantine fault tolerance was maintained for half of the nodes.

## 4 Related Work

For the authentication process, Chen et al. [16] proposed a blockchain-based trust transfer scheme using different trusted third parties in the case of cross-domain access. For data sharing, Liu et al. [17] proposed a solution when sharing data between different departments and they describe how to protect data integrity during the transaction. Information sharing between nodes is done using private blockchains.

Payeras-Capella et al. [18] presented two different schemes of an electronic delivery service to reduce the involvement of trusted third parties. The systems were based on a private and public electronic delivery system using blockchain and smart contracts to provide fair trade and reduce the role of trusted third parties.

## 5 Conclusion

This paper shows a way of securing an e-government platform by integrating blockchain technology instead of a centralized national identity registry. The blockchain makes it possible to fill the gaps in the delayed synchronization of data and their rapid non-compliance with reality. By using an adapted consensus algorithm (pBFT), the identity blockchain implemented in order to be able to replace the existing NIR efficiently.

For more granularity, it would be appropriate to split the blockchain into two: one for natural persons and the other for legal persons; the economic sphere may have other stakeholders in the form of a validator such as the Security and Public Order TTP for example. The two blockchains would be operated jointly. As is the case with the implemented blockchain, it is possible to reuse existing code between similar transactions such as the birth of an individual and the creation of a business. Another perspective would be the implementation of smart contracts, thus allowing the automatic dissemination of changes in the state of physical or legal entities to the authorized actors of our e-government platform.

## References

1. Fishenden, J., Thompson, M.: Digital government, open architecture and innovation: why public sector IT will never be the same again. J. Public Adm. Res. Theory 23, 977–1004 (2013)
2. Agbozo, B.K.A.E.: The role of e-government systems in ensuring government effectiveness and control of corruption. R-Economy 5(2) (2019)
3. U.D.O.E.A.S.A. Division for Public Institutions and Digital Government. E-Government (2022). https://publicadministration.un.org/egovkb/en-us/About/UNeGovDDFramework
4. Grigalashvili, V.: E-government and E-governance: various or multifarious concepts 5(11) (2022)
5. Putra, D.A., et al.: Tactical steps for e-government development. Int. J. Pure Appl. Math. 119(115) (2018)

6. Goede, M.: E-Estonia: the e-government cases of Estonia, Singapore, and Curaçao. Arch. Bus. Res. **7**(12), 25 (2019)
7. Khern, N.C.: Digital Government, Smart Nation: Pursuing Singapore's Tech Imperative (2019). https://www.csc.gov.sg/articles/digital-government-smart-nationpursuing-singapore%27s-tech-imperative
8. Berbar, A., Belkhir, A.: Identification in the service of national solidarity. In: Proceedings of the 4th International Conference on Smart City Applications, pp 1–6, Morocco (2019)
9. Kolb, J., Abdelbaky, M., Katz, R., Culler, D.E.: Core concepts, challenges, and future directions in blockchain: a centralized tutorial. ACM Comput. Surv. **53**(1), 1–39 (2020)
10. Francesca Fallucchi, M.G.: Blockchain, State-of-the-Art and Future Trends. Catania, IT (2021)
11. Executive Decree No. 10–210 of 16 September 2010 Establishing the Unique National Identification Number, Official Journal of the People's Democratic Republic of Algeria, n°154, 3–4, September 19 (2010)
12. Xiao, Y., Zhang, N., Lou, W., Hou, Y.T.: A survey of distributed consensus protocols for blockchain networks (2020)
13. Tomić, N.Z.: A review of consensus protocols in permissioned blockchains. J. Comput. Sci. Res. **3**(12) (2021)
14. Abdennouri, T., Saidi, T.: e-Government Platform: Application to National Solidarity, Master Thesis, Faculty of Electronics and Informatics, USTHB, N °161 (2016)
15. Governement at a glance, annex B : classification of the functions of government (COFOG), OECD (2011)
16. Chen, Y., Dong, G., Bai, J., Hao, Y., Li, F., Peng, H.: Trust enhancement scheme for cross domain authentication of PKI system. In: Proceedings of the International Conference on Cyber-Enabled Distributed Computing and Knowledge Discovery (CyberC), Guilin, China, pp. 103–110 (2019)
17. Liu, L., Piao, C., Jiang, X., Zheng, L.: Research on governmental data sharing based on local differential privacy approach. In: Proceedings of the 2018 IEEE 15th International Conference on e-Business Engineering (ICEBE), Xi'an, China, pp. 39–45 (2018)
18. Payeras-Capella, M.M., Mut-Puigserver, M., Cabot-Nadal, M.A.: Blockchain-based system for multiparty electronic registered delivery services. IEEE Access **7**, 95825–95843 (2019)
19. Nielsen, J.: Usability engineering (1993). ISBN10 0125184069

# Application of Machine Learning Forecasting Model for Renewable Generations of Adrar's Power System

S. Makhloufi[✉], M. Debbache, S. Diaf, and R. Yaiche

Centre de Développement des Energies Renouvelable, BP. 62, Route de l'Observatoire, Bouzareah, Algiers, Algeria

{s.makhloufi,m.debbache,s.diaf,r.yaiche}@cder.dz

**Abstract.** Uncertainties and intermittency of wind and solar power generations pose difficulties in power system operation. Thus, raises the importance of developing an accurate prediction model. This article proposes long short-term memory (LSTM) networks a complete tool to predict power output of two photovoltaic power plants and a wind farms installed in the isolated Adrar's power system. Meteorological variables such as, solar irradiation, ambient and cellule temperatures, relative humidity, wind speed, and pressure, are introduced into LSTM for predicting power outputs. The LSTM is trained on the collected time series of real recorded dataset. The RMSE accuracy of LSTM networks forecasting methods is used to prediction performance of the proposed methods. The LSRTM can predict the power output of PV plants and wind turbine with a lowest RMSE of 12.48 kW is successively obtained for the wind turbine.

**Keywords:** Long short-term · Photovoltaic power · Wind power prediction · Isolated power system

## 1 Introduction

Due to the ubiquitous, plentiful of renewable energy sources (RESs) in Algeria principally in wind and solar (W&S) resources, they considered the most promising energy alternatives in future energy mix. The total installed capacity reached by the end of 2019 in Algeria was 344.1 MW, while a wind farm of 10.2 MW and 53 MW of total capacity of photovoltaic (PV) power generations feeds the isolated Adrar-InSalah power system [1].

In fact, many issues may be emerged by the intermittent and variability of W&S resources [2, 3]. For example, fast change in demand and production balance may threaten the stability of a weak power system [2–5]. Therefore, large-scale integration of W&S energy into weak power systems and micro-grids are expected to significantly increase the importance of developing a complete accurate prediction tool that gathers renewable energy.

M. Hatti (Ed.): IC-AIRES 2023, LNNS 984, pp. 86–96, 2024.
https://doi.org/10.1007/978-3-031-60629-8_9

In this context, a series of researches have been conducted to perform W&S power prediction [6–18]. One of the most recent popular techniques in W&S forecasting is artificial intelligence approaches [3]. In photovoltaic (PV) power forecasting problem, Ref. [19] reviewed contemporary forecasting techniques proposed for PV power plants. Furthermore, different approaches categorized into physical, statistical, artificial intelligence, ensemble and hybrid approaches are critiqued and compared. The authors conclude that artificial neural networks models are best for forecasting short-term PV power. The large variations in historic W&S power output can affect performance of a single machine learning approach. One of the solutions is to hybrid this approach. Yu Feng et al. [20] proposed the combination of particle swarm optimization and extreme learning machine (PSO-ELM) models to predict daily PV power output. Their proposed approach can particularly be used in areas where in-situ measurements are unavailable. Dewangan et al. [21] examined different combined-forecast methods such as mean, median, linear regression, non-linear regressions, and supervised machine-learning algorithms, which are applied to three solar plants in Australia. Mishra et al. [22] proposed Long-Shot-Term-Memory network (LSTM) based deep learning (DL) technique and wavelet transformation for short-term solar power prediction. Ref. [23] proposed a hybrid deep neural network for PV power forecasting. Their approach based on a combination of extended short-term memory networks (LSTM) and convolutional neural networks (CNN). The proposed model includes the energy of electrons and photons as well climatic conditions as input data.

In the same manner, a series of researches have been conducted to perform wind power prediction. Aiming to improve accuracy of short-term wind power prediction by LSTM model, Shahid et al. [24] proposed to encompass the dynamic behavior of temporal data using Wavelet kernels. Similar, Li Han et al. [25] adopted Variational Mode Decomposition (VMD) method to improve LSTM prediction. In their model, the VMD decomposes wind power signal to the three components (long-term, fluctuation and random). Ref. [26] proposed an extensive statistical analysis and resampling routines, to predict short-term wind power. A hybrid forecasting model based on LSTM-enhanced forget-gate network (LSTM-EFG) developed in Ref. [27]. The parameters of the model are optimized using cuckoo search optimization algorithm, where after used to forecast the subseries data that are extracted using ensemble empirical mode decomposition. Authors of Ref. [28] employed state-of-the-art temporal convolutional networks (TCN) to extract temporal dynamics of the wind turbines and an orthogonal array tuning method based on the Taguchi design of experiments utilized to optimize the hyper-parameters of the proposed TCN model. Then, deep learning used to predict wind power curves of a wind farm. Zhou et al. [29] proposed multi-objective prediction intervals based on deep neural networks.

In fact, due to the inherent stochastic nature of the weather data that varies from region to another, forecast performances of LSTM model are still in improvement. Therefore, assessment on real application must be generalized. Considering the specific location of the isolated Adrar-InSalah power system and penetration level of the installed capacity of W&S generations, this article assess the concurrency of LSTM approach to predict W&S power outputs.

## 2  Long-Short Term Memory (LSTM)

A LSTM network based on memory cells is a subset of machine deep learning. The LSTM network allows making future predictions based on history dataset.

In fact, deep learning refers to artificial neural networks with more than one layer of neurons. The deep learning as its name implies permitting a "deep thinker" and solving several complex problems.

LSTMs are developed to avoid the dependency on old information with the ability to forget things. Both standard recurrent neural network (RNN) and LSTM have a repeating structure, but in LSTM, each element has four layers that decides what old information pass on to the next layer.

Four main elements compose an LSTM: input matrix, forget matrix, output matrix, and neuron. The input matrix determines the importance of new information, while the output matrix determines the target information to be predicted. The forget matrix can modulate the memory cells self-recurrent connection, allowing the cell to remember or forget its previous state.

A fundamental LSTM approach employs the following equations [30]:

$$f_1 = \partial\left(W_f \cdot [h_{t-1}, x_t] + b_f\right) \tag{1}$$

$$i_{t1} = \partial\left(W_i \cdot [h_{t-1}, x_t] + b_i\right) \tag{2}$$

$$\widehat{C}_t = \varphi\left(Wc \cdot [h_{t-1}, x_t] + b_c\right) \tag{3}$$

$$C_t = f_t \otimes C_{t-1} + i_t \otimes \tau_t \tag{4}$$

$$O_t = \delta\left(W_0 \cdot [h_{t-1}, x_t] + b_0\right) \tag{5}$$

$$h_t = O_t \otimes \varphi(C_t) \tag{6}$$

where i, o, and f are the input, output, and forget matrices respectively, W and b are the weight and bias, respectively, ht − 1 is the memory cells at the previous sequence step, σ is the element-wise application of the sigmoid function, ɸ is the element-wise application of the tanh function, and C is the cell state.

After updating the memory cell at current time t, the output Yt + p is calculated as follows [30]:

$$Y_{t+p} = W_d \cdot h_t + b_d \tag{7}$$

Figure 1 illustrate the four elements of LSTM approach.

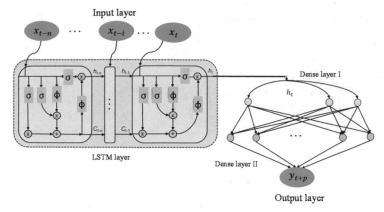

**Fig. 1.** Elements of LSTM approach [30].

# 3   Data Description

In this study, renewable power forecasting for a weak power system supplied Adrar-InSalah region in southwestern of Algeria is studied. Adrar-InSalah power system consists of 53 MW of PV power plants and a wind farms of 10.2 MW of capacity. Two PV power plants installed in Timimoun with a total capacity of 10 MW and Reggane with a total capacity of 5 MW and the wind farm installed in Kabertene are chosen to perform our approach.

The period spans from January 01 to March 30, 2018. The data are daily recorded with the interval of 15 min for the PV power plants and 10 min for the wind farm. The parameters of each plant are the average values calculated between 15 min and 10 min for PV power plants and for wind turbine respectively.

The database provided by the PV plants and wind turbine includes PV power outputs and the corresponding recorded meteorological data are given as following:

**Parameter of PV power plants**

- **Inputs**

  - Air temperature (AT) and Cell temperature (Tcel), given in °C
  - Inclined solar irradiance (ISI), given in W/m2
  - Relative humidity (H), given in %
  - Air pressure (P), given in Hpa
  - Wind speed (Ws), given in m/s

- **Output**

  - Active power output (Pout), given in kW

**Parameter of wind power turbine**

- **Inputs**

  - Ambient temperature (AT), given in °C
  - Alternator bearing temperature (AltBT), given in °C
  - Alternator temperature (AltT), given in °C
  - Alternator speed (AS), given in tr/mn
  - Wind speed (Ws), given in m/s

- **Output**

  - Active power output (Pout), given in kW

It is well known that the power output of the PV power plants are strongly depended on the solar irradiation. Generally, solar irradiation are available between 7 am to 18 pm. Therefore, for the PV power plant, we have selected this time period to calculate the matrix correlation. Matrices correlations of parameters are presented in Figs. 2, 3, 4. As can be seen in Fig. 2, there is a strong correlation between the power output and temperature of the wind generator.

We note that to protect the wind turbines from high temperature, they must be shutdown at a temperature ambient of 45 °C.

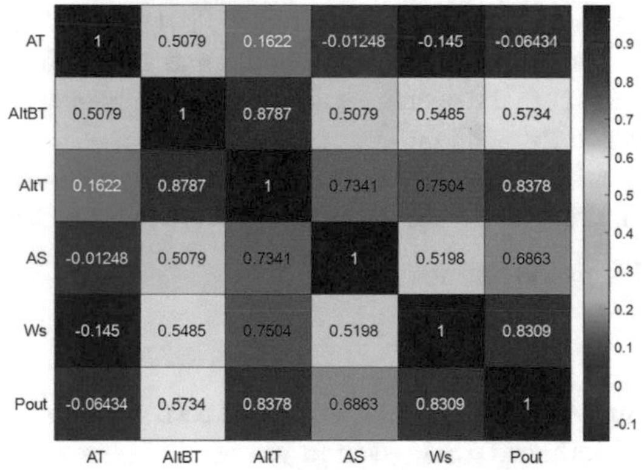

**Fig. 2.** Matrix correlation of recorded parameters of Kabertane's WT N03.

Figures 5, 6, 7 present active power generated by the PV plants and the wind turbine of a rated power 850 kW. Figures of power output of the PV plants show that they are shutdown in several times during high availability in solar irradiation. The bleu zones represent the low power, where the sudden variation in the colour, in other words, value corresponds to the disconnection of the PV plants to maintain the operation of the isolated power stable and avoid a total blackout. However, the disconnection of renewable energy under high availability of solar and wind resources may false the training learning.

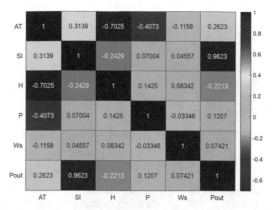

**Fig. 3.** Matrix correlation of recorded parameters of Reggane's PV power.

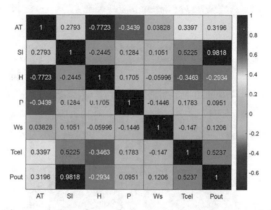

**Fig. 4.** Matrix correlation of recorded parameters of Timimoune's PV power.

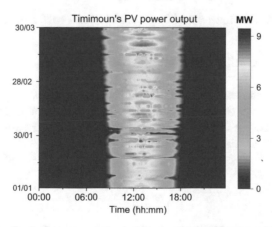

**Fig. 5.** Active power output of Timimoune's PV power plant.

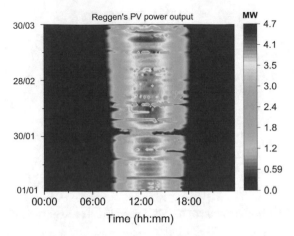

**Fig. 6.** Active power output of Reggane's PV power plant.

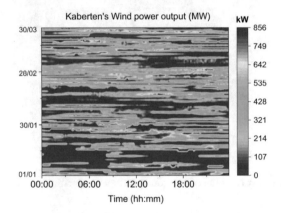

**Fig. 7.** Active power output of Kabertans's wind turbine.

## 4   Simulation Results and Discussion

In order to perform the ability of the deep leaning approach based on long short-term memory (LSTM) network-model, available in Matlab toolbox, to forecast the three-day-ahead power output of the PV plants and wind turbine, Root Mean Square Error (RMSE) is calculated to evaluate the accuracy of forecasting results.

$$RMSE = \left( \frac{1}{N} \sum_{t=1}^{N} (S_t - P_t)^2 \right)^{1/2} \tag{8}$$

where St represents the forecasted result at the time point t; Pt is the actual value at the time point t and N is the total number of Pt.

Setting parameters of the LSTM introduced in Matlab are given below:

- Number of Hidden Units = 200
- Maximum Epochs = 2000
- Gradient Threshold = 1
- Initial Learn Rate = 0.005
- Learn Rate Schedule: piecewise
- Learn Rate Drop Period = 125
- Learn Rate Drop Factor = 0.2

The data divided into training and test sets. The training set consists of the first 86 days, and the test set consists of the remaining three days. Figure 8 illustrates the predicted and measured power outputs and the calculated error between them. In Fig. 8, the total realized and the predicted powers are illustrated and showing excellent prediction results, with a very low error reached a maximum of 54 kW.

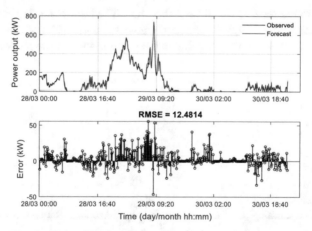

**Fig. 8.** Three-day-ahead-forecasting results of wind turbine.

Results of the two PV power plants are presented in Figs. 9 and 10. As can be seen, the predicted power follow the curvature of the output of the wind farm, as well the PV plants.

For Reggane's PV plant, acceptable deviations observed during the high solar irradiation times. This perfumes the LSTM model for predicting power output under unpredictable meteorological conditions.

The maximum error deviation reached 300 kW and 1000 kW for Reggane's and Timimoune's PV plants.

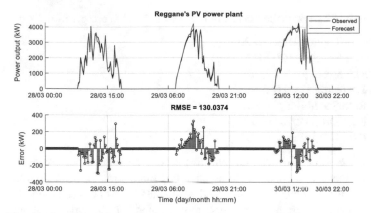

**Fig. 9.** Three-day-ahead-forecasting results of Reggane's PV power plant.

From the above results, we can see that the LSTM approach is able to give good accuracy for renewable energy forecasting. Under strict variation of the meteorological conditions, the proposed LSTM highlights the performance of the proposed model.

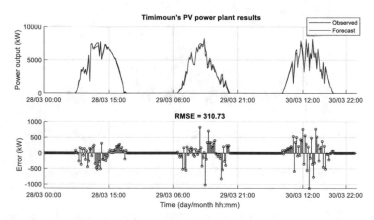

**Fig. 10.** Three-day-ahead-forecasting results of Timimoun's PV power plant.

## 5 Conclusion

This article proposes LSTM approach available in Matlab toolbox for PV and wind power forecasting. The performance of the proposed approach was tested on three operational power plants installed in the isolated Adrar-InSalah power system, which allow supplying electricity in the localities of Reggane, Timimoune, and Kabertene. The figures of the power outputs of these plants during January 01, to March 30, 2018 show high variability and disconnection during many times. Therefore, perform LSTM approach for power prediction of an intermittent and variable renewable energy under such condition permits well performing the LSTM approach.

The results show minor errors with very low RMSE below than 12 for wind turbine. Therefore, the LSTM approach available in Matlab toolbox can be used to predict accurately the renewable energy.

In the future, others PV power plants installed in the same isolated power system supplying the localities of Adrar, InSalah, Aoulef, and Kabertane, and under different operation conditions for all seasons should be investigated. Furthermore, multivariable correlation methods, such as Pearson, Kendall, and Spearman should be analyzed before starting any training process. This analysis is useful to accelerate deep learning and neglect unnecessary variables.

# References

1. Makhloufi, S., Debbache, M., Boulahchiche, S. : Long-term forecasting of intermittent wind and photovoltaic resources by using adaptive neuro fuzzy inference system (ANFIS). In: Proceedings of the 2018 International Conference on Wind Energy and Applications in Algeria (ICWEAA), pp. 1–4. Algiers, Algeria (2018)
2. Ding, M., Xu, Z., Wang, W., Wang, X., Song, Y., Chen, D.: A review on China's large-scale PV integration: progress, challenges and recommendations. Renew. Sustain. Energy Rev. **53**, 639–652 (2016)
3. Lim, Y.S., Tang, J.H.: Experimental study on flicker emissions by photovoltaic systems on highly cloudy region: a case study in Malaysia. Renew. Energy **64**, 61–70 (2014)
4. Lee, B.H.: A study on simplified robust optimal operation of micro-grids considering the uncertainty of renewable generation and loads. Trans. Korean Inst. Electr. Eng. **66**, 513–521 (2017)
5. Fermín, R., Alice, F., Ainhoa, G.: Predicting solar energy generation through artificialneural networks using weather forecasts for microgrid control. Renew. Energy **126**, 855–864 (2018)
6. Ogliari, E., Dolara, A., Manzolini, G., Leva, S.: Physical and hybrid methods comparison for the day ahead PV output power forecast. Renew. Energy **113**, 11–21 (2017)
7. Peiyuan, C., Troels, P., Bak-Jensen, B., Chen, Z.: ARIMA-based time series model of stochastic wind power generation. IEEE Trans. Power Syst. **25**(2), 667–676 (2010)
8. Cyril, V., Ted, S.: Statistical parameters as a means to a priori assess the accuracy of solar forecasting models. Energy **90**, 671–679 (2015)
9. Fermín, R., Alice, F., Ainhoa, G.: Predicting solar energy generation through artificial neural networks using weather forecasts for microgrid control. Renew. Energy **126**, 855–864 (2018)
10. Paulescu, M., Brabec, M., Boata, R., Badescu, V.: Structured, physically inspired (gray box) models versus black box modeling for forecasting the output power of photovoltaic plants. Energy **121**, 792–802 (2017)
11. Lorenz, E., Hurka, J., Heinemann, D., Beyer, H.G.: Irradiance forecasting for the power prediction of grid-connected photovoltaic systems. IEEE J. Sel. Top Appl. Earth Obs. Remote Sens. **2**(1), 2–10 (2009)
12. Antonanzas, J., Osorio, N., Escobar, R., Urraca, R., Martinez-de-Pison, F.J., Antonanzas-Torres, F.: Review of photovoltaic power forecasting. Sol. Energy **136**, 78–111 (2016)
13. Das, U.K., et al.: Forecasting of photovoltaic power generation and model optimization: a review. Renew. Sustain. Energy Rev. **81**, 912–928 (2018)
14. Cardell, J., Anderson, L., Tee, C.Y.: The effect of wind and demand uncertainty on electricity prices and system performance. IEEE PES T D 2010, 1–4 (2010)
15. Ghiasi, M., Esmaeilnamazi, S., Ghiasi, R., Fathi, M.: Role of renewable energy sources in evaluating technical and economic efficiency of power quality. Technol. Econ. Smart Grids Sustain. Energy **5**(1), 1 (2020)

16. Albadi, M., El-Saadany, E.: Overview of wind power intermittency impacts on power systems. Electric Power Syst. Res. **80**(6), 627–632 (2010)
17. Wang, J., Zhong, H., Lai, X., Xia, Q., Wang, Y., Kang, C.: Exploring key weather factors from analytical modelling toward improved solar power forecasting. IEEE Trans. Smart Grid **10**, 1417–1427 (2019)
18. Rafique, S.F., Zhang, J.H.: Energy management system, generation and demand predictors: a review. IET Gener. Trans. Dis. **12**(3), 519–530 (2018)
19. Ahmed, R., Sreeram, V., Mishra, Y., Arif, M.D.: A review and evaluation of the state-of the-art in PV solar power forecasting: techniques and optimization. Renew. Sustain. Energy Rev. **124**, 109792 (2020)
20. Feng, Y., Hao, W., Li, H., Cui, N., Gong, D., Gao, L.: Machine learning models to quantify and map daily global solar radiation and photovoltaic power. Renew. Sustain. Energy Rev. **118**, 109393 (2020)
21. Dewangan, C.L., Singh, S.N., Chakrabarti, S.: Combining forecasts of day-ahead solar power. Energy **202**, 117743 (2020)
22. Mishra, M., Dash, P.B., Nayak, J., Naik, B., Swain, S.K.: Deep learning and wavelet transform integrated approach for short-term solar PV power prediction. Measurement **166**, 108250 (2020)
23. Chang, K., Omer, A.A.I., Chang, K.C., Wang, H., Lin, Y.T., Nguyen, T.: Sung solar PV power forecasting approach basedon hybrid deep neural network. In: Hassanien, A.-E., et al. (Eds.): AMLTA 2021, AISC 1339, pp. 125–133 (2021)
24. Shahid, F., Zameer, A., Mehmood, A., Zahoor Raja, M.A.: A novel wavenets long short term memory paradigm for wind power prediction. Appl. Energy **269**, 115098 (2020)
25. Han, L., Jing, H., Zhang, R., Gao, Z.: Wind power forecast based on improved long short term memory network. Energy **189**, 116300 (2019)
26. De Caro, F., De Stefani, J., Bontempi, G., Vaccaro, A., Villacci, D.: Robust assessment of short-term wind power forecasting models on multiple time horizons. Technol. Econ. Smart Grids Sustain. Energy **5**, 19 (2020)
27. Devi, A.S., Maragatham, G., Boopathi, K., Rangaraj, A.G.: Hourly day-ahead wind power forecasting with the EEMD-CSO-LSTM-EFG deep learning technique. Soft. Comput. **24**, 12391–12411 (2020)
28. Mek, R., Alaeddini, A., Bhaganagar, K.: A robust deep learning framework for short-term wind power forecast of a full-scale wind farm using atmospheric variable. Energy **221**, 119759 (2021)
29. Zhou, M., Wang, B., Guo, S., Watada, J.: Multi-objective prediction intervals for wind power forecast based on deep neural networks. Inf. Sci. **550**, 207–220 (2021)
30. He, C., Patel, N., Kobilarov, M., Iordachita, I.: Real time prediction of sclera force with LSTM neural networks in robot-assisted retinal surgery. Appl. Mech. Mater. **896**, 183–194 (2020)

# MPPT and Optimization

# Using the PSO Algorithm to Optimize a Self-tuning PID-Type Fuzzy Controller for Indoor Temperature

Ahmed Bennaoui[1]([⊠]), Slami Saadi[2], Hossam A. Gabbar[3], and Aissa Ameur[1]

[1] Faculty of Technology, University of Amar Telidji, 03000 Laghouat, Algeria
{a.bennaoui,a.ameur}@lagh-univ.dz
[2] Faculty of Sciences and Technology, Ziane Achour University, BP 3117, Djelfa, Algeria
[3] Faculty of Engineering and Applied Science, Ontario Tech University, Oshawa, Canada
hossam.gaber@ontariotechu.ca

**Abstract.** In this work, we developed the fuzzy PID controller to regulate the indoor temperature where we used the PSO technique to optimize and select the values of the fuzzy PID controller to regulate the indoor temperature. The structure of the fuzzy PID control consists of two main elements: a classic conventional PID control to control the indoor air temperature, and a FLC that regulates the values of the PID controller according to the current indoor conditions. The results obtained demonstrate that using the FLC with the PSO technique leads to proper values for the PID controller,and which achieves the desired output of the system and give better performance (fast response, achieve lower the ISE and the IAE) by compared to the fuzzy PID controller without the PSO algorithm.

**Keywords:** PID · PSO · FLC · Fuzzy-PID

## 1 Introduction

A self-tuning fuzzy PID controller is a variant of the classical PID controller used in control engineering. PID stands for Proportional-Integral-Derivative, is a control algorithm commonly used to tune dynamic systems [1–3]. The main difference between a self-tuning fuzzy PID controller and a conventional PID controller is the use of fuzzy logic to automatically tune the parameters of the PID controller [4–6]. In a typical PID controller, the proportional, integral, and derivative coefficients must be set manually by a control engineer, which can be a complex and tricky task, especially when the system is subject to variations or changes in behavior. The self-tuning fuzzy PID controller uses fuzzy logic techniques to automatically adjust the PID coefficients based on real-time system responses. Fuzzy logic makes it possible to model uncertain or ill-defined behaviors and to use linguistic rules to adapt the parameters of the PID accordingly.. The performance of the fuzzy logic con-troller is influenced by the choice of the shapes and parame-ters of the membership functions. Some researchers have performed studies using optimization algorithms to tune the membership function parameters of the fuzzy

M. Hatti (Ed.): IC-AIRES 2023, LNNS 984, pp. 99–110, 2024.
https://doi.org/10.1007/978-3-031-60629-8_10

logic controller (FLC), as mentioned in [7–10]. The objective of this study is to use the PSO algorithm to op-timize and select the membership parameters of the fuzzy PID controller for indoor temperature [11]. This paper is organized as follows: part 2. The particle swarm optimization (PSO) method part 3. The Fuzzy PID Controller With PSO Algorithm. Part 4. Simulation and Results. Part 5. Conclusion.

## 2 Particle Swarm Optimization (PSO) Method

The PSO draws inspiration from this model and applies it to solving optimization problems. Within the framework of the PSO, each optimization problem is represented by a particle, which is comparable to a bird evolving in a search space. Each particle has a fitness value determined by an optimization function, as well as a speed that influences its direction and the distance traveled. The particles track the best solution found so far, called Pbest (best individual solution), as well as the best solution found by the entire population, called Gbest (best global solution). The initialization of the PSO is done by creating a swarm of random particles, representing random initial solutions. At each iteration, the particles are updated based on two extremes. The first is the best solution

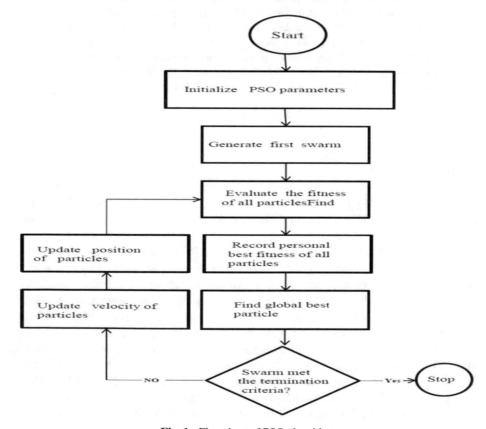

**Fig. 1.** Flowchart of PSO algorithm

found individually by each particle (Pbest), while the other is the best solution found by the entire population (Gbest). It is also possible to use the best solutions among the neighbors of a particle, instead of the whole population, thus creating local extremes. The advantages of the PSO method over other optimization methods are as follows [12, 13, 14]: Ease of implementation and programming, Easy concept and Mainly provides a better solution and faster convergence. The flowchart is represented in Fig. 1.

## 3 The Fuzzy PID Controller with PSO Algorithm

We suggested the use of the PSO algorithm to determine the optimal values of the membership functions of the inputs and outputs of the fuzzy PID controller for indoor temperature Fig. 2.

$$\text{We are taken as the fitness function} = \text{ISE} = \int_{0}^{+\infty} [e(t)]^2 dt \tag{1}$$

$$\int [0] \tag{2}$$

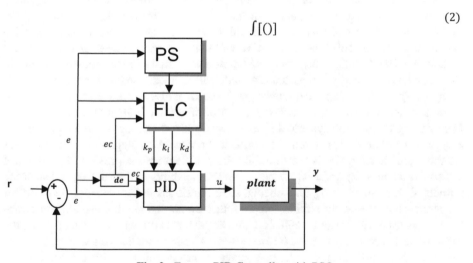

**Fig. 2.** Fuzzy - PID Controller with PSO

Where:
The transfer function of the indoor temperature change [11]:

$$\frac{T_i(s)}{Q_{h1}(s)} = \frac{0.028}{(506.52) * s + 1} \tag{2}$$

where: $T_i$ is the indoor temperature, $Q_{h1}$ Is the work rate of the heater,

we propose: the temperature difference between the indoor and outdoor temperature is: $r(k) = 5^0$ C

## 3.1  Fuzzy Logic System Design (Rules and MFS)

In an article written in 1961, Lofti A. Zadeh mentions that a new technique is needed, a special type of mathematics, that considers multivalued logical values, but it was not until 1965 that he published the first article on fuzzy logic. Fuzzy logic1 (FL) is a multi-valued logic that allows (by means of membership sets) a more practical way of approaching problems as seen in the real world. Unlike binary (yes/no) information, fuzzy logic emulates reasoning ability and makes use of approximate data to find precise solutions. A fuzzy system can be configured to map inputs to outputs, for the same purpose as any other computing system [15, 16, 17, 18, 19]. Basically, it consists of 3 stages: fuzzification, rule evaluation and defuzzification. Fussification, is a translation process to obtain the fuzzy representation from the current or crisp values (for example, temperature), for which it uses the membership functions. Rule evaluation, or fuzzy inference, is the way of producing fuzzy numerical responses from linguistic rules applied to the input fuzzy values. Defuzzification is a translation process that allows obtaining a representative numerical value of all the outputs from the fuzzy information produced by the evaluation of rules. Fuzzy logic finds application in scenarios where the involved process is highly complex, and precise mathematical models for nonlinear processes are not adequately available., and when definitions and knowledge involve a high degree of uncertainty. On the other hand, the use of fuzzy logic should be avoided when a mathematical model is sufficient to actually solve the problem, when the problems are linear, or when there is no solution.This technique has been used quite successfully in the industry, mainly in Japan, and is increasingly being used in a multitude of fields. The first time it was used in a major way was in the Japanese subway, with excellent results. Air conditioning control systems Automatic focus systems in cameras Optimization of industrial control systems Handwriting recognition systems Improvement in the efficiency of the use of fuel in engines expert knowledge systems computer technology Advantages and disadvantages As the main advantage, it is worth noting the excellent results provided by a control system based on fuzzy logic: This system has the capability to deliver fast and accurate outputs, thereby minimizing fundamental state transitions in the physical environment it operates. It possesses the skill to foresee future events and maintain stable conditions in the physical environment it controls. However, the main drawback lies in deciding the appropriate membership function for fuzzy sets, as it can be challenging to determine the impact of the quantifiers in our natural language on the said function. Wrong specification of membership function in the system could lead to a probable failure of the entire system.

A fuzzy controller in Fig. 3 is characterized by inputs and outputs which are correlated by membership functions (MF).

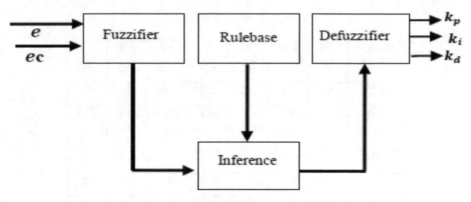

**Fig. 3.** The fuzzy logic controller

In this work we used the two-input fuzzy sets are defined in Fig. 4. [11], and three outputs fuzzy set are defined in (Fig. 5.). [11] and we used fuzzy rules in tables flowing: Tables 1, 2 and 3.

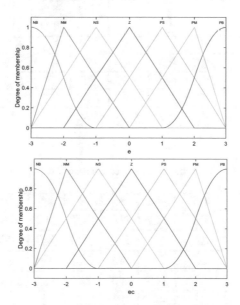

**Fig. 4.** The fuzzy sets of the FLC inputs prior to tuning.

**Table 1.** Fuzzy rule base of kp

| e↓de→ | PB | PM | PS | Z | NS | NM | NB |
|---|---|---|---|---|---|---|---|
| NB | Z | Z | PS | PM | PM | PB | PB |
| NM | NS | Z | PS | PS | PM | PB | PB |
| NS | NS | NS | Z | PS | PM | PM | PM |
| Z | NM | NM | NS | Z | PS | PM | PM |
| PS | NM | NM | NS | NS | Z | PS | PS |
| PM | NB | NM | NM | NM | NS | Z | PS |
| PB | NB | NB | NM | NM | NM | Z | Z |

**Table 2.** Fuzzy rule base of ki

| e↓de→ | PB | PM | PS | Z | NS | NM | NB |
|---|---|---|---|---|---|---|---|
| NB | Z | Z | PS | NM | NM | NB | NB |
| NM | Z | Z | PS | NS | NM | NB | NB |
| NS | PS | PS | Z | NS | NS | NM | NB |
| Z | PM | PM | NS | Z | NS | NM | NM |
| PS | PB | PM | NS | PS | Z | NS | NM |
| PM | PB | PB | NM | PM | PS | Z | Z |
| PB | PB | PB | NM | PM | PS | Z | Z |

**Table 3.** Fuzzy rule base of kd

| e↓de→ | PB | PM | PS | Z | NS | NM | NB |
|---|---|---|---|---|---|---|---|
| NB | PS | NM | NB | NB | NB | NS | PS |
| NM | Z | NS | NM | NM | NB | NS | PS |
| NS | Z | NS | NS | NM | NM | NS | Z |
| Z | Z | NS | NS | NS | NS | NS | Z |
| PS | Z | Z | Z | Z | Z | Z | Z |
| PM | PB | PS | PS | PS | PS | NS | PB |
| PB | PB | PS | PS | PM | PM | PM | PB |

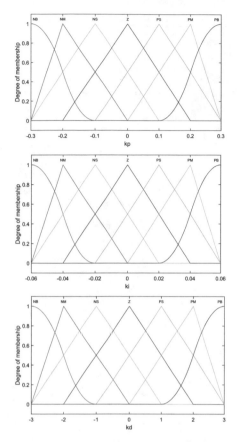

**Fig. 5.** The fuzzy sets of the FLC outputs prior to tuning.

## 4   Simulation and Results

Following the simulation process, we got this results:

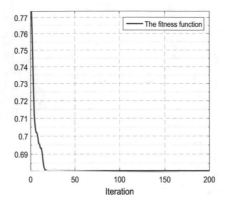

**Fig. 6.** Iterative convergence curve

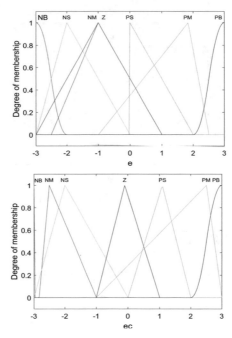

**Fig. 7.** The inputs fuzzy sets of FLC after tuning by PSO

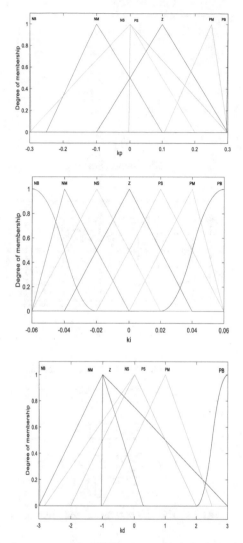

**Fig. 8.** The fuzzy sets of the FLC outputs after being tuned using PSO.

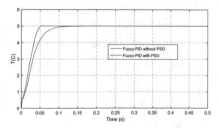

**Fig. 9.** Output response

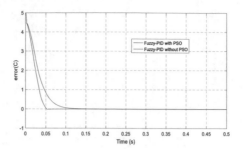

**Fig. 10.** Tracking error(c)

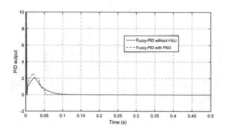

**Fig. 11.** The PID output

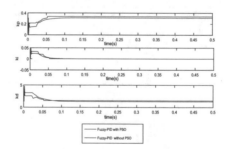

**Fig. 12.** The PID parameters auto tuning

**Table 4.** Comparative analysis of performance: Fuzzy − PIDwithPSO Controller *and* Fuzzy − PIDwithoutPSOController

|  | ISE | *IAE* |
|---|---|---|
| *Fuzzy − PIDwithPSO* | 0.6809 | 0.2175 |
| *Fuzzy − PIDwithoutPSO* [11] | 0.8762 | 0.2976 |

# 5 Conclusion

In this study, we used the PSO algorithm to determine the optimal parameters of the fuzzy-PID controller membership functions for indoor temperature management. The simulations results (Figs. 6, 7, 8, 9, 10, 11, 12 and Table 4)demonstrated that the combination of fuzzy logic and PSO algorithm achieves suitable parameters for the PID controller, which achieves the target output of the system. Fuzzy- PID control with PSO has excellent performance, including fast response, no overshoot, and reduced square integrals of errors (ISE) and absolute errors (IAE) compared to fuzzy- PID control alone. Thus, the fuzzy-PID controller proposed with the PSO strategy guarantees the desired control performance for indoor temperature management. Finally,in future work we will propose use Intelligent PID controller based on Type 2 fuzzy fuzzy logic control to regulate the indoor temperature.

# References

1. Soyguder, S., Karakose, M., Alli, H.: Design and simulation of self-tuning PID-type fuzzy adaptive control for an expert HVAC system. Elsevier. Expert Syst. Appl. **36**(3.1), 4566–4573 (2009)
2. Haissig, C.M., Woessner, M.: An adaptive fuzzy algorithm for domestic hot water. Temperature control of a combi boiler. ASHRAE Trans. **106**(2) (2000)
3. Huang, S., Nelson, R.M.: A PID-law-combining fuzzy controller for HVAC applications. ASHRAE Trans. **97**(2), 768–774 (1991)
4. Xu, J.X., Hung, C.C., Liu, C.C.C.: Parallel structure and tuning of a fuzzy PID controller. Automatica **36**, 673–684 (2000)
5. Zhao, Z.Y., Tomizuka, M.: Fuzzygain scheduling of PID Controllers. IEEE Trans. Syst. Man Cybern. **23**(5), 1392–1398 (1993)
6. Xu, J.X., Hung, C.C., Liu, C.: Tuning and analysis of a fuzzy PI controller based on gain and phase margins. IEEE Trans. Syst. Man Cybern. **28**(5), 685–691 (1998)
7. Jianghui, C., Yongsheng, Z., Chongzhu, W.: Research on optimization of fuzzy membership function based on ant colony algorithm. IEEE Chinese Control Conf., 1209–1213 (2006)
8. Botzheim, J., Cabrita, C., Koczy, L.T., Ruano, A.E.: Estimating fuzzy membership functions parameters by the Levenberg-Marquardt algorithm. In: 2004 IEEE International Conference on Fuzzy Systems (2004). https://doi.org/10.1109/FUZZY.2004.1375431
9. Ullah, A., Li, J., Hussain, A., Shen, Y.: Genetic optimization of fuzzy membership functions for cloud resource provisioning. In: IEEE Symposium Series on Computational Intelligence (SSCI) (2016). https://doi.org/10.1109/SSCI.2016.7850088
10. Bennaoui, A., Ameur, A., Saadi, S., Bennaoui, A.: Moth-flame optimizer algorithm for optimal of fuzzy logic controller for nonlinear system. In: Hatti, M. (eds) Advanced Computational Techniques for Renewable Energy Systems. IC-AIRES. Lecture Notes in Networks and Systems, vol. 591. Springer, Cham (2023).https://doi.org/10.1007/978-3-031-21216-1_72
11. Yang, S.: Intelligent PID controller based on fuzzy logic control and neural network technology for indoor environment quality improvement. PhD thesis, University of Nottingham (2014)
12. Li, W., Song, W., Chun, W., Li, L., Huanhuan, M.: Electric vehicle charging strategy under time-of-use electricity price mechanism based on particle swarm optimization algorithm. In: 2023 4th International Conference on Computer Engineering and Application (ICCEA), Hangzhou, China, pp. 270–273 (2023). https://doi.org/10.1109/ICCEA58433.2023.10135375

13. Bansal, J.C.: Particle Swarm Optimization. In: Bansal, J.C., Singh, P.K., Pal, N.R. (eds.) Evolutionary and Swarm Intelligence Algorithms. SCI, vol. 779, pp. 11–23. Springer, Cham (2019). https://doi.org/10.1007/978-3-319-91341-4_2

14. Kennedy, J., Eberhart, R.: Particle swarm optimization. In: Proceedings of the IEEE International Conference on Neural Networks, Perth, Australia, pp. 1942–1948 (1995)

15. Mahalakshmi, G., Brijesh, M., Gokul Raj, M., Gowchik, S., Barathkavi, A.: Design of fuzzy logic controller for solar PV fed BLDC motor with zeta converter for precision controlled system. In: 2023 4th International Conference on Signal Processing and Communication (ICSPC), Coimbatore, India, pp. 111–115 (2023). https://doi.org/10.1109/ICSPC57692.2023. 10126041.

16. Jabbar, R.I., et al.: A modified P&O-MPPT technique using fuzzy logic controller for PV systems. In: 2023 IEEE IAS Global Conference on Emerging Technologies (GlobConET), London, United Kingdom, pp. 1–7 (2023). https://doi.org/10.1109/GlobConET56651.101 49991

17. Arslankaya, S.: Comparison of performances of fuzzy logic and adaptive neuro fuzzy inference system (ANFIS) for estimating employee labor loss. J. Eng. Res., 100107 (2023)

18. Lee, C.C.: Fuzzy logic in control systems: fuzzy logic controller-Part I/II. IEEE Trans. Syst. Man Cybern. **20**, 404–435 (1990)

19. Zadeh, L.A.: The concept of a linguistic variable and its application to approximate reasoning. Inf. Sci. **8**, 199–249 (1975)

# Kalman MPPT Controller for Home Turbine Connected to the Utility Grid

A. Abbadi[1](✉), F. Hamidia[1], M. R. Skender[2], A. Morsli[1], and F. Bettache[3]

[1] Electrical Engineering Department, Research Laboratory in Electrical Engineering and Automatic (LREA), University of Médéa, Médéa, Algeria
amel.abbadi@yahoo.fr
[2] Electrical Engineering Department, Renewable Energy and Materials Laboratory (REML), University of Médéa, Médéa, Algeria
[3] Electrical Engineering Department, Laboratory of Mechanics, Physics and Mathematical Modeling, University of Médéa, Médéa, Algeria

**Abstract.** This article aims to present an MPPT (Maximum Power-Point Tracking) approach based on the Kalman filter principle. This technique ranks among the methods that do not require a mathematical model. It is very interesting because the prediction of the future state is made by means of the present and the estimated past. This approach is used to optimize the performance of a wind energy conversion system (WECS) connected to residential applications. This MPPT controller was developed in two variants: with and without a speed sensor. The main objective of this work is to test the feasibility of the proposed approaches (Kalman MPPT) to extract the maximum power from the WECS and to analyse the results obtained during the variation of the wind speed (for small and large changes). The total harmonic distortion (THD) of residential current is used as a comparison parameter between the proposed method and the disturbance and observation technique.

**Keywords:** MPPT controller · Kalman filter · wind energy conversion system · maximum power · tracking algorithm

## 1 Introduction

To mitigate or even overcome the impact of energy crises and environmental problems, we are moving towards the production of electricity from renewable sources, such as wind, photovoltaic (PV), biomass, etc. Wind Energy Conversion System (WECS) is one of the best sources of renewable energy because of the advantages it has such as abundant availability of energy sources; which is also non-polluting and low cost.

The most widely used components in wind energy conversion systems are permanent magnet synchronous generators (PMSG) feeding three-phase rectifiers followed by grid-connected inverters [1–4].

As is known, the continuous extraction of wind energy is highly dependent on the variation in wind speed. Thus, to extract the maximum power from wind energy, several

M. Hatti (Ed.): IC-AIRES 2023, LNNS 984, pp. 111–118, 2024.
https://doi.org/10.1007/978-3-031-60629-8_11

MPPT techniques have been proposed [2, 3, 5–7]. These MPPT regulators are generally subdivided into two large groups. The first requires prior knowledge of the parameters of the turbine and the MSAP machine to calculate the maximum operating point [8]. The second group searches for the optimum using power and/or rotational speed increments based either on iterative techniques [9] or on prediction and correction as for the Kalman filter [10].

Interested in the advantages offered by the prediction-correction type MPPT controller, this article presents a model-free MPPT controller based on a Kalman filter. In this article, the Kalman filter is used to extract the maximum power from the wind energy conversion system. The proposed MPPT controller automatically varies the reference voltage Vdc of the inverter.

This article is structured as follows. The configuration of the wind energy conversion system is detailed in Sect. 2. Section 3 describes the Kalman MPPT technique. Simulations are performed in Sect. 4 to validate the efficiency of the proposed MPPT controller. Finally, in Sect. 5, conclusions are drawn.

## 2 Configuration of the Residential Wind Energy Conversion System

Figure 1 shows the proposed system's block diagram. The wind energy conversion system is composed of a permanent magnet synchronous generator, a rectifier stage, an inverter control module, a load (Home), and a large-scale interconnected grid.

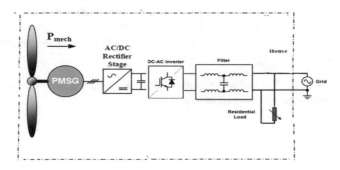

**Fig. 1.** WECS (wind energy conversion system) is associated with residential applications.

The wind turbine's extracted power can be determined through [11]:

$$P_{Turbine} = \rho \pi R^2 C_p(\lambda, \beta) v^3 \big/ 2 \qquad (1)$$

where, $\rho$ is the air density (kg/m$^3$), $R$ is the rotor blade radius (m), $v$ is the wind speed (m/s), and $C_p$ is the power coefficient of the turbine which is a function of two variables: the blade pitch angle $\beta$ (in degrees) and the tip speed ratio $\lambda$. The tip-speed ratio is obtained as:

$$\lambda = \omega_m R \big/ v \qquad (2)$$

where $\omega_m$ is the blades angular velocity (in rad/sec). $T_m$ is the mechanical torque generated by the wind turbine. It results by calculation as follows:

$$T_m = \frac{1}{2}\rho A C_p(\lambda, \beta)v^3 \frac{1}{\omega_m} \tag{3}$$

The power coefficient $C_p$ $(\lambda, \beta)$ used in this study is expressed as [11]:

$$C_p = c_1\left(\frac{c_2}{\lambda_i} - c_3\beta - c_4\right)e^{-\left(\frac{c_5}{\lambda_i}\right)} + c_6\lambda; \quad \frac{1}{\lambda_i} = \frac{1}{\lambda + 0.08\beta} - \frac{0.035}{\beta^3 + 1} \tag{4}$$

The coefficient $c_1$ to $c_6$ are: $c_1 = 0.5176$, $c_2 = 116$, $c_3 = 0.4$, $c_4 = 5$, $c_5 = 21$ and $c_6 = 0.0068$.

## 3  Kalman MPPT

### 3.1  Kalman Filter

Rudolf Kalman gave his name to the Kalman filter. It is an approach involving delivering a suitable answer in the sense of least squares iteratively. This strategy has particular attraction since it predicts the future state using the present and estimated past, even in the absence of a formal model.

The Kalman filter consists of two phases: prediction and correction.

**The Prediction Phase:**  This phase estimates the current state by using the state reflected in the preceding instant.

$$x_k^- = A x_{k-1} + B u_{k-1} \tag{5}$$

where: $x_k^-$ is the estimated state in the predicted iteration $k$ from the previous iteration. $x_{k-1}$ is the previous state estimate. $u_{k-1}$ is the control process of the $k-1$ iteration process. $A$ is a linear model. $B$ a constant that depends on the model that is used in the control process [12].

The predicted error covariance matrix is given by:

$$H_k^- = A H_{k-1}A^T + Q \tag{6}$$

where $H_k^-$ and $H_{k-1}$ are the estimated error covariance matrices in the predicted iteration k and k-1 respectively. $Q$ represents the model error.

**The Correction Phase:**  The observations of the present instant are utilized to adjust the anticipated state in order to obtain a more precise estimate during the correction phase. To do so, we begin by calculating the K gain.

$$K_k = H_k^- C^T \left(C H_k^- C^T + R\right)^{-1} \tag{7}$$

Afterward, the estimate $x_k$ is updated through output $y_k$.

$$x_k = x_k^- + K_k\left(y_k - C x_k^-\right) \tag{8}$$

Finally, the covariance error matrix is updated from the equation:

$$H_k = (I - H_k C) H_k^-$$ (9)

where: $K_k$ is the Kalman gain. $R$ Covariance of noise. $y_k$ the measurement. C a constant related to the Kalman Filter system and the observed space [12].

## 3.2 MPPT Kalman Filter

MPPT techniques have been employed in solar panels since the 1970s. We chose to develop our Kalman MPPT controller considering that the power-speed characteristics of wind turbines are similar to the power-voltage characteristic curve of solar panels.

In our case, the Kalman filter is designed to find the maximum power point. The reference voltage is the output of this filter. To predict the MPP point, it suffices to analyze the P-ω curve of Fig. 2

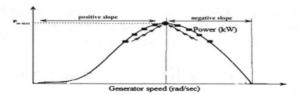

**Fig. 2.** Characteristic curve of the wind turbine: power versus speed

As in [10], we consider that A = 1 and B = M. Thus, the equations for the prediction step are:

$$V_k^- = V_{k-1} + M \frac{\Delta P}{\Delta \omega}$$ (10)

$$H_k^- = H_{k-1} + Q$$ (11)

where: $V_k^-$ is the estimated voltage value using the Kalman filter in the k. M equivalent to B and it is considered a scaling factor. $\Delta P / \Delta \omega$ is the slope of the curve P-ω and it is considered as the equivalent to the control $u_{k-1}$ [10].

The correction step process is made it as follows.

The $K$ gain, according to [10], is calculated using $C = 1$ and the covariance error is as

$$K_k = H_k^- \left( H_k^- + R \right)^{-1}$$ (12)

The voltage value is calculated (corrected) by taking into account the voltage generated by the turbine (measured):

$$V_k = V_k^- + K_k \left( V_{turbine,k} - V_k^- \right)$$ (13)

The updated covariance error matrix is:

$$H_k = (1 - K_k)H_k^-$$ (14)

Once the MPP value is reached, $V_k = V_{k\,ref\_mpp}$.

In this MPPT control scheme, we should use a speed sensor. It will be more interesting if we can design an MPPT controller without speed sensor; using only the measurements: current and voltage generated by our PMSG.

Since the output voltage of the rectifier is proportional and linear to the rotor speed of the PMSG [13], we can adopt the following approach:

$$V_k^- = V_{k-1} + M \frac{\Delta P}{\Delta V}$$ (15)

## 4 Simulation Result

To test the efficiency of the MPPT controller using the Kalman filter with the two approaches (with and without speed sensor: Fig. 3.a and Fig. 3.b), the wind energy conversion system connected to residential applications was simulated under different scenarios.

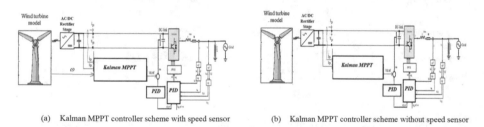

(a)  Kalman MPPT controller scheme with speed sensor          (b)  Kalman MPPT controller scheme without speed sensor

**Fig. 3.** Proposed Kalman MPPT controllers.

Two rectangular wind speed profiles were generated, the first varying from 6.5 m/s to 7 m/s (Fig. 4.a) and the second varying from 8 m/s to 11 m/s (Fig. 4.b) with variable wind speed steps. Figures 4 and 5 show the wind speed, rotor speed, and power coefficient as a result of the two scenarios, with and without speed sensor.

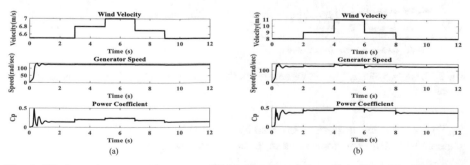

**Fig. 4.** Wind speed, rotor speed, power coefficient for two wind speed profiles with speed sensor

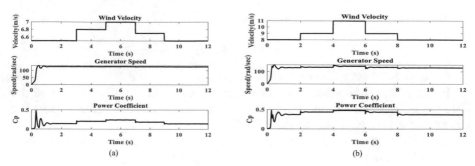

**Fig. 5.** Wind speed, rotor speed, power coefficient for two wind speed profiles without speed sensor

Figures 6 and 7 show the residential load voltage and current for the two suggested wind speed profiles using both approaches (with and without speed sensor). For both wind speed profiles, the proposed MPPT techniques displayed good tracking performance.

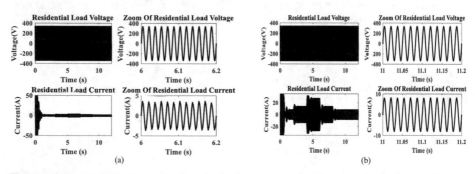

**Fig. 6.** Inverter voltage, inverter current for both proposed wind speed profiles with speed sensor

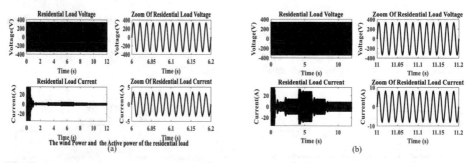

**Fig. 7.** Inverter voltage, inverter current for both proposed wind speed profiles without speed sensor

Figure 8 depicts the power generated by the wind turbine as well as the active power generated by the inverter for the two suggested wind speed profiles using both approaches

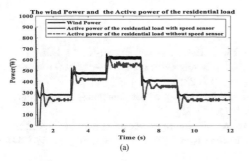 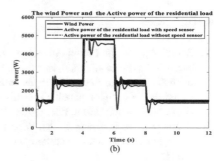

(a)                                                              (b)

**Fig. 8.** Power generated by the wind turbine and active power generated by the inverter for both proposed wind speed profiles with and without speed sensor

(with and without speed sensor). The proposed MPPT techniques performed similarly for both wind speed profiles.

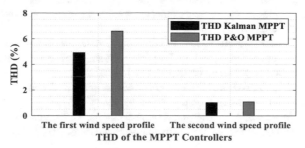

THD of the MPPT Controllers

**Fig. 9.** The total harmonic distortion of the residential load current for the both wind speed profiles with Kalma MPPT controller and P&O MPPT controller

Residential load currents exhibited the same total harmonic distortion (THD) for both wind speed profiles with and without the speed sensor. Figure 9 depicts the total harmonic distortion (THD) of the residential load current using two MPPT controllers: the Kalman MPPT controller and the P&O MPPT controller for the two wind speed profiles. The proposed MPPT controller clearly provided lower total harmonic distortion (THD) than the P&O MPPT controller.

## 5 Conclusion

A Kalman filter is applied as an MPPT controller in this paper to extract the maximum power from a grid-connected wind energy conversion system. This MPPT controller was designed in two ways (with and without a speed sensor). It should be highlighted that the MPPT controllers proposed are model-free MPPT controllers. The wind speed was simulated using two profiles, one with modest fluctuations and one with big fluctuations. The simulation results show that both wind profiles are adequately handled by the suggested controllers (with and without a speed sensor). The results of the simulations in the two proposed scenarios show that the THD is less than 5%, as required by the IEEE 519 standard.

# References

1. Raouf, A., Basem Tawfiq Albassioni, K., Eldin, E.T., Youssef, H., El-Kholy, E.E.: Wind energy conversion systems based on a synchronous generator: comparative review of control methods and performance. Energies **16**(5), 1–22 (2023)
2. Gaied, H., et al.: Comparative analysis of MPPT techniques for enhancing a wind energy conversion system. Front. Energy Res. **10**, 1–15 (2022)
3. Zhang, X., Jia, J., Zheng, L., Yi, W., Zhang, Z.: Maximum power point tracking algorithms for wind power generation system: review. Comparison Anal. Energy Sci. Eng. **11**, 430–444 (2023)
4. Dehghan, S.M., Mohamadian, M., Varjani, A.Y.: A new variable-speed wind energy conversion system using permanent-magnet synchronous generator and Z-source inverter. IEEE Trans. Energy Conv. **24**(3), 714–724 (2009)
5. Abdullah, M.A., Yatim, A.H.M., Tan, C.W., Saidur, R.: A review of maximum power point tracking algorithms for wind energy systems. Renew. Sustain. Energy Rev. **16**(5), 3220–3227 (2012)
6. Linus, R.M., Damodharan, P.: Maximum power point tracking method using a modified perturb and observe algorithm for grid connected wind energy conversion systems. IET Renew Power Gen. **9**(6), 682–689 (2015)
7. Ahmed, B., Nurul, H., Yusoff, H.S.H., et al.: Paper review: maximum power point tracking for wind energy conversion system. In: 2nd International Conference on Electrical, Control and Instrumentation Engineering (ICECIE), pp. 1–6 (2020)
8. Kot, R., Rolak, M., Malinowski, M.: Comparison of maximum peak power tracking algorithms for a small wind turbine. Math. Comput. Simul **91**, 29–40 (2013)
9. Katche, M.L., Makokha, A.B., Zachary, S.O., Adaramola, M.S.: A comprehensive review of maximum power point tracking (MPPT) techniques used in solar PV systems. Energies **16**(5), 1–23 (2023)
10. Kang, B.O., Park, J.H.: Kalman filter MPPT method for a solar inverter. In: IEEE Power and Energy Conference at Illinois, pp. 1–5 (2011)
11. Patil, K., Mehta, B.: Modeling and control of variable speed wind turbine with permanent magnet synchronous generator. In: IEEE International Conference on Advances in Green Energy (ICAGE'14), pp. 1–7 (2017)
12. Costa, P.J., Dunyak, J.P., Mohtashemi, M.: Models, prediction, and estimation of outbreaks of infectious disease. Proc. IEEE SoutheastCon, 174–178 (2005)
13. Putri, R.I., Syamsiana, I.N., Rifa'i, M., Aditya, F.: Voltage control for variable speed wind turbine using buck converter based on PID controller. IOP Conf. Ser.: Mater. Sci. Eng. 1073 012048, 1–7 (2021)

# Variation of the Speed of a DC Motor Powered by the Photovoltaic Generator Using MPPT Control

Fatima Moulay[1,2]($\boxtimes$) and Assia Habbati[2]

[1] Irecom Laboratory, University of Technology, University of Djillali Liabés, Sidi bel Abbes, Algeria
fatimamoulay66@yahoo.fr
[2] Energarid Laboratory, University of Technology, Bechar, Algeria

**Abstract.** This paper presents the analysis, modeling and control model of the electrical part of a photovoltaic generation system, connected to the DC motor by a Boost converter using MPPT control. Initially, we carried out a detailed study of the direct current motor with its characteristics, then we developed a mathematical model of the Photovoltaic Module where it is necessary to take meteorological data (irradiance and temperature) as input variables, the current, voltage or power as output variables. The model predicts the behavior and characteristics of the PV module based on the equivalent circuit of the mathematical model using the Matlab/Simulink platform under different temperatures and solar radiation readings. Then, a basic circuit of the boost converter is designed with constant DC source voltage, in the third step, a comparative study was also carried out for the converter connected to the PV system directly with the converter connected to the load (the motor has direct current) powered by the pulse generator. Therefore, the design of the model with the contribution of the boost converter, which responds to the use of systems aimed at increasing the input voltages and providing the loads with the desired outputs, leads us to suggest the efficiency and the use of this model.

**Keywords:** Modelling · PV system · duty pulses generator · irradiance · DC motor · Boost Converter · MPPT

## 1 Introduction

Thus, a rigorous study is necessary to make the best choice and the most efficient with the lowest possible cost [1]. The performance of a PV system is highly dependent on weather conditions, such as solar radiation, temperature and wind speed [2]. This energy conversion takes place through a so-called photovoltaic (PV) cell based on a physical phenomenon called the photovoltaic effect, which consists of producing an electromotive force when the surface of this cell is exposed to light. The voltage generated may vary depending on the material used to manufacture the cell. The association of several PV cells in series/parallel gives rise to a photovoltaic generator (GPV) which has a nonlinear current-voltage (I-V) characteristic presenting a point of maximum power [3]. Today, the

M. Hatti (Ed.): IC-AIRES 2023, LNNS 984, pp. 119–129, 2024.
https://doi.org/10.1007/978-3-031-60629-8_12

global photovoltaic (PV) industry relies heavily on the needs of remote areas for reliable and inexpensive power supply. In a large number of applications, photovoltaic is simply the most cost-effective solution. Examples of such applications include isolated systems supplying cottages or remote residences, utilities and the military, water pumping on farms, and emergency call stations on campuses or highways. The objective of this work is to simulate a photovoltaic system which supplies a direct current motor and this through a booster chopper adaptation stage [5]. This paper presents the analysis, modeling and control model of the electric part of a PV generation system connected to the utility load (DC motor) by a boost converter illustrate by the Fig. 1.

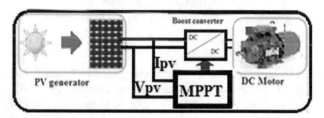

**Fig. 1.** Synoptic diagram of the PV system

## 2   DC Motor

### 2.1   Modeling of the DC machine

The modeling of the motor + load assembly is feasible from the basic equations of the direct current machine and the fundamental principle of dynamics [12]:

$$u(t) = E + R_a i(t) + L_a \frac{di(t)}{dt} \tag{1}$$

$$\begin{cases} C_e = K_t i(t) \\ E = K_e \Omega(t) \end{cases} \tag{2}$$

$$Ce = J \frac{d\Omega(t)}{dt} + f\Omega(t) + Cr \tag{3}$$

$C_e$ Motor torque
$C_r$ Resistant torque
E Back electromotive force
V Input voltage
$\Omega$ Angular speed
$R_a$, $L_a$ Armature resistance and inductance
J Rotating inertial measurement of motor
f Damping coefficient

Motor parameters: Ra = 0.6 $\Omega$; La = 0.02 H; Kt = 1.25; Ke = 1.25; J = 0.03; f = 0.06.

Load: Cr = 100 Nm (application in steady state).

Command step 200 rad/sec.

## 2.2  Modeling of the Three-Phase Source

It is assumed that the rectifier is supplied by a balanced voltage three-phase network.
$V_m = 220$ V.
f = 50 Hz.

## 2.3  Modeling of the Rectifier

And if we neglect the encroachment effect, the output voltage of the rectifier will be defined as following:

$$U_{red}(t) = Max\ [U_a(t).U_b(t).U_c(t)] - Min\ [U_a(t).U_b(t).U_c(t)] \tag{4}$$

## 2.4  Modeling of the Filter

One can pull the filter of "Butterworth" from the Simulink library/Signal Processing Blockset/Filtering/Filter Designs/
Lf = 400 mH.
Cf = 600 µF.

## 2.5  Static Converter (chopper) and Pulse Generator

With the pulses thus produced, we attack the trigger of the transistor; we notice that when we vary the voltage Vcom the pulses will change in their turn, which causes either the increase or the decrease in the supply voltage of the DC motor therefore a variation speed. The controller forces this speed to follow the set point.
*Time interval:* [0 1/ (50*30)]. **Exits**: [0 1]

## 2.6  Choice and the Expression of the PI Regulator

For the regulation of our speed, we chose the PI regulator (proportional integral regulator) because it improves the precision of the loop (zero static error).The general structure of a proportional integral (PI) regulator is composed of the proportional function and the integral function placed in parallel [13].
    The PI regulator is modeled by the transfer function G(s), with:

$$G(s) = \frac{sK_p + K_i}{s} = K_p + \frac{1}{s}K_i \tag{5}$$

$K_p = 150$ and $K_i = 12$.

## 2.7  Global block diagram of the system studied under Simulink

In this part of the work, we studied the direct current machine (DC MOTOR) and we chose the model that best corresponds to our application by modeling it with these own mathematical equations and we even integrated a speed regulator given by the Fig. 2.

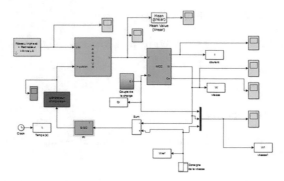

**Fig. 2.** Global block diagram of the system studied under Simulink

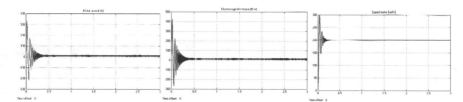

**Fig. 3.** DC motor output characteristics

## 2.8 Study of the Direct Current Motor Without Resistive Torque

At the Fig. 3, the motor made an inrush current at start-up to bring its speed to its nominal speed, thereafter it stabilizes in steady state.

- In steady state the motor turns at a constant speed (the motor runs off-load), at this moment the torque value is very low so the armature current is also low: $I_a = 6.5$ A and in the transient state $I_a = 300$A.
- The electromagnetic torque evolves in the same way as with the armature current because we have $C_{em} = K_f *I_a = 8.125$ Nm and in the transient state $C_{em} = K_f *I_a = 1.25*300 = 375$ Nm.
- The speed increases as a function of time then it stabilizes at a constant value, at this instant the resistive torque is negligible: $W_r = 200$ rd/s

## 2.9 DC motor with resistive torque

This puts into action a resistant torque at the instant "t = 1.5 s" which is equal to "Cr = 100 N.m".

Steady state, from the Fig. 4, obtained, the following values are taken values:

Armature current Ia = 95.5 A.

Electromagnetic torque Cem = 1.25. 95.7 = 119.37 N.m

Rotational speed Wr = 190.01 rd/s and it stabilized at 200 rd/s after 0.2 s and this is thanks to the PI regulator inserted in the motor control loop.

Puts into action a resistant torque from the moment "t = 1.5 s" to "t = 2s" which is equal to "Cr = 100 N.m".

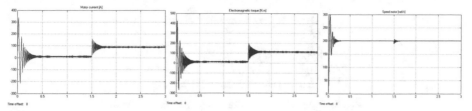

**Fig. 4.** The output characteristics of the dc motor after the application of the first resistive torque

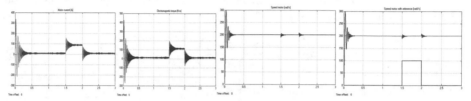

**Fig. 5.** The output characteristics of the dc motor after the application of the second resistive torque

In the presence of the disturbance which is the application of the resistive torque from t = 1.5s to t = 2s, we notice the increase in the armature current as well as the electromagnetic torque but for the speed, it remained constant and it follows the set point value which is 200 rd/sec and with a small disturbance at the time of the application of the resistive torque and this thanks to the chosen PI regulator or its values Kp = 150 and Ki = 12 were chosen by trial and error.

A resisting torque has been applied to the machine to illustrate its influence on current and speed. In the same way as the system alone, we note that when the machine is connected to direct current, the system operates on the same principle with an increase in voltage and power due to an increase in the number of modules.

According to the Fig. 5, which represents the output speed of the motor, a resistive torque Cr = 10 N.m has been applied at time t = 1.5s to t = 2 s, we see that when a torque is applied, the speed will drop, which will change the operation of the machine (DC MOTOR). It is for that we used a speed regulator (PI) to be able to bring the speed of the machine (DC MOTOR) to its reference speed whatever the load applied.

## 3    Modelling of a Photovoltaic Module

PV cells are connected together to form photovoltaic module or photovoltaic panel. The equivalent circuit of a solar cell is shown in Fig. 6. There are mathematical models of the characteristic current–voltage of the photovoltaic cell [14]. However, the most practical by their relative simplicity in the calculation are the single exponential model (a single diode), in which, the obscurity current (diode current), representing uniquely the saturation current resulting of the diffusion phenomenon, and the two-exponential model (two diodes) representing both the components, diffusion and recombination.

The equivalent circuit of the general model which consists of a photo current, a diode, a parallel resistor expressing a leakage current, and a series resistor describing an internal

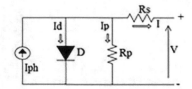

**Fig. 6.** Equivalent circuit of the photovoltaic module

resistance to the current flow, is shown in Fig. 6. The voltage-current characteristic equation of a solar cell is given as:

$$I = Iph - I_0\left[exp\left(\frac{e(V + Rs\,I)}{\lambda KTc}\right) - 1\right] - \frac{V + Rs\,I}{Rp} \tag{10}$$

V The voltage at the terminals of the module,
λ Coefficient characterizing the power variation as a function of temperature,
K The Boltzmann constant,
e The electron charge,
I0 The saturation current.

Rs and Rp Resistors represent the losses of metal contacts and leakage of the PN junction respectively Fig. 6,gives the PV module Matlab/SIMULINK model [15]. The mathematical model of the characteristic I (V) represented by Eq. 10, made it possible to draw the curves of the variation in I–V and P–V characteristics. The evolution of the I − V characteristic as a function of temperature shows that the current increases very rapidly when the temperature rises and generates a less pronounced decrease in the open circuit voltage given by the Fig. 7.

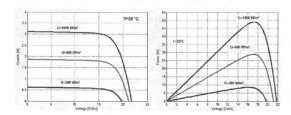

**Fig. 7.** I–V, P–V characteristics for different irradiation levels for T constant.

## 4   Boost Converter

We took the simple example of a continuous source which feeds a boost converter. We plug the circuit with the values of the 2813 µH inductor and the 2200 µF capacitor (Fig. 8). Voltage Vs = 100 V, V0 = 220V, R = 10 Ω and switching frequency 50 KHz.

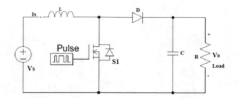

**Fig. 8.** Circuit diagram of the Boost converter

## 5   Modeling and Simulation of the PV System with the Converter

In the Fig. 9, we take the PV module shown in Fig. 8 and we connect the boost DC-DC converter with the same parameters calculated for the example and for the same objective which is to have a boost input voltage of 100 V and a voltage boost output of 220 V or load voltage for our case we took a resistive load of 50 ohms.

After simulation, the results are given by Fig. 11, for constant irradiance at a value of 1000 W/m².

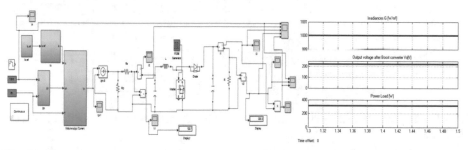

**Fig. 9.** PV and boost converter with impulsion generator and the output characteristic of the PV and the Boost converter

## 6   Simulation of a PV-Boost Panel with DC Motor and MPPT Control

### 6.1   Block diagram

We take the diagram of Fig. 9, we replace the resistive load by the direct current motor and the pulse generator by the MPPT control.

The MPPT command must have a high level of simplicity and a reasonable cost. In addition, regarding its performance, the MPPT control must have good dynamic and static behavior to ensure rapid and precise adaptation to climatic changes. The most commonly encountered classical methods are commonly referred to respectively as Perturbation and Observation (P&O) and Incremental Conductance (Inc Cond). Two sensors are generally necessary to measure the voltage and the current of the PV generator from which the power is calculated. These methods can be implemented on a microcontroller, which can keep in memory the previous values of the voltage and the current from the PV panel [11–14].

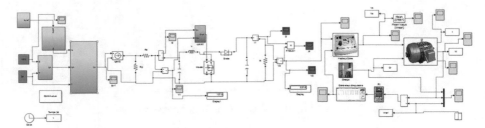

**Fig. 10.** The PV system simulation block powering a DC motor (Simulink / Matlab)

## 6.2 Simulation and results of the DC motor with constant excitation

In Fig. 11, the output of the boost converter is varied, to have an output voltage equal to 220 V, to power the motor. The regulator forces the speed to follow the setpoint.

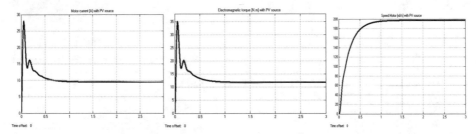

**Fig. 11.** Current, electromagnetic torque and speed characteristic as a function of time of the DC MOTOR

The motor made an inrush current at start-up to bring its speed to its nominal speed, thereafter it stabilizes in steady state.

In steady state the motor turns at a constant speed (the motor runs off-load), at this moment the torque value is very low so the armature current is also low: Ia = 9.45 A and in in transient state Ia = 28 A.

The electromagnetic torque evolves in the same way as with the armature current because we have Cem = Km* Ia = 1.25*9.45 = 11.81 N.m and in in transient state Cem = Km* Ia = 1.25* 28 = 35 N.m

The speed increases as a function of time then it stabilizes at a constant value, at this instant the resistive torque is negligible (Wr = 200 rd/s).

The motor uses a high current at start-up to bring its speed to a nominal speed, thereafter it stabilizes in a steady state and the motor runs at a constant speed (the motor operates at no load). The speed increases with time and then stabilizes at a constant value.

## 6.3 Result and Interpretation

The Fig. 12, represent the power of the generator as well as the voltage and the output power of the DC-DC converter.

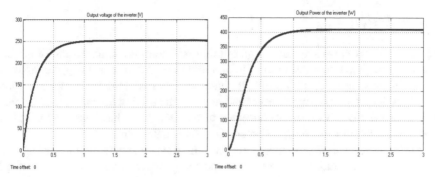

**Fig. 12.** DC-DC Converter Output Voltage

In our case, we varied the boost parameters to obtain the desired voltage, but in reality, we had to increase the number of panels. The voltage increased after using four panels in series, which allowed us to connect the machine to a direct current.

### 6.4 Influence of irradiance and temperature

In order to analyze the reaction of the system following a change in the irradiance, we proceed to a change from G = 1000 w/m² to G = 600 W/m² and T = 25 °C.

It is remarkable that the value of the voltage has a direct and proportional link to the intensity of the irradiance; it decreases with the decrease of the latter. The same for the power since it depends on the voltage with a decrease in the ripples.

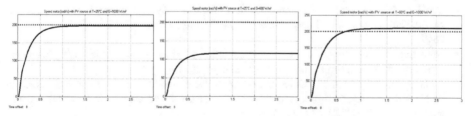

**Fig. 13.** Speed/time characteristic of the DC MOTOR with T = 25 °C and G = 1000 W/m² and G = 600 W/m² (T = 60 °C and G = 1000 W/m that powers the DC motor)

Figure 13 shows that the irradiance has a direct influence on the speed of the DC motor.

To see as a function of time the reaction of the system to the change in temperature, we apply a change from T = 25 C° to T = 60 C°. We notice,that the temperature has a direct influence on the output voltage of the photovoltaic generator, the output voltage before and after the converter drops quite sharply when the temperature increases.

# 7 Conclusion

The strength of this work is the design of the model with the contribution of the boost converter, which is answered in the use of systems aimed at increasing the input voltages and supplying loads with desired outputs. In this study, the fundamental circuit is considered for its simplicity of design and to make the system less complex. A detailed analysis of the Boost Converter connected to the DC motor has been performed. The basic boost converter was designed according to the calculated parameters.

In this paper, we have presented the modelling of system elements and the simulation for the different conditions of irradiance and temperature. The performance of a PV generator is strongly influenced by climatic conditions, particularly solar irradiation and the temperature of the PV module. In order to extract the maximum power available at the terminals of the generator and to transfer it to the load, the technique conventionally used is to use an adaptation stage, between the PV generator and the load. This stage plays the role of interface between the two elements by ensuring, through a control action, the transfer of the maximum power supplied by the generator so that it is as close as possible to maximum power. The solution frequently adopted is the incorporation of a static converter which acts as a source-load adapter controlled directly by the PWM technique. The choice of conversion structure depends on the load to be supplied.

# References

1. Ortega, E., Aranguren, G., Saenz, M.J., Gutierrez, R., Jimeno, J.C.: Study of photovoltaic systems monitoring methods. In: 2017 IEEE 44th Photovoltaic Specialist Conference (PVSC) (2017). https://doi.org/10.1109/pvsc.2017.8366523
2. Van Sark, W., Louwen, A., Tsafarakis, O., Moraitis, P.: PV System monitoring and characterization. Photovoltaic Solar Energy, 553–563 (2017). https://doi.org/10.1002/9781118927496.ch49
3. Ngoc Son, N., The Vinh, L.: Parameter estimation of photovoltaic model, using balancing composite motion optimization. Acta Polytech. Hung. **19**(11) 2022
4. Muangkote, N., Sunat, K., Chiewchanwattana, S., Kaiwinit, S.: An advanced onlooker-ranking-based adaptive differential evolution to extract the parameters of solar cell models. Renew. Energy **134**, 1129–1147 (2019)
5. Farah, A., Belazi, A., Benabdallah, F., Almalaq, A., Chtourou, M., Abido, M.A.: Parameter extraction of photovoltaic models using a comprehensive learning Rao-1 algorithm. Energy Convers. Manag. **252**, 115057 (2022)
6. Fan, Y., Wang, P., Heidari, A.A., Chen, H., Mafarja, M.: Randomre selection particle swarm optimization for optimal design of solar photovoltaic modules. Energy **239**, 121865 (2022)
7. Pardhu, B.S.S.G., Kota, V.R.: Radial movement optimization based parameter extraction of double diode model of solar photovoltaic cell. Sol. Energy **213**, 312–327 (2021)
8. Hamied, A., Mellit, A., Zoulid, M.A., Birouk, R.: IoT-based experimental prototype for monitoring of photovoltaic arrays. In: Conference: 2018 International Conference on Applied Smart Systems (ICASS) (2018). https://doi.org/10.1109/icass.201
9. Houssein, E.H., Zaki, G.N., Diab, A.A.Z., Younis, E.M.G.: An efficient Manta Ray Foraging Optimization algorithm for parameter extraction of three-diode photovoltaic model. Comput. Electr. Eng. **94**, 107304 (2021)

10. Sallam, K. M., Hossain, M. A., Chakraborty, R. K., Ryan, M. J.: An improved gaining-sharing knowledge algorithm for parameter extraction of photovoltaic models. In: Energy Convers. Manag. **237**, 114030 (2021)
11. Luu, T. V., Nguyen, N. S.: Parameters extraction of solar cells using modified JAYA algorithm. Optik (Stuttg). **203**(1), 164034 (2020)
12. Ben Messaoud R.: Extraction of uncertain parameters of single and double diode model of a photovoltaic panel using Salp Swarm algorithm. Meas. **154**(6), 107446 (2020)
13. Liang, J., et al.: Evolutionary multi-task optimization for parameters extraction of photovoltaic models. Energy Convers. Manag. **207**, 112509 (2020)
14. Hashim, F.A., Houssein, E.H., Hussain, K., Mabrouk, M.S., AlAtabany, W.: Honey Badger Algorithm: New metaheuristic algorithm for solving optimization problems. Math. Comput. Simul **192**, 84–110 (2022)

# MOTH-FLAME-MPPT Algorithms of PV Panels Under PSC

Hamidia Fethia[1]([✉]), Abbadi. Amel[1], Skender. Mohamed Redha[2],
Morsli Abdelkader[1], Bettache Farouk[3], and Salhi Fatima[1]

[1] Research Laboratory in Electrical Engineering and Automatic, Medea University,
Medea, Algeria
fehamidia@gmail.com
[2] Renewable Energy and Materials Laboratory, Medea University,
Medea, Algeria
[3] Mechanics, Physics and Mathematical Modeling Laboratory, Medea University,
Medea, Algeria

**Abstract.** Photovoltaic solar panels are designed to provide the maximum desired power. But unexpected shading effects from clouds, buildings, tree branches and dust significantly reduce the performance of these panels. Shading is one of the factors that influences the reduction of the output power of photovoltaic modules. When a photovoltaic array receives partial shading during operation, its performance becomes difficult to predict due to the behavior of the nonlinear modules. The power-voltage curves are characterized by multiple local peaks and a single global peak.

Meta-heuristic algorithms are optimization methods that present satisfactory performance in terms of exploration, exploitation, avoidance of local optima and convergence. This paper discusses the application of Mot-flame algorithm compared to different optimization method for finding the maximum power point. These MPPT algorithms (Maximum Power Point Tracking) are based on meta-heuristic optimization for photovoltaic systems under partial shading conditions. The paper investigates and compares the tracking capability of MPPT-metaheuristic algorithms was carried out under strong partial shading conditions (PSC), considering the number of evaluations, tracking time, efficiency and successful rate.

**Keywords:** Meta-heuristic Algorithm · Partial Shading Condition · Maximum Power Point Technique · Global and Local peak

## 1 Introduction

Man has always used the riches of his environment to be able to survive and to satisfy their needs. Much of the world's energy production comes from fossil sources. The consumption of these sources results in greenhouse gas emissions and therefore an increase in pollution. In recent decades, our planet has seen an increase in the rate of greenhouse gas emissions which has been the result also of considerable advancement in technology and industry. This has had the adverse consequence of climatic upheaval and notable natural disasters. Resorting to the development of renewables energies then

© The Author(s), under exclusive license to Springer Nature Switzerland AG 2024
M. Hatti (Ed.): IC-AIRES 2023, LNNS 984, pp. 130–138, 2024.
https://doi.org/10.1007/978-3-031-60629-8_13

appeared as the ultimate solution to this problem. Interests in clean energy technologies have been growing in recent years [1, 2]. Renewable energies, in particular photovoltaic solar energy, are clean, silent, available and free energy. In addition, the photovoltaic (PV) array itself has a fixed structure with low maintenance cost. This is why, PV source is mostly well suited for use in rural zones. Photovoltaic solar panels are very sensitive to shade. Full or partial shading conditions have a significant impact on power generation capacity and can lead to losses. The shading comes from the environment of the building. Mountains, trees, clouds, dust or dead leaves. Most modules are now equipped with bypass diodes to reduce the effect due to shading (and protect the cells) but these effects remain considerable.

Photovoltaic panels have a specific electrical characteristic which is given by the manufacturer in the form of curves. These curves generally represent the evolution of the current and the power compared to the voltage of the panel. The electrical characteristic of the panel is non-linear in nature and has a particular point called the "Maximum Power Point" (MPP). This point is the optimum operating point for which the panel operates at its maximum power. Photovoltaic energy is highly dependent on climatic conditions and site location, which makes the position of the MPP variable over time and therefore difficult to locate. In partially shaded conditions, when the photovoltaic system does not receive uniform insolation, the characteristic curve in this case become more complex, with display of a single global peak and numerous local peaks. The global maximum point corresponds to the maximum power while the other points correspond to much lower powers.

MPPT technique, as its name suggests, tracks the MPP over time and thus allows us to get the maximum power that the panel is able to provide. The MPPT aims to improve and optimize the operation of photovoltaic systems. Technically, the MPPT uses an interface between the panel and the load which is usually a power conversion device as Boost converter, the most classical used algorithms are P&O, Constant Voltage, INC[3, 4], and those based on artificial intelligence methods as fuzzy logic [5–7], neural network [8], neural fuzzy method [9] and metaheuristic approach as PSO [10, 11] proposed by James Kennedy and Russell Eberhart [12], ACO[13] proposed by Marco Dorigo [14], ABC[15] proposed Karaboga [16], Bat [17] proposed by X.Yang [18], GA[19] proposed by John Holland [20], MFA [21] and WOA [22] developed by Mirjalili [23, 24] etc., In this work, we will propose MFA and we will compare it with a PSO and GWO MPPT-Meta-heuristic algorithms.

This work is divided into five sections as follows. Section 2 introduces the electrical characteristics of photovoltaique panel. The Sect. 3 proposes and discuss the application of three Meta-Heuristic-MPPT algorithms for PV systems as MFA, GWO and PSO. Section 4 covers the obtained simulation of the implemented algorithm under partial shading algorithm. Finally we conclude by conclusion in Sect. 5.

## 2 Photovolatic Panel

Thus, the simplest equivalent circuit of a solar cell [25, 26] consists of a diode which represents the P-N junction of the cell, and a constant current source whose current amplitude depends on the intensity of the radiation. Two resistances are taken into consideration for a more exact description, parallel resistance $R_{sh}$ characterizing the leakage current and series resistance $R_s$ representing the various contact resistances (Fig. 1).

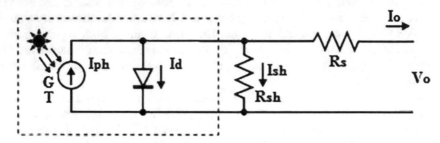

**Fig. 1.** Model of photovoltaic cell

In practice, the parallel resistance $R_{sh}$ is very high (in the order of mega Ohms) and the series resistance $R_s$ is very low (in the order of a few milli-ohms).

With such an equivalent electric circuit, we can write

$$I = I_{ph} - I_0 \left[ e^{\left( \frac{V + I.R_S}{a.V_{th}} \right)} - 1 \right] - \frac{V + I.R_S}{R_{sh}} \tag{1}$$

## 3 Maximum Power Point Tracking by Moth-Flame Algo

In order to extract, at each instant, the maximum power available at the terminals of the PV generator and to transfer it to the load, an adaptation stage is used. This block plays the role of interface between the source and the load. It ensures, through a control action (duty cycle), the transfer of the maximum power supplied by the generator.

The adapter commonly used in PV is a static converter (DC/DC power converter), the type of this static converter used in our work, is Boost converter. The conversion structure is chosen according to the load to be supplied and increase the voltage of the PVG. The proposed configuration consists of five modules connected in series under partial shading condition (800–1000-200–500-500W/m²).

Classical methods as P&O are very effective when the characteristic P-V have a single optimal point (absence of shading). Otherwise, the application of these algorithms to locate the maximum power point in the presence of shading conditions (dynamic regime) gives unsatisfactory results, the algorithm has difficulty searching more precisely the MPP, it converges to first LMPP (local maximum). To deal with the problem of existence of several local maxima, we consider here the implementation of an MPPT control based on the intelligent metaheuristic techniques. The Mothflame Algorithm (MFA) and Grey Wolf Optimization Algorithm (GWO) [26, 27] was proposed by Mijalili. in 2015 and 2014 and PSO is proposed by J.Kennedy and R.Eberhart [12] in 1995.

The MFO approach is inspired by the nocturnal navigation mechanism used by butterflies in nature. The procedure for developing the algorithm is quite simple. This

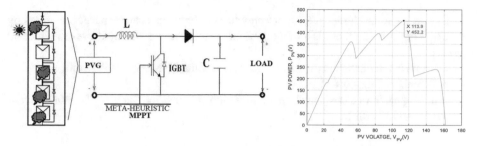

**Fig. 2.** PV-DC Load under partial shading 800–1000-200–500-500W/m$^2$

is presented in the MFO flowchart. The first step is to generate random moths in the neighborhood or solution space. Then the fitness value (Maximum power) for each butterfly is calculated and the best position $G_{best}$ obtained is labeled by the flame. Then the update process takes place, after which the process will be repeated until a point where the process end criteria are met, the model candidate solution has been assumed to be the moth and the problem variables to resolve take a position. The Fig. 3 presents the flowcharts of these metaheuristic algorithms by applying to get MPPT.

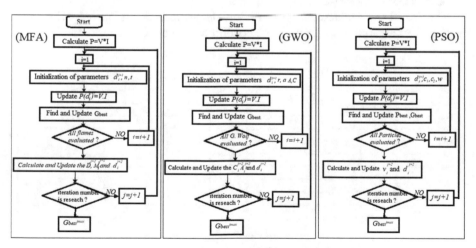

**Fig. 3.** Flow-diagram of proposed MPPT-Meta-Heuristic methods

Initialization of the positions of the Moths which consists in creating M

$$M_{ij} = rand(n, dm).(ub_i - lb_i) + lb_i \qquad (2)$$

Update of the positions of the moths

$$M_{ij} = S(M_{ij}, F_{ij}) = Ds_i * e^t * \cos(2\pi t) + F_{ij} \qquad (3)$$

$$Ds_i = |M_{ij} - F_{ij}| \qquad (4)$$

S: it is the logarithmic spiral function which defines the search space. S is given by the Eq. 3 considering that t is a random number uniformly distributed over the interval [−1,1]

Ds: this is the distance between the position of the $i^{th}$-Moth and $j^{th}$-Moth

n: Number of moths $\epsilon$ [1,…, n]

dm: Dimension of the search space in which the position vectors of moth (search-space size $\epsilon$ [1,…, dm])

## 4   Matlab Simulation Results

Simulation is a powerful tool for evaluating the theoretical performance of a system. Indeed, the latter can be tested by using the followed PV Module parameters: $I_{sc} = 8.21A$, $P_{max} = 200W$, $V_{oc} = 32.9V$, $R_s = 0.2172\Omega$.

All simulation results was represented with PV (5 panels) under partial shading (800–1000-200–500-500 w/m²), the modules are connected in serie as shown Fig. 2. By using Metaheuristic algorithms it can be observed that the proposed algorithms start searching the maximum power (Fig. 4), and after 300 iterations as shown Fig. 6, the proposed methods can get the maximum values, the obtained GMPP are nearly equal to the desired GMPP value (452.2W) except SSA.

The Figs. 5 and 7 represents the Power and duty respsonses of these algorithms and the three best algorithms under the same conditions weather and algorithms.

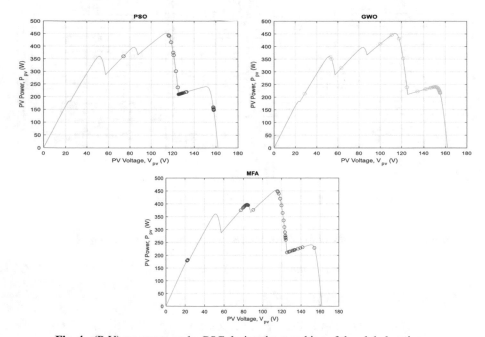

**Fig. 4.** (P-V) responses under PSC during the searching of the global optimum

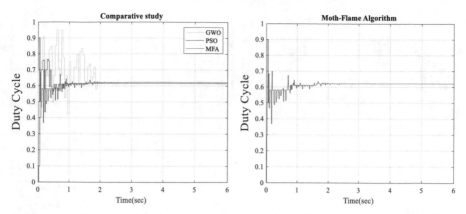

**Fig. 5.** Duty Cycle response under PSC of the proposed MPPT-Metaheuristic Algorithms

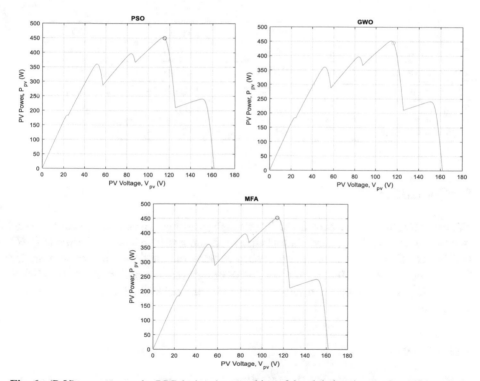

**Fig. 6.** (P-V) responses under PSC during the searching of the global optimum after 300 iterations

According to these results, we summarize the performance of Metaheuristic algorithms in the following table.

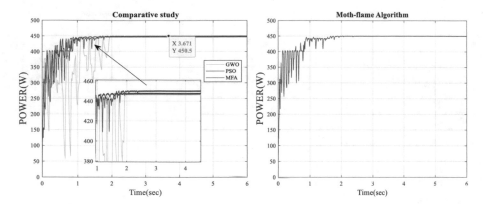

**Fig. 7.** Power response of metaheuristic-MPPT algorithm under PSC (800–1000-200–500-500 w/m$^2$)

|  | PSO | MFA | GWO |
|---|---|---|---|
| Tracking Efficiency | 99,00% | 99,55% | 99,33% |
| Convergence Tracking Time | 01.50s | 01.64s | 01.60s |
| Convergence | To GMPP | To GMPP | To GMPP |
| Iteration | 300 | 300 | 300 |
| Population number | 40 | 40 | 40 |

## 5 Conclusion

The proposed work has presented and discussed the use of very promising meta-heuristic optimisation method named Mothfame algorithm and it will be latter compared to (PSO and GWO), these algorithms were proposed to track the maximum power point under PSC (800–1000-200–500-500 w/m$^2$) of Photovoltaic system supplied DC load via Boost converter, Matlab simulation results present that the proposed MFA offer to us a very satisfactory results.

## References

1. Huang, C., Wang, L., Long, H., Luo, X., Wang, J.-H.: A hybrid global maximum power point tracking method for photovoltaic arrays under partial shading conditions. Optik **180**, 665–674 (2020)
2. Mat, Z.B.A., Madya, Kar, Y.B., Abu Hassan, S.H.B., Talik, N.A.B.: Proton Exchange Membrane (PEM) and Solid Oxide (SOFC) fuel cell based vehicles-a review. In: 2nd IEEE International Conference on Intelligent Transportation Engineering (2017)
3. Mishra, J., Das, S., Kumar, D., Pattnaik, M.: Performance comparison of P&O and INC MPPT algorithm for a stand-alone PV system. In: Innovations in Power and Advanced Computing Technologies (i-PACT) (2019)

4. Elgendy, M.A., Zahawi, B., Atkinson, D.J.: Assessment of perturb and observe MPPT algorithm implementation techniques for PV pumping applications. IEEE Trans. Sustain. Energy **3**, 21–33 (2012)
5. Chiu, C.S.: T-S fuzzy maximum power point tracking control of solar power generation systems. IEEE Trans. Energy Convers. **25**(4), 1123–1132 (2010)
6. Altin, N.: Interval type-2 fuzzy logic controller based maximum power point tracking in photovoltaic systems. Adv. Electric. Comput. Eng. **13**(3) (2013)
7. Hamidia, F., Abbadi, A., Tlemçani, A.: Improved pumping system supplied by double photovoltaic panel. Rev. Roum. Sci. Techn. Électrotech. Énerg. **64**(1), 87–93 (2019). Bucarest
8. Lin, W.M., Hong, C.M., Chen, C.H.: Neural-network-based MPPT control of a stand-alone hybrid power generation system. IEEE Trans. Power Electron. **26**(12), 3571–3581 (2011)
9. Al-Majidi, S.D., Abbod, M.F., Al-Raweshidy, H.S.: Maximum power point tracking technique based on a neural-fuzzy approach for stand-alone photovoltaic system. In: 55th International Universities Power Engineering Conference (2020)
10. Javed, S., Ishaque, K.: A comprehensive analyses with new findings of different *PSO* variants for *MPPT* problem under partial shading. Ain Shams Eng. J. **13**(5), 101680 (2022)
11. Hamidia, F., Abbadi, A., Tlemçani, A.: Improved hybrid pumping system with storage battery-based particle swarm algorithm. Rev. Roum. Sci. Techn. Électrotech. Énerg. **66**(1), 243–248 (2021), Bucarest
12. Kennedy, J., Eberhart, R.: Particle swarm optimization. In: Proceeding of International Conference on Neural Network (ICNN 1995), pp. 1942–1948. IEEE (1995)
13. Titri, S., Larbes, C., Youcef Toumi, K., Benatchba, K.: A new MPPT controller based on the Ant colony optimization algorithm for Photovoltaic systems under partial shading conditions. Appl. Soft Comput. **58**, 465–479 (2017)
14. Dorigo, M., Stützle, T.: The ant colony optimization metaheuristic: algorithms, applications, and advances. In: Handbook of Metaheuristics, Chapter 9, pp. 250–285. Springer, Boston (2003). https://doi.org/10.1007/0-306-48056-5_9
15. Benyoucef, A.S., Chouder, A., Kara, K., Silvestre, S., Aitsahed, O.: Artificial bee colony based algorithm for maximum power point tracking (MPPT) for PV systems operating under partial shaded conditions. Appl. Soft Comput. **32**, 38–48 (2015)
16. Karaboga, D., Basturk, B.: Artificial bee colony (ABC) optimization algorithm for solving constrained optimization problems. In: Foundations of Fuzzy Logic and Soft Computing, pp. 789–798. Springer, Heidelberg (2007). https://doi.org/10.1007/978-3-540-72950-1_77
17. Kaceda, K., Larbesa, C., Ramzanb, N., Bounabia, M., Dahmanea, Z.E.: Bat algorithm based maximum power point tracking for photovoltaic system under partial shading conditions. Sol. Energy **158**, 490–550 (2017)
18. Yang, X.-S., He, X.: Bat algorithm: literature review and applications. Int. J. Bio-Inspired Comput. **5**(3), 141–149 (2013)
19. Ramaprabha, R., Gothandaraman, V., Kanimozhi, K., Divya, R., Mathur, B.L.: Maximum power point tracking using GA-optimized artificial neural network for Solar PV system. In: Proceedings of the 1st International Conference on Electrical Energy Systems (ICEES); Newport Beach, pp. 264–268, pp. 141–149 (2011)
20. Sastry, K., Goldberg, D., Kendall, G.: Genetic Algorithms. In: Search Methodologies, Chapter 4, pp. 97–125. Springer, Boston (2005). https://doi.org/10.1007/0-387-28356-0_4
21. Shi, J.Y., et al.: Moth-flame optimization-based maximum power point tracking for photovoltaic systems under partial shading conditions. J. Power Electron. **19**(5), 1248–1258 (2019)
22. Mirjalili, S.: Moth-flame optimization algorithm: novel nature-inspired heuristic paradigm. Knowl.-Based Syst. **89**, 228–249 (2015)
23. Ebrahim, M.A., Osama, A., Kotb, K.M., Bendary, F.: *Whale* inspired algorithm based *MPPT* controllers for grid-connected solar photovoltaic system. Energy Procedia **162**, 77–86 (2019)

24. Mirjalili, S., Lewis, A.: The whale optimization algorithm. Adv. Eng. Softw. **95**(51–67), 2016 (2016)
25. .Prabhu, N., Balraj, R., Ayyubh, S.: Controller design for hybrid power systems. Int. J. Eng. Sci. Res. Technol. **4**(1), 318–324 (2015)
26. .Hamidia, F., Abbadi, A.: MPPT based Grey wolf Optimization. In: ICAIRES, Algeria, pp. 57–65. Springer, Cham (2021). https://doi.org/10.1007/978-3-030-92038-8_6
27. Mirjalili, S., Mirjalili, S.M., Lewis, A.: Grey wolf optimizer. Adv. Eng. Softw. **69**, 46–61 (2014)

# Improving Wind Turbine Performance in Residential Grid-Connected System Operating at Variable Wind Conditions with WOA-MPPT Based Strategy

Nadia. Douifi[1]([✉]), Amel. Abbadi[2], and Fethia. Hamidia[2]

[1] Laboratory of Advanced Electronic Systems (LSEA), Electrical Engineering Department, Faculty of Technology, University of Medea, Medea, Algeria
douifi.nadia@gmail.com

[2] Electrical Engineering and Automatic Laboratory (LREA), Electrical Engineering Department, Faculty of Technology, University of Medea, Medea, Algeria

**Abstract.** The utilization of renewable energy sources like wind power has been gaining attention as an alternative source of energy. The maximum power point tracking technique is essential for enhancing the performance of wind energy systems. In this context, this paper presents an MPPT control strategy based on whale optimization algorithm for a domestic wind turbine grid-connected system. The system comprises a small wind turbine driving a permanent magnet synchronous generator, connected to the grid through a rectification stage and single-phase inverter. The proposed algorithm utilizes a variable DC reference signal to adjust the inverter's output voltage to extract maximum power from the wind turbine. Simulation results demonstrate that the WOA MPPT-based algorithm can effectively regulate the operating point of the wind turbine, allowing it to operate at its maximum efficiency under varying wind conditions.

**Keywords:** Maximum power point tracking · Permanent magnet synchronous generator · Small scale wind turbine · Whale optimization algorithm

## 1 Introduction

Energy shortage has emerged as a critical issue for many countries around the world. As a result, households in some nations have been restricted from controlling their daily electricity usage, and the cost of energy from power utilities is increasing. To mitigate these challenges and reduce energy bills and carbon dioxide emissions, small renewable generators have become a popular solution [1]. These generators not only sell excess electricity back to the national grids but also serve as backup sources during power cuts. Among various renewable generators, wind turbines (WTs) are considered one of the most efficient and clean technologies available [2].

Small wind turbines (SWTs) are often utilized in distributed energy systems, both in residential and urban environments where the installation of larger wind turbines is

© The Author(s), under exclusive license to Springer Nature Switzerland AG 2024
M. Hatti (Ed.): IC-AIRES 2023, LNNS 984, pp. 139–147, 2024.
https://doi.org/10.1007/978-3-031-60629-8_14

not feasible [3]. SWTs are versatile and can be installed in various locations including rooftops, poles, and towers. They can be used for both off-grid and grid-connected power systems, offering a sustainable and reliable source of energy that can help to reduce the dependency on traditional utility sources [4].

Wind energy conversion systems (WECSs) can utilize four main types of electric power generators, including synchronous generators, induction generators, doubly-fed induction generators, and permanent magnet synchronous generators (PMSGs). Among these, PMSGs are a promising technology for wind energy generation, offering high efficiency, reliability, and control performance, as well as lower maintenance costs [5]. The multiple pole design of PMSGs enables slow-speed operation and gearless wind energy conversion, which further enhances their efficiency and reliability, making them a popular choice in the industry, particularly for small wind turbines.

The quantity of energy collected from the wind is determined by the Tip-Speed Ratio (TSR), which is the ratio of the rotational speed of the blades to the wind speed. Each turbine has an ideal TSR that maximizes extracted power. Since wind speed is constantly changing, a variable rotational speed is necessary to maintain the optimal TSR at all times [6, 7]. Hence, similar to photovoltaic systems, maximum power point tracking (MPPT) control schemes are included in modern WECSs to operate the wind turbines at their optimal point and maximizing the extracted energy for all wind speed conditions [8].

There exists a considerable amount of literature on MPPT algorithms, with a particular focus on their application in wind energy systems. Some of the commonly used techniques include Perturb and Observe (P&O) [9], Fuzzy Logic Control (FLC) [10], Model predictive control (MPC) [11] and Optimal Torque Control (OTC) [12]. Traditional MPPT techniques can cause oscillations, reducing efficiency and ability of maximum power tracking limitations under transient environments. However, researchers are developing advanced variations of these techniques, as well as new algorithms such as artificial intelligent techniques, to address these challenges and improve the effectiveness of MPPT. Taking into account the effectiveness, adaptability, exploration-exploitation balance, nature-inspired approach, and ease of implementation offered by the Whale Optimization Algorithm (WOA), it emerges as a highly advantageous selection among various metaheuristic methods for MPPT in wind systems.

In this context, this paper introduces a sensorless MPPT controller that utilizes the WOA to extract optimal power from a 5 kW residential SWT grid connected system. To this aim, the suggested strategy automatically adjusts the VDC reference signal of the inverter at each iteration toward the optimal operating point in a way that guarantees extracting the maximum available power from the wind. The rest of the paper is organized as follows: Sect. 2 describes the modelling of the wind energy system, while Sect. 3 explains the WOA-MPPT based algorithm and its implementation in the system. The simulation results are presented and discussed in Sect. 4, finally conclusion is presented in Sect. 5.

## 2   System Modelling

Figure 1 depicts the schematic configuration of the PMSG-based residential wind conversion system. The major components of the model include WT directly coupled with PMSG, three phase diode rectifier to convert the AC voltage into DC, an inverter with a voltage and current regulators, an MPPT controller, residential load and residential power network. The proposed system is designed to ensure that the Wind Energy Conversion System operates at maximum power output while maintaining the stability and quality of the power supplied to the distribution network and the residential load.

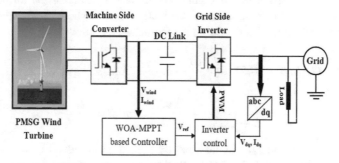

**Fig. 1.** Schematic layout of the modelled residential SWT grid-connected system.

### 2.1   Characteristics of the Wind Turbine

Kinetic energy of the wind is converted by the wind turbine into mechanical energy which subsequently powers a generator to produce electrical energy. The amount of kinetic energy generated by a moving object is typically determined by its velocity and mass as expressed in the following equation:

$$E_{kin} = \frac{1}{2}mV_w{}^2 \tag{1}$$

where in this case $V_w$ is defined as the wind velocity, the mass $m$ as the air mass and can be obtained by:

$$m = \rho(A.d) \tag{2}$$

where $\rho$ is the air density, $A$ and $d$ are the area swept by the WT blades and distance travelled by the wind, respectively.

The wind turbine's mechanical power ($P_m$) can be defined as the amount of kinetic energy generated over a given period of time. Therefore, $P_m$ can be represented mathematically as:

$$P_m = \frac{E_{kin}}{t} = \frac{1}{2}\rho AV_w{}^3 \tag{3}$$

Equation (3) represents the theoretical maximum power that can be harnessed by a wind turbine. However, the actual power output of the turbine is influenced by its efficiency, which depends on the radius of turbine $R$, air density and the turbine power coefficient ($C_p$). Hence, the actual mechanical power generated by the WT can be expressed as [13]:

$$P_m = \frac{1}{2}\rho\pi R^2 C_p(\lambda, \beta) V_w{}^3 \tag{4}$$

The power coefficient is a measure of the aerodynamic efficiency of a wind turbine. $C_p$ is a nonlinear function that is commonly modeled based on the turbine's unique characteristics. These characteristics are typically related to both the tip speed ratio ($\lambda$) and the blade pitch angle ($\beta$). $C_p$ ($\lambda$, $\beta$) is formulated as follow [14]:

$$C_p(\lambda, \beta) = 0.5176\left(\frac{116}{\lambda_i} - 0.4\beta - 5\right)exp^{\frac{-21}{\lambda_i}} + 0.0068\lambda \tag{5}$$

$$\frac{1}{\lambda_i} = \frac{1}{\lambda + 0.08\beta} - \frac{0.035}{\beta^3 + 1}, and\ \lambda = \frac{\omega_m \times R}{V_w} \tag{6}$$

$\omega_m$ is the rotor's mechanical angular velocity measured in rad/s. Figure 2a depicts the $C_p$-$\lambda$ characteristics curves. According to the figure, at a fixed pitch angle, $C_p$ value mainly depends on $\lambda$ and it reaches its maximum value at the optimal tip speed ratio ($\lambda_{opt}$) [15]. Figure 2b illustrates the $P$-$\omega_m$ characteristic curves for various wind speeds. It can be observed that when the speed of the wind changes, it affects the rotor speed and the amount of power produced by a wind turbine [16].

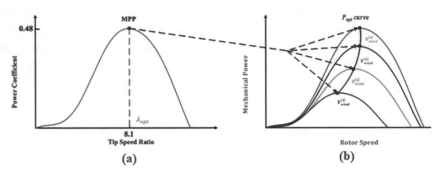

**Fig. 2.** $C_P$ - $\lambda$ and $P$ - $\omega_m$ characteristic curves of the WT [17].

At a certain wind speed, there is an optimal rotor speed ($\omega_{opt}$) that maximizes the output power of the turbine. Therefore, a control method known as MPPT is implemented to ensure WT is operating at Maximum power at different wind speeds. An effective MPPT algorithm should be able to track changes in wind speed and adjust the rotor speed to maintain the turbine at the optimal operating conditions $\lambda_{opt}, C_{p\text{-}opt}$ and $\omega_{opt}$.

## 3  WOA-MPPT Based Algorithm

The Whale Optimization Algorithm (WOA) is a metaheuristic optimization algorithm inspired by the feeding behavior of humpback whales, which hunt for a school of fish despite their large size. It was proposed by Seyedali Mirjalili and Andrew Lewis in 2016. The algorithm is designed to solve optimization problems by using a combination of random search and local search to explore the search space and converge towards the global optimal solution [18].

In this paper, the proposed WOA optimization algorithm is utilized to search for the optimal operating point of the wind turbine in order to maximize power output under varying wind speeds. To achieve this, the voltage and current of the small wind turbine are measured and serve as inputs to the MPPT controller. The MPPT controller tracks changes in wind speed and regulates the VDC reference signal ($Vref$) of the inverter's VDC controller accordingly. WOA-MPPT based algorithm utilizes the particle position to represent the $Vref$ and the objective function is designed to maximize the turbines power. The WOA algorithm updates $Vref$ iteratively based on the three main hunting behaviors of humpback whales, which are: searching for prey using (7) when $|A| \geq 1$, surrounding prey using (9) when $|A| < 1$, and attacking prey using (12).

$$Vref_{t+1} = Vref_{rand} - A.D_{rand} \ if \ |A| \geq 1 \ \& \ p < 0.5 \tag{7}$$

$$D_{rand} = \left| C.Vref_{rand} - Vref_t \right| \tag{8}$$

$$Vref_{t+1} = Vref^* - A.D \quad if \ |A| < 1 \ \& \ p < 0.5 \tag{9}$$

$$D_{rand} = \left| C.Vref^* - Vref_t \right| \tag{10}$$

$$A = 2.a.r - 2, a = 2 - 2\left(\frac{t}{t_{max}}\right), and \ c = 2.r \tag{11}$$

$$Vref_{t+1} = Vref^* + D'.e^{bl}.cos(2\pi l) \ if \ p \geq 0.5 \tag{12}$$

$$D' = \left| Vref^* - Vref_t \right|, l = (a_2 - 1) * rand + 1, and \ a_2 = -1 - \frac{t}{t_{max}} \tag{13}$$

where: factor $A$ is a parameter that varies randomly between $[-2, 2]$, variable $l$ is chosen randomly from the range of $[-1, 1]$, $t$ represents the current iteration, $Vref^*$ represents best value obtained and $t_{max}$ is the maximum number of iterations. Parameter $b$ is set to 1, and random variable $r$ is uniformly distributed between $[0, 1]$.

The WOA algorithm iteratively updates particle position as described in the flowchart shown in Fig. 3 until a stopping criterion is met or maximum iterations are reached.

To distinguish between power variations caused by wind changes and those caused by previous perturbations, the condition presented in (14) must be verified. If the condition is met, the particles must be re-initialized to search for a new maximum power point (MPP).

$$\left| \frac{P^{t+1} - P^t}{P^t} \right| > \Delta P \tag{14}$$

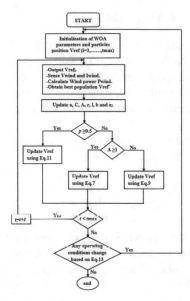

**Fig. 3.** Flowchart of the proposed WOA-MPPT algorithm.

## 4 Simulation Results

MATLAB/SIMULINK was utilized to model and simulate the residential SWT grid-connected system operating at 230 V/60 Hz. The simulation considered varying wind speed conditions to evaluate the performance of the studied WOA-MPPT strategy.

A wind speed profile ranging from 8 to 11 m/s was used in the wind turbine model, with both large and small step sizes. Although step changes in wind speeds do not represent actual wind speed changes, they provide a worst-case scenario for assessing the tracking algorithm's performance. The simulation results are presented in Figs. 4, 5, 6, 7, and 8.

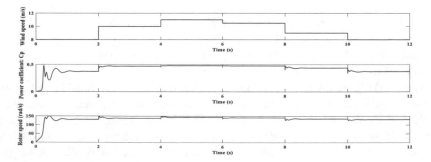

**Fig. 4.** Machine side results for the wind speed profile.

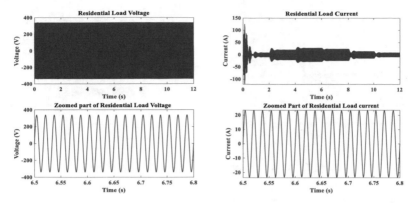

**Fig. 5.** Residential load voltage and current with a zoomed part.

The simulation results demonstrate that the WOA-MPPT based wind turbine system was able to effectively optimize the generator speed and tune the wind power coefficient in response to changes in wind velocity of the proposed profile as seen in Fig. 4.

Figures 5, 6, 7 and 8 show that simulations results have produced satisfactory performance regarding the optimization of MPPT for the wind turbine, the effective transfer of power from the turbine to the grid, and the high quality of the transferred power (reduce harmonic distortion, improve power factor, and stabilize the output voltage of the DC bus).

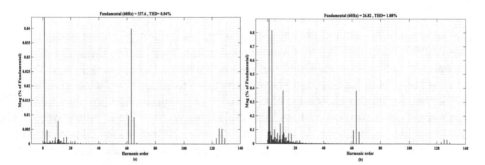

**Fig. 6.** Total harmonic distortion (THD) of the Residential load voltage (a) and current (b) for the proposed wind speed profile.

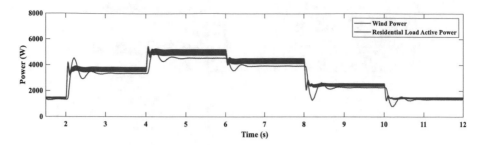

**Fig. 7.** Active power injected to the residential load.

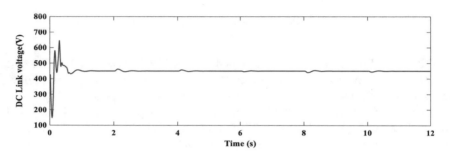

**Fig. 8.** DC link voltage.

## 5   Conclusion

The paper presents a WAO-MPPT based algorithm for achieving optimal power operation of a grid-connected residential small wind turbine system. The simulation results demonstrate that the algorithm performs well under varying wind conditions, and it is capable of extracting the maximum power from the wind turbine at any wind speed without the need for precise knowledge of the wind or rotor speed. Furthermore, electrical energy quality was also achieved, with the total harmonic distortion of both current and voltage meeting the norms set by IEEE. Overall, the results proved the effectiveness of the proposed MPPT control strategy in improving the performance and efficiency of the wind turbine.

## References

1. Zentani, A., Almaktoof, A., Kahn, M.T.E.: DC-DC boost converter with P&O MPPT applied to a stand-alone small wind turbine system. In: 2022 30th Southern African Universities Power Engineering Conference (SAUPEC), pp. 1–5. IEEE (2022)
2. Aubrée, R., Auger, F., Macé, M., Loron, L.: Design of an efficient small wind-energy conversion system with an adaptive sensorless MPPT strategy. Renew. Energy **86**, 280–291 (2016)
3. Khan, A., Memon, S., Said, Z.: Predictive permanent magnet synchronous generator based small-scale wind energy system at dynamic wind speed analysis for residential net-zero energy building. Int. J. Solar Thermal Vacuum Eng. **3**(1), 29–49 (2021)

4. Zammit, D., Spiteri Staines, C., Micallef, A., Apap, M.: MPPT with current control for a PMSG small wind turbine in a grid-connected DC microgrid. In: Wind Energy Exploitation in Urban Environment: TUrbWind 2017 Colloquium 1, pp. 205–219. Springer, Heidelberg (2018)

5. Kumar, S., Ali, I., Siddiqui, A.S.: Control of wind energy connected single phase grid with MPPT for domestic purposes. In: 2022 2nd International Conference on Intelligent Technologies (CONIT), pp. 1–6. IEEE (2022)

6. Ali, Y.A., Ouassaid, M.: Sensorless MPPT controller using particle swarm and grey wolf optimization for wind turbines. In: 2019 7th International Renewable and Sustainable Energy Conference (IRSEC), pp. 1–7. IEEE (2019)

7. Han, K., Chen, G.Z.: A novel control strategy of wind turbine MPPT implementation for direct-drive PMSG wind generation imitation platform. In: 2009 IEEE 6th International Power Electronics and Motion Control Conference, pp. 2255–2259. IEEE (2009)

8. Balbino, A.J., Nora, B.D.S., Lazzarin, T.B.: An improved mechanical sensorless maximum power point tracking method for permanent-magnet synchronous generator-based small wind turbines systems. IEEE Trans. Indust. Electron. **69**(5), 4765–4775 (2021)

9. Mousa, H.H., Youssef, A.R., Mohamed, E.E.: State of the art perturb and observe MPPT algorithms based wind energy conversion systems: a technology review. Int. J. Electr. Power Energy Syst. **126**, 106598 (2021)

10. Salem, A.A., Aldin, N.A.N., Azmy, A.M., Abdellatif, W.S.: A fuzzy logic-based MPPT technique for PMSG wind generation system. Int. J. Renew. Energy Res. **9**(4), 1751–1760 (2019)

11. Srivastava, A., Bajpai, R.S.: Model predictive control of grid-connected wind energy conversion system. IETE J. Res. **68**(5), 3474–3486 (2022)

12. Deng, X., Yang, J., Sun, Y., Song, D., Yang, Y., Joo, Y.H.: An effective wind speed estimation based extended optimal torque control for maximum wind energy capture. IEEE Access **8**, 65959–65969 (2020)

13. Hu, L., et al.: Sliding mode extremum seeking control based on improved invasive weed optimization for MPPT in wind energy conversion system. Appl. Energy **248**, 567–575 (2019)

14. Mahmoud, M.M., Aly, M.M., Salama, H.S., Abdel-Rahim, A.M.M.: Dynamic evaluation of optimization techniques–based proportional–integral controller for wind-driven permanent magnet synchronous generator. Wind Eng. **45**(3), 696–709 (2021)

15. Qais, M.H., Hasanien, H.M., Alghuwainem, S.: A grey wolf optimizer for optimum parameters of multiple PI controllers of a grid-connected PMSG driven by variable speed wind turbine. IEEE Access **6**, 44120–44128 (2018)

16. Mousa, H.H., Youssef, A.R., Mohamed, E.E.: Variable step size P&O MPPT algorithm for optimal power extraction of multi-phase PMSG based wind generation system. Int. J. Electr. Power Energy Syst. **108**, 218–231 (2019)

17. Lima, G.F.D., Kremes, W.D.J., Siqueira, H.V., Aliakbarian, B., Converti, A., Illa Font, C.H.: A three-phase phase-modular single-ended primary-inductance converter rectifier operating in discontinuous conduction mode for small-scale wind turbine applications. Energies **16**(13), 5220 (2023)

18. Qais, M.H., Hasanien, H.M., Alghuwainem, S.: Enhanced whale optimization algorithm for maximum power point tracking of variable-speed wind generators. Appl. Soft Comput. **86**, 105937 (2020)

# Improved Power Quality with Active Shunt Power Filter Based on MPPT Firefly Controller of a Wind Turbine

A. Abbadi[1]([✉]), F. Hamidia[1], M. R. Skender[2], A. Morsli[1], and F. Bettache[3]

[1] Electrical Engineering Department, Research Laboratory in Electrical Engineering and Automatic (LREA), University of Médéa, Médéa, Algeria
amel.abbadi@yahoo.fr

[2] Electrical Engineering Department, Renewable Energy and Materials Laboratory (REML), University of Médéa, Médéa, Algeria

[3] Electrical Engineering Department, Laboratory of Mechanics, Physics and Mathematical Modeling, University of Médéa, Médéa, Algeria

**Abstract.** This paper proposes an MPPT controller based on the Firefly algorithm applied to a wind energy conversion system (WECS) integrated with a shunt active power filter (SAPF) to solve power quality problems. Indeed, the current electrical distribution system involves the use of non-linear loads that cause quality problems in electrical power systems. In addition, the penetration of renewable energy sources, especially wind turbines, is increasing in the power grids to meet the ever-increasing energy demand. Thus, the studied system consists of a wind turbine based on a permanent magnet synchronous generator (PMSG), a rectifier, and a grid-connected three-phase voltage source inverter. The wind energy conversion system is controlled using the Firefly algorithm to track the maximum power point (MPPT) in the wind system. The dynamic performance of SAPF is optimized using the direct power control technique. These controllers aim to minimize total harmonic distortion (THD) produced by non-linear loads. Simulation results are provided to prove the effectiveness of the proposed system. The results show that the adopted control algorithms are effective in eliminating harmonic currents and injecting the available active power of the PMSG wind turbine into the load and/or power supply systems.

**Keywords:** Shunt active power filter · power quality · total harmonic distortion · wind energy conversion system · MPPT controller · Firefly algorithm

## 1 Introduction

Electricity demand is rising every day and the availability of traditional energy sources (gas, oil and coal, etc.) is decreasing, hence the need to use alternative and durable energy sources. Among the various renewable energy sources, wind power is one of the most important alternative energies as it is freely available and environmentally friendly [1, 2].

© The Author(s), under exclusive license to Springer Nature Switzerland AG 2024
M. Hatti (Ed.): IC-AIRES 2023, LNNS 984, pp. 148–156, 2024.
https://doi.org/10.1007/978-3-031-60629-8_15

Power quality difficulties arise as a result of the employment of modern power electronic devices and nonlinear loads. These problems, which include harmonic currents, reactive power loads, and unbalance, have far-reaching negative consequences for utilities and customers [3, 4].

The most common problem is the harmonic currents produced by nonlinear loads. These reduce the efficiency of the electrical network and thus damage the connected electrical and electronic equipment.

The shunt active power filter (SAPF) is a device that injects the compensation currents and thus cancels the harmonics produced [5, 6].

The SAPF-based wind energy conservation system (WECS) is used to improve power quality and enhance bi-directional energy flow, as well as maximum energy extraction under varying wind speeds [7, 8].

As known, the maximum energy can be extracted from the WECS using the MPPT controller. MPPT has a variety of control techniques like enhanced hill-climbing MPPT technique [9], fuzzy logic [10], neural network [11], etc...Few works have been reported in the literature on Bio-inspired MPPT approaches for the WECS [12].

Bio-inspired MPPT algorithm offer several advantages, the most important are: there is no need for mathematical modeling of the process to optimize the control, their convergence speed is appreciable, and they are robust to load changes [13]. Interested in the benefits of bio-inspired MPPT controllers, this article introduces a sensorless wind turbine MPPT controller based on an optimization algorithm called Firefly.Generally, the Firefly algorithm (FA) is more commonly used as an MPPT controller for photovoltaic systems under partial shade conditions [14].

In this article, a firefly MPPT controller is proposed to extract the maximum power from the WECS by adjusting the duty cycle of the boost converter and thus optimize the performance of the SAPF in maintaining the THD according to the IEEE 519 standard.

This paper is organized as follows. The following is how this document is organized. Section 2 outlines the system under study. Section 3 represents the Firefly MPPT controller. Section 4 evaluates the efficacy of the proposed MPPT controller using simulations. Finally, in Sect. 5, the conclusions are provided.

## 2    Configuration of the System

Figure 1 depicts the structure of a shunt active power filter powered by a wind energy conversion system (WECS). This system includes a permanent magnet synchronous generator (PMSG) as a wind turbine coupled to a three-phase inverter connected to the

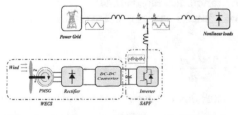

**Fig. 1.** Block Diagram of the shunt active power filter supplied by wind energy conversion system.

grid and a non-linear load via a shunt active power filter. The inverter transfers power from the wind turbine and at the same time compensates for harmonic currents.

## 2.1 Wind Energy Conversion System

A wind energy conversion system consists of a wind turbine that converts wind energy into mechanical energy. The wind turbine shaft is connected to the permanent magnet synchronous generator shaft via a gearbox. The configuration of this WECS is shown in Fig. 1.

### 2.1.1 Wind Turbine Model

The mechanical energy obtained from the wind turbine is given as follows [7]:

$$P_{Turbine} = \rho \pi R^2 C_p(\lambda, \beta) v^3 / 2 \tag{1}$$

where, $\rho$ is the air density (kg/m$^3$), $R$ is the rotor blade radius (m), $v$ is the wind speed (m/s), and $C_P$ is the power coefficient of the turbine, $\beta$ is the blade pitch angle (in degrees)and $\lambda$ the tip speed ratio. The tip-speed ratio $\lambda$ is expreced as:

$$\lambda = \omega_m R / v \tag{2}$$

where $\omega_m$ is the blades angular velocity (in rad/s).

$T_m$ is the mechanical torque output of the wind turbine. It is expressed as:

$$T_m = \frac{1}{2} \rho A C_p(\lambda, \beta) v^3 \frac{1}{\omega_m} \tag{3}$$

The power coefficient $C_p(\lambda, \beta)$ used in the present research is expressed as [13]:

$$C_p = c_1 \left( \frac{c_2}{\lambda_i} - c_3 \beta - c_4 \right) e^{-(c_5/\lambda_i)} + c_6 \lambda \quad ; \frac{1}{\lambda_i} = \frac{1}{\lambda + 0.08\beta} - \frac{0.035}{\beta^3 + 1} \tag{4}$$

The coefficient $c_1$ to $c_6$ are: $c_1 = 0.5176$, $c_2 = 116$, $c_3 = 0.4$, $c_4 = 5$, $c_5 = 21$ and $c_6 = 0.0068$.

### 2.1.2 PMSG Modeling

The main characteristics of wind turbine technology based on permanent magnet synchronous generators are mainly based on the absence of the gearbox (source of fault), the absence of rotor excitation, increased efficiency, smaller size, reduced maintenance costs, and near-low rotor losses. Its tension in the reference frame rotor d-q is expressed as follows [7]

$$v_{sd} = R_s i_{sd} + \frac{d\lambda_{sd}}{dt} - \omega_r \lambda_{sq} \quad ; \quad v_{sq} = R_s i_{sq} + \frac{d\lambda_{sq}}{dt} - \omega_r \lambda_{sd} \tag{5}$$

where $\lambda_{sd}$ and $\lambda_{sq}$ are stator flux linkage in d-q axis of rotor and are formulated as:

$$\lambda_{sd} = L_s i_{sd} + \psi_f \quad ; \quad \lambda_{sq} = L_s i_{sq} \tag{6}$$

where $\psi_f$ is the flux of the permanent magnet, $L_s$ is the stator inductance of PMSG. The electromagnetic torque $T_e$ of three-phase PMSG can be written as:

$$T_e = \frac{3}{2}p(\lambda_{sd}\, i_{sq} - \lambda_{sq}\, i_{sd})$$ (7)

where '$p$' is the number of pole pairs.

## 2.2  Shunt Active Power Filter

The active filter structure is made up of two parts: a power part and a control part. An inverter, a coupling filter, and an energy source constitute the power part. The control part takes responsibility for controlling the switching of the semiconductors which includes the inverter from the power part. It is feasible to generate harmonic signals at the inverter output to compensate for those present in the electrical network using appropriate control techniques.

# 3  Firefly MPPT Controller

Dr. Xin She Yang designed the Firefly Algorithm in 2007 to solve the issue of optimization [15]. FFA is an approach founded on firefly conduct. Fireflies often emit short-duration light with a distinct beat.

The following rules were adopted and merged to form the FFA:

a. All fireflies are unisex, which means that they are attracted to each other despite their gender.
b. The allure of fireflies is related to the intensity of the light generated. The fascination of fireflies diminishes with increasing distance. If none of the fireflies lights brighter than the others, the fireflies will move at random.
c. The configuration of the goal function influences or determines the light produced by fireflies.

Assume there are two fireflies, p and q, separated by rpq, and their positions are Xp and Xq. The Cartesian distance between the two fireflies is provided as follows:

$$r_{pq} = \|X_p - X_q\|$$ (8)

The relative brightness and attractiveness of each firefly are expressed as follows

$$I = I_0\, e^{-\gamma\, r_{pq}} \; ; \quad \beta(r) = \beta_0\, e^{-\gamma\, (r_{pq})^n} \, , n \geq 1$$ (9)

The following is an update to the position of the fireflies '$p$':

$$X_p^{t+1} = X_p^t + \beta(r)\, (X_p - X_q) + \alpha\left(rand - \frac{1}{2}\right)$$ (10)

In which case: p,q: firefly, r is the distance between fireflies, while rpq is the distance between firefly p and q. $X_p$: the location of the firefly p, $X_q$: firefly location q,: the firefly's attraction,$I$: the firefly's radiance, $\gamma$: absorption parameter, $n$: integer higher than one, and $\alpha$: regulate the randomness.

The performance of the FA is exploited for the design of an MPPT controller to search for the optimal MPP in different working conditions.

In the implementation of this algorithm in MPPT, the position of the firefly (the regulated variable) represents the duty cycle that controls the DC-DC converter, and the objective function is chosen as the wind output power (Fig. 2).

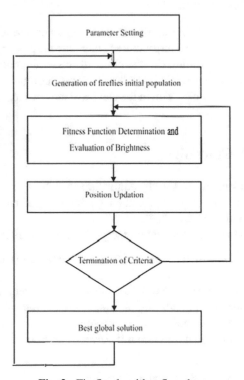

**Fig. 2.** Firefly algorithm flowchart

**The proposed method works as follow:**

**STEP 1**: **Parameter Setting**: Determine the FA constants, namely termination criterion, population size N, βo, γ, n, and α. The location of the firefly acts as the duty cycle d of the dc-dc converter in this approach. The brightness of any given firefly is defined as the generated power Pwind corresponding to this firefly's position.

**STEP 2: Firefly Initialization:** The fireflies are placed in the permissible solution area between dmin and dmax, where dmin and dmax indicate the minimum and maximum values of the dc-dc converter's duty ratio.

**stage 3: Brightness Examination:** In this stage, the dc-dc converter is progressively operated in accordance with the position of each firefly (i.e., duty ratio). Each firefly's brightness or light intensity is considered the equivalent wind output power, Pwind, for each duty ratio. This procedure is done for all fireflies in the population.

**STEP 4: Update Firefly Position:** The firefly with the highest brightness remains in its current location, while the rest of the fireflies modify their positions.

**STEP 5: Criteria termination:** If the termination requirement is met, terminate the program; otherwise, proceed to step 3. When the displacement of all fireflies in consecutive steps reaches a predetermined minimum value, the optimization technique has ended. When the program is terminated, the dc-dc converter operates at the optimum duty cycle, which matches GMPP.

**STEP 6:** Restart the FA: if the wind changes its velocity.

# 4   Simulations Results

To test and validate the proposed control approach under nonlinear load conditions, a MATLAB/Simulink model integrating the SAPF and the WECS is developed..

We have adopted two cases. In the first one, the WECS is not connected. Figure 3 shows the source voltages, the source current Ias, the filter current Iaf, and the load current IaL before and after compensation by the shunt APF.

In Fig. 3, It is clear that the source voltage waveform is purely sinusoidal and balanced. The source current waveform was distorted before filtering, but once the active filter is inserted into the configuration (t = 0.2s), the waveform instantly becomes sinusoidal.

As depicted in the Fig. 4.a, the THD (Total Harmonic Distortion) is considerably reduced after the activation of the filter on the electrical network, which confirms the good quality of the filtering. As shown in Fig. 4.b, the source current satisfies the load requirement ($I_{aL}$).

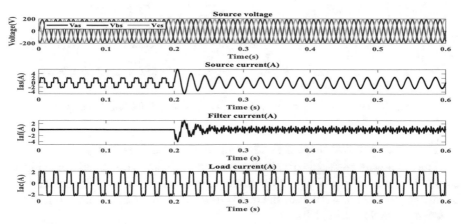

**Fig. 3.** The source voltages, the source current Ias, the filter current Iaf, and the load current IaL before and after compensation by the shunt APF.

In the second case, the shunt active power filter is supplied by the wind energy conversion system. We employed two wind speed profiles from the wind turbine, one at 6m/s and the other at 8 m/s.

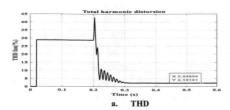 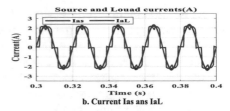

a.    THD                                      b. Current Ias ans IaL

**Fig. 4.  a.** Temporal pattern of the THD without and with SAPF of the source current Ias. **b.** Source current Ias,and the load current IaL with SAPF

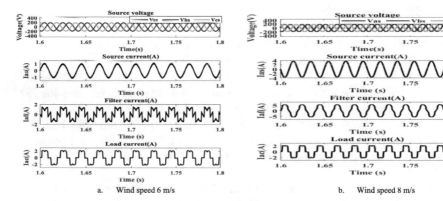

a.    Wind speed 6 m/s                    b.    Wind speed 8 m/s

**Fig. 5.** The source voltages, the source current Ias, the filter current Iaf, and the load current IaL for both wind speed.

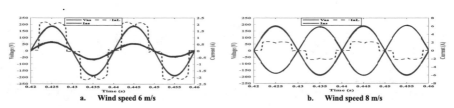

a.    Wind speed 6 m/s                    b.    Wind speed 8 m/s

**Fig. 6.** The source current Ias and the load current IaL for both wind speed.

The source's current shape is clearly sinusoidal at both speeds (Fig. 5).

The load current is supplied by both sources (WECS and grid), as shown in Fig. 6.a; however, based to Fig. 6.b, part of the wind energy satisfies the load demand and additional energy is fed into the grid, and this is justified by the fact that there is a phase π between *Vas* and *Ias* (Fig. 6.b).

As shown in the Fig. 7, the THD (Total Harmonic Distortion) complies with the IEEE 519 standard.

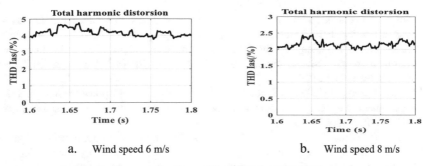

    a.   Wind speed 6 m/s               b.   Wind speed 8 m/s

**Fig. 7.** Temporal pattern of the THD for both wind speed.

## 5   Conclusion

This article has focused on the application of a Firefly MPPT controller to control a boost converter in a grid-connected wind energy conversion system. Direct power control to control a three-phase inverter is also used. The suggested control technique is applied to minimize harmonics while injecting available active power from the PMSG wind turbine into the load and/or network. The examination of the simulation results indicates that the proposed system is efficient and performs well. The direct power control technology performs well in terms of injecting active power generated by the PMSG wind turbine into distribution networks while simultaneously compensating for harmonics.

## References

1. Manwell, J.F., McGowan, J.G., Rogers, A.L.: Wind Energy Explained: Theory, Design and Application. Wiley, Chichester (2009)
2. Wagner, H.J., Mathur, J.: Introduction to Wind Energy Systems: Basics, Technology and Operation. Springer, Berlin/Heidelberg (2018)
3. Martinez, R., et al.: Techniques to locate the origin of power quality disturbances in a power system: a review. Sustainability **14**, 7428 (2022)
4. Thentral, T.M.T., Palanisamy, R., Usha, S., Bajaj, M., Zawbaa, H.M., Kamel, S.: Analysis of power quality issues of different types of household applications. Energy Rep. **8**, 5370–5386 (2022)
5. Hoon, Y., Mohd Radzi, M.A., Mohd Zainuri, M.A.A., Zawawi, M.A.M.: Shunt active power filter: a review on phase synchronization control techniques. Electronics **8**, 791 (2019)
6. Hoon, Y., Radzi, M.A.M., Hassan, M.K., Mailah, N.F.: Control algorithms of shunt active power filter for harmonics mitigation: a review. Energies **10**(12), 2038 (2017)
7. Kumar, R., Bansal, H.O.: Design and control of wind integrated shunt active power filter to improve power quality. In: IEEE 8th Power India International Conference (PIICON) (2018)
8. Hoseinpour, S., Barakati, M., Ghazi, R.: Harmonic reduction in wind turbine generators using a shunt active filter based on the proposed modulation technique. Int. J. Electr. Power Energy Syst. **43**(1), 1401–1412 (2012)
9. Muhammad, S., Kazmi, R., Goto, H., Guo, H.: A novel algorithm for fast and efficient speed-sensorless maximum power point tracking in wind energy conversion systems. IEEE Trans. Ind. Electron. **58**(1), 29–36 (2011)

10. Galdi, V., Piccolo, A., Siano, P.: Designing an adaptive fuzzy controller for maximum wind energy extraction. IEEE Trans. Energy Convers. **23**(2), 559–569 (2008)
11. Wei, C., Zhang, Z., Qiao, W., Qu, L.: Reinforcement learning-based intelligent maximum power point tracking control for wind energy conversion systems. IEEE Trans. Ind. Electron. **62**(10), 6360–6370 (2015)
12. Le, X.C., Duong, M.Q., Le, K.H.: Review of the modern maximum power tracking algorithms for permanent magnet synchronous generator of wind power conversion systems. Energies **16**(1) 402 (2023), pp. 1–25
13. Zafar, M.H., Khan, N.M., Mirza, A.F., Mansoor, M.: Bio-inspired optimization algorithms based maximum power point tracking technique for photovoltaic systems under partial shading and complex partial shading conditions. J. Clean. Prod. **309**, 127279 (2021), pp. 1–18
14. Bengourina, M.R., Rahli, M., Saadi, S., Hassaine, L.: PSO based direct power control for a multifunctional grid connected photovoltaic system. Int. J. Power Electron. Drive System (IJPEDS) **9**(2), 610–621 (2018)
15. Sagonda, A.F., Folly, K.A.: A comparative study between deterministic and two meta heuristic algorithms for solar PV MPPT control under partial shading conditions. Syst. Soft Comput. **4**, 1–22 (2022)

# Optimal Design of an Off-Grid Photovoltaic-Battery System for UAV Charging in Wildlife Monitoring

F. Fodhil[1(✉)] and O. Gherouat[2]

[1] Research Centre in Industrial Technologies (CRTI), P.O. Box 64, 16014 Cheraga, Algiers, Algeria
fadfod1@yahoo.fr

[2] Department of Automation, Faculty of Electrical Engineering, Université des Sciences et de la Technologie Houari Boumediene (USTHB), BP 32, El Alia, 16111 Bab Ezzouar, Algiers, Algeria

**Abstract.** This paper aims to determine the most efficient design for an off-grid photovoltaic-battery system, which plays a critical role in powering a charging station for Unmanned Aerial Vehicles (UAVs) used in environmental monitoring, particularly for aerial surveys of wildlife. The study focuses on designing an off-grid PV-Battery system that provides sustainable and reliable energy for the UAVs' charging needs. The main objective of the optimization is to minimize the levelized cost of energy (LCOE). Based on the power consumption profile of the case study, the optimal system successfully fulfills the electrical demand with only a slight unmet load of 1.09%. Furthermore, a sensitivity analysis is performed to examine the effect of $SOC_{min}$ on the behavior of the optimal system.

**Keywords:** Unmanned Aerial Vehicles (UAVs) · photovoltaic-battery system · charging station · optimal design

## 1  Introduction

Unmanned aerial vehicles (UAVs) have gained widespread utilization across diverse missions owing to their adaptability, cost-efficiency, and capacity to undertake tasks without endangering human lives. These versatile UAVs have found applications in an array of fields and industries, encompassing surveillance, reconnaissance (ISR) operations, search and rescue missions, disaster response efforts, precision agriculture, environmental monitoring, infrastructure inspection, journalism and filmmaking endeavors, and humanitarian aid missions. To ensure the uninterrupted operation of these UAVs, a reliable and efficient charging infrastructure is of paramount importance [1, 2]. By incorporating renewable energy into UAVs charging stations, such as through the development and enhancement of PV-powered charging stations, we establish a sustainable and eco-friendly solution that aligns with environmental conservation and clean energy initiatives. This approach offers a viable and pragmatic way to meet the energy needs of

M. Hatti (Ed.): IC-AIRES 2023, LNNS 984, pp. 157–165, 2024.
https://doi.org/10.1007/978-3-031-60629-8_16

drone operations while reducing their environmental impact. Several researches in the literature focused on the design and optimization of these charging stations especially for cellular networks [3–6]. In [3], the main goal of the paper was to effectively control the energy usage of a cellular network based on UAVs, specifically designed to offer coverage for rural and economically underserved communities. The key emphasis was on optimizing the energy storage in both the UAVs and charging sites, accomplished through strategically scheduling UAV missions across various locations and time intervals. In [4], the authors conducted an optimization to determine the ideal size of an off-grid PV-battery energy system utilized for powering a UAV-based telecommunication infrastructure. In [5], a PV-battery power system was proposed for supplying UAV-assisted base stations in urban areas. The study involves estimating the power consumption profile of UAV batteries and conducting sensitivity analysis to assess the effects of PV system size and battery capacity on the overall system performance. In [6] the authors introduced a PV grid-connected system designed to power UAVs used for providing wireless access via small cells (SC) in hotspot locations. The model was evaluated in a realistic scenario using actual cellular data and a UAV power consumption method. In another context, Ali et al. [7] focused on a mobile PV charging system equipped with an automated battery replacement (ABR) technology. The aim is to support continuous drone missions for combating harmful Fall Armyworm (FAW) insects without resorting to chemical methods.

This paper presents an optimized design for an off-grid PV-battery system that acts as the power source for UAVs charging station. The proposed optimized design is intended for environmental monitoring purposes. The system is meant to power UAVs used in conducting aerial surveys and environmental observations, supporting the monitoring and tracking of wildlife. The design takes into account important factors such as solar resource availability, battery capacity, and load demand for the charging station. To find the best solution, HOMER Pro software is employed. Additionally, the study includes a sensitivity analysis to evaluate the impact of the minimum state of charge ($SOC_{min}$) on the performance of the optimal system.

## 2   Materials and Methods

To determine the most efficient design for an off-grid photovoltaic-battery system utilized in powering UAVs charging station, the study employs HOMER Pro software, considering input data related to the energy consumption model, meteorological conditions, and the techno-economic details of the system elements. This section encompasses a description of designing the proposed system included the consideration of the PV and battery models, net present cost (NPC) and the levelized cost of energy (LCOE). Figure 1 illustrates the basic schematic of the charging system.

### 2.1   PV Model

The hourly output $E_{PV}(t)$ (W) the PV generator is mathematically expressed as follows [8]:

$$E_{PV}(t) = d_{pv}.E_{STC}.\frac{I_h(t)}{I_{STC}}.\left[1 + \frac{\alpha}{100}(Temp_C(t) - Temp_{STC})\right] \qquad (1)$$

**Fig. 1.** The core illustration of an off-grid PV-battery charging station designed for UAVs.

where $d_{pv}$ represents the derating factor of the PV module. $I_h(t)$ (kWh/m²) signifies the solar radiation received on the inclined surface of the PV panels during the time interval of hour t. $E_{STC}$ (W$_p$) denotes the panel's output power under Standard Test Conditions (STC). $I_{STC}$ (kWh/m²) refers to the incident radiation under Standard Test Conditions (STC). $Temp_{STC}$ represents the PV cell temperature during STC ($Temp_{STC}$= 25 °C) and $Temp_C(t)$ (°C) stands for the PV cell temperature at the current time step, and it can be computed as follows [9]:

$$Temp_C(t) = Temp_a(t) + \left(\frac{NOCT - 20}{0.8}\right) \cdot \frac{I_{h_{year}}(t)}{1kWh/m^2} \tag{2}$$

where $Temp_a(t)$ denotes the ambient temperature (°C), and NOCT is the nominal operation cell temperature (°C).

## 2.2 Battery Model

The equation to compute the battery bank's power availability, $P_B(t)$ at hour t is as follows [10]:

$$P_B(t) = P_B(t-1).(1-\sigma) + (P_{PV}(t) - P_{load}(t)).\eta_{bat} \tag{3}$$

when:

$P_{PV}(t) - P_{load}(t) > 0$ and $P_B(t-1) < P_{Bmax}$ The charging process.
$P_{PV}(t) - P_{load}(t) < 0$ and $P_B(t-1) \geq P_{Bmin}$ The discharging process.

$$P_{B_{low}} \leq P_B(t) \leq P_{B_{up}} \tag{4}$$

The power availability of the battery bank at the previous hour $(t-1)$ is denoted by $P_B(t-1)$. The term $\sigma$ denotes the self-discharge rate of the battery bank. The variable $P_{load}(t)$ represents the load requirement during hour t. $\eta_{bat}$, is the battery efficiency. Moreover, $P_{B_{low}}$ and $P_{B_{up}}$ represent the lower and upper limits of allowable energy within the battery bank, respectively. Additionally, the upper threshold of state of charge ($SOC_{max}$) and the lower threshold of state of charge ($SOC_{min}$) for the battery bank during discharging are defined as follows:

$$SOC_{max} = C_B N_{bat} \tag{5}$$

$$SOC_{min} = SOC_{max}(1 - DOD_{max}), \tag{6}$$

$N_{bat}$ represents the overall quantity of batteries, $C_B$ represents the nominal capacity of each individual battery, and $DOD_{max}$ (%) represents the maximum allowable depth of discharge.

## 2.3  Net Present Cost (NPC) and Levelized Cost of Energy (LCOE)

NPC represents the summation of all costs and revenues occurring over the project's lifetime. To compute the total NPC of a system, the following equation is utilized [11]:

$$NPC = \frac{TC_{annual}}{CRF(j, N)} \tag{7}$$

where $TC_{annual}$ denotes the total annualized cost. CRF is the factor of capital recovery, defined as follows:

$$CRF(j, N) = \frac{j(1+j)^N}{(1+j)^N - 1} \tag{8}$$

where N stands for the number of years, j represents the annual real interest rate, and is computed using the subsequent formula:

$$j = \frac{k - f}{1 + f} \tag{9}$$

In this context, k signifies the nominal interest rate, and f represents the annual inflation rate.

LCOE, which stands for levelized cost of energy, represents the average cost of the system's useful electricity measured in $/kWh, and it can be determined using the subsequent equation [12]:

$$LCOE = \frac{TC_{annual}}{E} \tag{10}$$

where $TC_{annual}$ is the yearly total cost, and E is the overall electricity consumption in kWh/year.

# 3   Case Study

The UAVs are used for environmental monitoring through aerial surveys to observe and track wildlife. To design the off-grid PV system that powers the UAVs charging station, it is essential to estimate the daily energy requirement. The Li-Po battery of the UAV lasts approximately 20 min, and the UAV's missions continue for at least 6 h each day. The load profile is determined based on the mission duration and seasonal variations of daylight. Figure 2 shows the monthly load profile. The average daily consumption is 11.34 kWh/day.

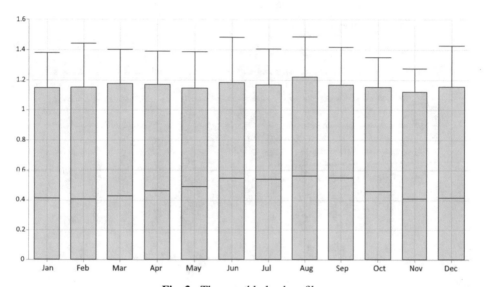

**Fig. 2.**  The monthly load profile.

## 3.1   Site Description

The charging station is situated on Chréa mountain, within the city of Blida, Algeria. The meteorological data for this location is depicted in Fig. 3. The solar resource utilized in this study was sourced from the NASA Surface Meteorology and Solar Energy (SSE) database. The global horizontal solar radiation at the site varies from 2.1 kWh/m$^2$/day to 7.2 kWh/m$^2$/day, with an annual average of 4.85 kWh/m$^2$/day, a clearness index of 0.58 and an annual average temperature of 18.06 °C.

## 3.2   Input Data

To simulate the system within HOMER, essential data including meteorological information, load profile, and the technical and economic details of system components are necessary. Comprehensive technical and economic information regarding the system elements can be found in Table 1.

**Table 1.** The technico-economic details of the system.

| Inputs | Value |
| --- | --- |
| Duration of the project (years) | 25 |
| PV module lifetime (years) | 25 |
| Battery lifetime (years) | 10 |
| Battery efficiency (%) | 95 |
| Nominal discount rate (%) | 8 |
| PV module cost ($/kW$_p$) | 1370 |
| Battery cost ($/kWh) | 830 |

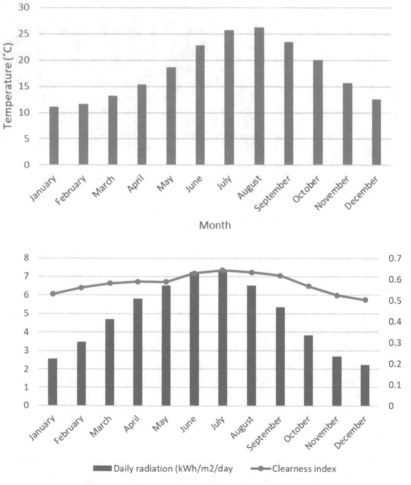

**Fig. 3.** Meteorological characteristics of the site.

# 4  Results and Discussion

Figure 4 presents the schematic diagram of the off-grid PV-battery system components and configuration. The inputs for the proposed approach include the characteristics of the PV module, battery, and load requirement. The SunPower X21-335-BLK PV module with 335 Wp and LGChem RESU10 (9.8kWh) battery are considered for the simulation. In cases where the PV energy generation surpasses the immediate hourly load demand, any excess energy is directed into the battery bank, while taking into consideration the current state of charge of the batteries. On the contrary, if the PV output is insufficient to fulfill the load demand, the battery bank discharges its stored energy to aid the PV in meeting the load requirements, with due consideration to the batteries' state of charge. To determine the optimal solution, the technical constraints are evaluated, and the total cost is calculated for each combination of PV system rated power and battery capacity. The simulation was conducted with an annual discount rate of 8% and a project lifetime of 25 years. Additionally, the system performance parameters were configured with a minimum battery state of charge ($SOC_{min}$) set at 20%. The optimal solution that best meets our constraints while minimizing both the NPC and LCOE is (7.27 kW PV, 9.8 kW Battery), Table 2 presents a summary of the various outcomes obtained from the optimal solution.

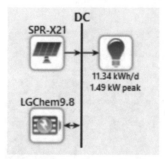

**Fig. 4.** The off-grid PV-battery system components and configuration.

**Table 2.** The techno-economic results of the optimal system.

| Technical | | Economic | |
|---|---|---|---|
| PV Array (kW) | 7.27 | NPC ($) | 24402.28 |
| Battery (kWh) | 9.8 | LCOE ($/kWh) | 0.457 |
| PV production (kWh/year) | 12383 | Capital cost ($) | 17453.8 |
| Battery autonomy (hour) | 15.4 | Replacement cost ($) | 6626 |
| Battery expected life (years) | 10 | O&M cost ($) | 1221 |
| Capacity shortage (%) | 1.09 | salvage ($) | −898.34 |

A load satisfaction rate of 1.09% was achieved with the optimized grid-connected PV system, and the resulting LCOE was 0.457 $/kWh. The dominant cost component was the battery, accounting for 54.2% of the NPC, while the PV generator represented 45.79%. The initial capital cost of the optimal system amounted to 17453.8 $, making up 71.52% of the NPC. The operating cost was estimated at 1221 $, and the replacement cost was 6626 $, accounting for 27.15% of the NPC. On the other hand, the salvage cost was evaluated at −898.34 $, providing a net benefit to the system.

A sensitivity analysis was conducted to examine the impact of variations in the minimum state of charge ($SOC_{min}$) on the performance of the hybrid system, focusing specifically on the levelized cost of energy (LCOE), PV generator size, and battery size. The results, depicted in Fig. 5, demonstrate a gradual increase in the system's LCOE from 0.6 $/kWh to 0.613 $/kWh as the minimum $SOC_{min}$ was raised from 5% to 21%. This increase was primarily influenced by a slight rise in the PV generator size, while the battery size remained constant at 19.6 kWh. However, at a minimum SOC of 22%, there was a significant change in the system's configuration. The PV generator size increased from 4 kW to 7.37 kW, and the battery size decreased to 9.8 kWh and remained at this level. As a result, these adjustments led to a notable decrease in the LCOE from 0.613 $/kWh to 0.461 $/kWh. Furthermore, with a further increase in the minimum SOC from 22% to 30%, the LCOE slightly rose from 0.461 $/kWh to 0.470 $/kWh, primarily due to a slight increase in the PV generator size.

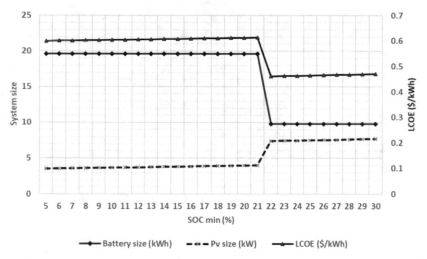

**Fig. 5.** Effect of the $SOC_{min}$ on the LCOE and the optimal system size.

## 5  Conclusion

This paper delves into the design and optimization of an off-grid PV-battery system used as a charging station for UAVs, specifically for environmental monitoring purposes. The optimized configuration of this system resulted in an annual PV production of 12383 kWh, with a minimal capacity shortage of only 1.09%. Additionally, a sensitivity analysis was performed to investigate the effects of the minimum state of charge ($SOC_{min}$) on the performance of the system.

The findings revealed that $SOC_{min}$ strongly impacts both the cost of energy and the optimal design size of the system. Moreover, The study suggests that the off-grid PV-battery system has significant potential as a practical and sustainable solution for powering UAVs, notably extending the operational endurance of UAVs during their mission coverage, especially in rural and remote regions.

# References

1. Ucgun, H., Yuzgec, U., Bayilmis, C.: A review on applications of rotary-wing unmanned aerial vehicle charging stations. Int. J. Adv. Robot. Syst. **18**(3), 17298814211015863 (2021)
2. Mohsan, S.A.H., Othman, N.Q.H., Khan, M.A., Amjad, H., Żywiołek, J.: A comprehensive review of micro UAV charging techniques. Micromachines **13**(6), 977 (2022)
3. Amorosi, L., Chiaraviglio, L., Galan-Jimenez, J.: Optimal energy management of UAV-based cellular networks powered by solar panels and batteries: formulation and solutions. IEEE Access **7**, 53698–53717 (2019)
4. Javidsharifi, M., Pourroshanfekr Arabani, H., Kerekes, T., Sera, D., Spataru, S.V., Guerrero, J.M.: Optimum sizing of photovoltaic-battery power supply for drone-based cellular networks. Drones **5**(4), 138 (2021)
5. Javidsharifi, M., Arabani, H.P., Kerekes, T., Sera, D., Guerrero, J.M.: PV-powered base stations equipped by UAVs in urban areas. In: 2022 IEEE 96th Vehicular Technology Conference (VTC2022-Fall), pp. 1–4 (2022)
6. Chiaraviglio, L., D'andreagiovanni, F., Choo, R., Cuomo, F., Colonnese, S.: Joint optimization of area throughput and grid-connected microgeneration in UAV-based mobile networks. IEEE Access **7**, 69545–69558 (2019)
7. Ali, E., Fanni, M., Mohamed, A.M.: A new battery selection system and charging control of a movable solar-powered charging station for endless flying killing drones. Sustainability **14**(4), 2071 (2022)
8. Dufo-López, R., Cristóbal-Monreal, I.R., Yusta, J.M.: Stochastic-heuristic methodology for the optimisation of components and control variables of PV-wind-diesel-battery stand-alone systems. Renewable Energy **99**, 919–935 (2016)
9. Diaf, S., Notton, G., Belhamel, M., Haddadi, M., Louche, A.: Design and techno-economical optimization for hybrid PV/wind system under various meteorological conditions. Appl. Energy **85**(10), 968–987 (2008)
10. Wang, H., Zhang, D.: The stand-alone PV generation system with parallel battery charger. In: 2010 International Conference on Electrical and Control Engineering, pp. 4450–4453 (2010)
11. Dalton, G.J., Lockington, D.A., Baldock, T.E.: Feasibility analysis of renewable energy supply options for a grid-connected large hotel. Renew. Energy **34**(4), 955–964 (2009)
12. Beccali, M.A.R.C.O., Brunone, S., Cellura, M., Franzitta, V.: Energy, economic and environmental analysis on RET-hydrogen systems in residential buildings. Renew. Energy **33**(3), 366–382 (2008)

# Efficient Solar Energy Harvesting and Power Management for Electric Vehicles Using MPPT Under Partial Shading Conditions and Battery/Super-Capacitor Integration

Ali Hamil$^{(\boxtimes)}$, Fethia Hamidia, and Amel Abbadi

LREA Laboratory, Yahia Feres University, Medea, Algeria
alihamil15@gmail.com

**Abstract.** This paper presents a comprehensive study on the implementation of power management strategies in electric vehicles equipped with solar panels. The research focuses on the integration of Maximum Power Point Tracking (MPPT) algorithms along with efficient battery/super-capacitor management techniques to optimize the use of renewable solar energy. The study addresses the challenges of partial shading on the solar panels and proposes innovative solutions to enhance energy harvesting and storage in electric vehicles. The aim of this research is to improve the overall energy efficiency and performance of electric vehicles by harnessing solar energy effectively. Through the use of MPPT, the system continuously tracks the maximum power output from the solar panels, even under partial shading conditions, ensuring optimal energy utilization. The paper discusses the methodology employed for power management, including the integration of battery and super-capacitor systems. It emphasizes the significance of efficient energy distribution between the two energy storage devices, ensuring seamless power flow and extended vehicle range.

**Keywords:** Electric Vehicles · Solar Energy · Power Management · MPPT · Battery · Supercapacitor · Partial Shading · Energy Efficiency

## 1 Introduction

The rise of electric vehicles (EVs) [1–3] and the growing interest in renewable energy technologies have led to significant advancements in sustainable transportation solutions. In particular, solar energy presents a promising source of clean and renewable energy that may be incorporated into electric cars to increase their general efficiency and lower their carbon footprint. This paper presents a comprehensive study focused on the implementation of power management strategies in electric vehicles equipped with solar panels. The primary objective of this research is to optimize the utilization of renewable solar energy through the integration of Maximum Power Point Tracking (MPPT) algorithms and efficient battery/super-capacitor management techniques. The dynamic nature of driving conditions, including partial shading on solar panels, presents

M. Hatti (Ed.): IC-AIRES 2023, LNNS 984, pp. 166–172, 2024.
https://doi.org/10.1007/978-3-031-60629-8_17

challenges for the effective utilization of solar energy in electric vehicles. Therefore, this study addresses these challenges and proposes innovative solutions to enhance energy harvesting and storage under various operating conditions. Through the use of MPPT algorithms, the power management system continuously tracks the maximum power output from the solar panels, even under partial shading conditions, ensuring optimal energy utilization. Additionally, the seamless integration of battery and super-capacitor systems in the power management process aims to extend the electric vehicle's range while maintaining a smooth power flow. In the subsequent sections, we will detail the methodologies, results, and implications of our research, aiming to foster a greener and more efficient future for electric vehicles.

## 2 Optimization of Solar Energy Using MPPT Under Partial Shading Conditions

Solar energy has emerged as a promising renewable energy source for electric vehicles (EVs), offering a clean and sustainable alternative to traditional fossil fuels. Integrating solar panels into EVs allows for on-the-go energy generation; however, the presence of partial shading poses challenges to efficient solar energy utilization. To address this issue, the implementation of Maximum Power Point Tracking (MPPT) algorithms is crucial [4, 5]. MPPT enables continuous monitoring and tracking of the maximum power output from the solar panels, even in partial shading conditions. By dynamically adjusting the operating point, MPPT optimizes energy capture, improving the overall efficiency of solar energy utilization in EVs.

To maximize solar energy utilization in partial shade circumstances, a number of MPPT approaches, including perturb and observe (P&O), incremental conductance, meta-heuristic algorithms and neural network-based algorithms, have been developed. These methods help mitigate the adverse effects of shading, ensuring that the solar panels operate at their peak efficiency. The integration of MPPT technology in electric vehicles not only increases energy efficiency but also reduces dependence on conventional grid charging. By effectively harnessing solar energy, EVs can extend their driving range and reduce carbon emissions, contributing to a more sustainable and eco-friendly transportation system.

In this study, the MPPT algorithm based on Particle Swarm Optimization (PSO) [6] was employed to optimize solar energy utilization in the presence of partial shading. The figure presented below illustrates the methodology used, demonstrating the effectiveness of the PSO-based MPPT approach in maximizing power capture from the solar panels.

Particle Swarm Optimization (PSO) is a nature-inspired optimization algorithm based on the social behavior of birds flocking or fish schooling. In PSO, a group of particles (representing potential solutions) moves through the solution space to find the optimal solution. Each particle has a position and velocity, and it evaluates its position's fitness based on an objective function. The particles collaborate by adjusting their velocities according to their own best position found so far (pbest) and the best position among all particles (gbest). During each iteration, particles update their positions by moving towards more promising regions of the solution space. This process continues until a stopping criterion is met, leading to the discovery of the optimal solution. In the context of optimizing solar energy utilization in electric vehicles under partial shading, PSO-based MPPT approaches can efficiently find the maximum power point, enhancing solar energy utilization and improving overall system efficiency (Fig. 1).

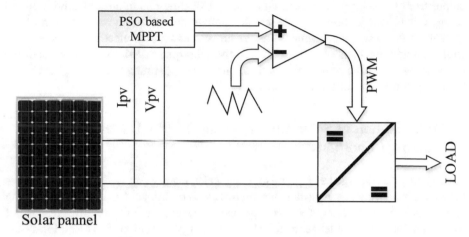

**Fig. 1.** Configuration of the proposed PV system

## 3   Power Management Battery/Super-capacitor

In the context of optimizing solar energy utilization in electric vehicles, the integration of both batteries and super-capacitors in the power management system [7–9] is essential. Each energy storage system serves a specific purpose, and their combination allows for a more efficient and reliable utilization of the harvested solar energy.

Batteries are known for their high capacity, making them well suited for storing significant amounts of energy over an extended period. They are ideal for meeting the long-term energy demands of electric vehicles, such as powering the vehicle during extended trips or continuous driving. However, batteries have limitations in terms of rapid charge and discharge rates. During high-demand situations, such as sudden accelerations or uphill drives, batteries may struggle to deliver the required power instantaneously. This is where super-capacitors come into play. Super-capacitors excel in providing rapid bursts of power and have an exceptional charge and discharge rate compared to batteries. During peak power demands, super-capacitors can rapidly release energy to support the batteries, ensuring smooth and responsive acceleration and power delivery. Their ability to handle high currents makes them particularly valuable in compensating for sudden spikes in power demand, which is critical for enhancing the overall performance of electric vehicles. The power management system intelligently allocates energy storage between batteries and super-capacitors based on real-time conditions. During steady-state driving or low power demands, batteries take the lead in providing sustained energy, optimizing their long-term utilization and lifespan. When sudden power demands occur,

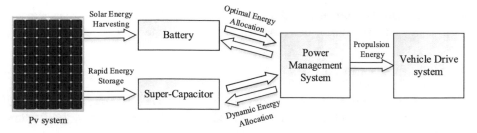

**Fig. 2.** Power Management Battery/Supercapacitor System for Solar-Electric Vehicles: Energy Flow and Optimization

the power management system utilizes the super-capacitors to supplement the batteries, preventing overloading and ensuring seamless power flow [10, 11] (Fig. 2).

This complementary energy storage approach offers several advantages for electric vehicles. It enhances the overall power management system's efficiency and extends the life of both batteries and super-capacitors by optimizing their use. Additionally, it ensures a smooth and responsive driving experience, enhancing vehicle performance and user satisfaction.

In our case, we have utilized PID controllers to effectively manage the power distribution between the battery and super-capacitor in the power management system. The Proportional-Integral-Derivative (PID) controllers play a pivotal role in dynamically regulating the energy flow based on real-time data and setpoint values, enabling us to optimize the utilization of solar energy in electric vehicles. As the solar energy input and power demands fluctuate during driving, the PID controllers continuously monitor the energy distribution and compare it to the desired setpoint, representing the optimal energy ratio between the battery and super-capacitor. The PID controllers allow us to achieve seamless energy flow, ensuring smooth transitions and minimal energy losses during power transfers.

The Proportional component (P) of the PID controller generates a control signal proportional to the error, which is the deviation between the setpoint and the actual energy distribution. This component enables us to respond promptly to any deviations, maintaining a balanced energy allocation between the battery and super-capacitor. The Integral component (I) of the PID controller accumulates past errors and calculates the integral of the error over time. This feature helps us address any steady-state errors or long-term imbalances in energy distribution, ensuring accurate and stable power management over extended driving periods. Additionally, the Derivative component (D) of the PID controller predicts future error trends based on the rate of change of the error. This capability allows us to proactively adjust the energy distribution to stabilize the system, preventing overshooting or oscillations and improving the overall efficiency of the power management process. The adaptability and responsiveness of the PID controllers are especially valuable in the dynamic context of solar energy utilization in electric vehicles. By continuously adjusting the power flow based on real-time conditions, the PID controllers optimize the energy utilization from the solar panel, contributing to enhanced vehicle efficiency and extended driving range. Overall, the implementation of PID controllers in our power management system has been instrumental in achieving efficient energy distribution between the battery and super-capacitor, thereby maximizing the benefits of solar energy for a sustainable and eco-friendly transportation solution.

## 4   Simulation and Results

In this section, we present the results of our simulation study and discuss the performance of the power management system with PID-based optimization of solar energy utilization in electric vehicles.

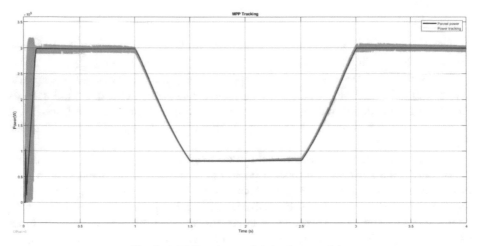

**Fig. 3.** MPPT under partial shading condition

Figure 3 showcasing MPPT under shading conditions provides valuable insights into the algorithm's effectiveness in optimizing solar energy utilization for electric vehicles. Despite challenging shading scenarios caused by passing clouds or obstructions, the MPPT algorithm adeptly tracks the maximum power point (MPP) of the solar panel. In the graph, we observe how the MPPT curve swiftly responds to changes in solar irradiance, adjusting the operating point to maintain the solar panel's MPP. This fast tracking capability ensures optimal power capture, contributing to enhanced energy harvesting. Compared to conventional fixed voltage or current-based charging methods, the MPPT algorithm significantly improves energy capture under shading conditions, leading to increased efficiency in the power management system. Moreover, the MPPT's adaptability and robustness are evident as it navigates between multiple local MPPs during shading events, avoiding being trapped at a local MPP. This dynamic behavior further enhances the system's efficiency and viability. Overall, the MPPT algorithm plays a crucial role in maximizing solar energy utilization in electric vehicles, resulting in extended driving range and reduced reliance on grid charging. Its ability to swiftly track the MPP and efficiently respond to changing conditions makes it an essential component in achieving sustainable and eco-friendly transportation (Fig. 4).

This figure demonstrates the performance of our power management system in accurately tracking and meeting the required power demand for the seamless operation of

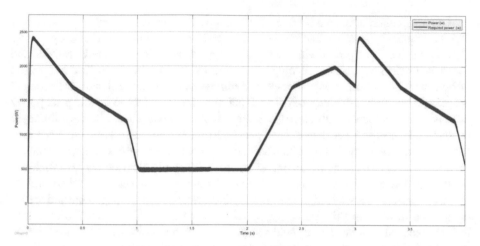

**Fig. 4.** Required Power and Managed Power graph

the electric vehicle. The graph compares the required power, representing the vehicle's power demand during various driving conditions, with the managed power delivered by the power management system. As shown in the figure, the managed power closely aligns with the required power, ensuring that the electric vehicle receives the necessary power to function optimally. This effective power tracking is crucial for maintaining the vehicle's performance, especially during dynamic driving scenarios such as acceleration, deceleration, and uphill driving. The power management system's ability to match the required power demand efficiently contributes to the vehicle's smooth operation, extended driving range, and overall energy efficiency. This close alignment between required power and managed power showcases the power management system's reliability and effectiveness in delivering a seamless and reliable driving experience while harnessing the benefits of renewable solar energy.

## 5  Conclusion

In this study, we have presented a comprehensive analysis of the power management system for electric vehicles equipped with solar panels and integrated with Maximum Power Point Tracking (MPPT) algorithms. The primary objective of this research was to optimize the utilization of solar energy and enhance the energy efficiency of electric vehicles, contributing to a more sustainable and eco-friendly transportation ecosystem.

Through the integration of advanced MPPT algorithms, we demonstrated the ability to efficiently track and capture the maximum power output from the solar panels, even under challenging shading conditions. The MPPT algorithm's adaptability and fast tracking response enable optimal energy harvesting, resulting in increased energy input to the power management system. The power management system's effectiveness in seamlessly distributing energy between the battery and super-capacitor further enhances the electric vehicle's performance. The Proportional-Integral-Derivative (PID) controllers play a vital role in dynamically regulating the energy flow, ensuring smooth transitions and minimal energy losses during power transfers.

Our experimental results revealed that the power management system effectively met the required power demand for the electric vehicle's optimal operation. By closely aligning the managed power with the required power, the system ensured a reliable and seamless driving experience, encompassing various driving scenarios. The integration of both battery and super-capacitor energy storage systems in the power management process proved to be advantageous. The super-capacitors' ability to handle peak power demands complemented the battery's energy storage capabilities, allowing for rapid bursts of power during acceleration and deceleration. The successful implementation of the power management system demonstrated its potential for real-world applications in electric vehicles. The system's adaptability, reliability, and enhanced energy efficiency contribute to extended driving range and reduced reliance on conventional grid charging.

In conclusion, our study highlights the significance of power management and MPPT integration in electric vehicles equipped with solar panels. The collective efforts to harness renewable solar energy hold the potential to revolutionize the transportation industry, leading us towards a greener and more sustainable future.

## References

1. Hannan, M.A., Azidin, F.A., Mohamed, A.: Hybrid electric vehicles and their challenges: a review. Renew. Sustain. Energy Rev. **29**, 135–150 (2014)
2. Ding, N., Prasad, K., Lie, T.T.: The electric vehicle: a review. Int. J. Electric Hybrid Vehicles **9**(1), 49–66 (2017)
3. Mo, T., et al.: Trends and emerging technologies for the development of electric vehicles. Energies **15**(17), 6271 (2022)
4. Figueiredo, S., Leão e Silva, R.N.A.: Hybrid MPPT technique PSO-P&O applied to photovoltaic systems under uniform and partial shading conditions. IEEE Latin Am. Trans. **19**(10), 1610–1617 (2021)
5. Motahhir, S.., El Hammoumi, A., El Ghzizal, A.: The most used MPPT algorithms: review and the suitable low-cost embedded board for each algorithm. J. Clean. Prod. **246**, 118983 (2020)
6. Yap, K.Y., Sarimuthu, C.R., Lim, J.M.-Y.: Artificial intelligence based MPPT techniques for solar power system: a review. J. Mod. Power Syst. Clean Energy **8**(6), 1043–1059 (2020)
7. Choi, M.-E., Seo, S.-W.: Robust energy management of a battery/supercapacitor Hybrid Energy Storage System in an electric vehicle. In: 2012 IEEE International Electric Vehicle Conference. IEEE (2012)
8. Pay, S., Baghzouz, Y.: Effectiveness of battery-supercapacitor combination in electric vehicles. In: 2003 IEEE Bologna Power Tech Conference Proceedings, vol. 3. IEEE (2003)
9. Kouchachvili, L., Yaïci, W., Entchev, E.: Hybrid battery/supercapacitor energy storage system for the electric vehicles. J. Power Sources **374**, 237–248 (2018)
10. Shin, D., et al.: Battery-supercapacitor hybrid system for high-rate pulsed load applications. In: 2011 Design, Automation & Test in Europe. IEEE (2011)
11. Prasad, T.N., et al.: Power management in hybrid ANFIS PID based AC–DC microgrids with EHO based cost optimized droop control strategy. Energy Rep. **8**, 15081–15094 (2022)

# New Hybrid Optimized MPPT Technique for PV Systems Under Partial Shadow Conditions

Salah Anis Krim, Fateh Krim$^{(\boxtimes)}$, and Hamza Afghoul

Department of Electronics, University of Setif-1, Setif, Algeria
{salahanis.krim,h_feroura}@univ-setif.dz, krim_f@ieee.org

**Abstract.** The maximum production of electricity is needed to satisfy the consumption from renewable energy source such as solar energy. Solar cell is more appropriate energy source for long usage. The irradiance level is measured for finding the maximum earned power from solar energy so that several maximum power point tracking (MPPT) techniques have been introduced. Thus the performance of photovoltaic (PV) system is increased by using Honey Badger Algorithm (HBA) and Coyote Optimizer Algorithm (COA). The HBA is extracted from the foraging behavior of honey badger which is used to get maximum power from the PV panel. Additionally COA optimization is used for making efficient and accurate output power by analyzing the placement of PV panel in the whole process based on the social behavior of coyotes. From these algorithms, the maximum power is obtained in any climate change situations and also the efficiency and accuracy are maintained using the proposed optimization techniques. The novelty of this study is to produce maximum power and also maintain the efficiency and accuracy of output throughout the process. The simulation is carried out under Matlab/Simulink environment. The key results analyzed the performance of the proposed technique, in terms of steady state behavior, tracking speed, tracking efficiency, and distortions in waveforms and was compared to the most recent and effective MPPT algorithms as FPA, GSA and PSO, under different solar radiance levels, and the comparison showed that is superior to them, a faster tracking speed, higher tracking efficiency, and lower oscillations around the steady state.

**Keywords:** PV system · HBA · COA · MPPT · Partial shading

## 1 Introduction

Electricity consumption is increasing, owing to the growth of technologies and population day by day, so that power generation is focused on renewable energy sources such as solar energy. This later is inexhaustible and affordable energy source that produces energy from sun light [1]. The disadvantages of traditional Perturb and Observe (P&O) algorithm can be reduced using the fractional short-circuit current (FSCC) method under different conditions. [2]. An adaptive perturbation size is achieved by multiplying 2D Gaussian function and Arctangent function then the PV system steady state is developed by duty cycle computed with variable perturbation frequency [3].

M. Hatti (Ed.): IC-AIRES 2023, LNNS 984, pp. 173–182, 2024.
https://doi.org/10.1007/978-3-031-60629-8_18

A high voltage gain DC-DC converter was implemented in P&O algorithm for low power applications [4]. The difference between successive powers with corresponding voltages was determined by analyzing the differential power algorithm to effectively track the maximum power [5]. To reduce the oscillation with high speed and good tracking accuracy, fractional order using fuzzy logic method was used [6]. The traditional variable step size is eliminated by changing the step size with the method of auto-scaling variable step-size [7]. In [8] it focuses on the generating speed by increasing the number of optimal blades in which impeller solidity is measured using the starting and stopping speed of blades.

The total harmonic distortion (THD) is reduced to improve the performance of the proposed PV system by integrating of conjoins three series connected full bridge inverters and a single half bridge inverter [9]. The main objective of this technology is to produce maximum power under different partial shading conditions (PSC) in which whale optimization and differential evolution algorithms are used to produce high quality output [10]. This method is used to produce maximum power by reducing the shading effect and atmospheric changes using voltage scanning technique within the short period of complex PSC [11]. This system is generated for low power applications that use single diode based PV systems to enhance accuracy and efficiency by a MPPT algorithm [12]. The maximum power depends on the radiance and temperature and the efficiency is achieved by a new P&O algorithm [13]. This system constantly generates the maximum power using array of PV cells with the MPPT algorithm of deep reinforcement and fuzzy mechanism [14].

The purpose of the proposed methodology is to improve the PV panel performance using a stability algorithm for producing maximum power; unlike conventional technologies where the maximum power is not stable for all types of loads. The remainder of this paper is organized as follows. In Sect. 2, MPPT tracker using proposed algorithm to enhance the performance of the system is explained. In Sect. 3, the key simulation results of the proposed technique are discussed and compared to existing algorithms as Flower Pollination Algorithm (FPA), Gravitational Search Algorithm (GSA) and Particle Swarm Optimisation (PSO), in Sect. 3. Section 4 ends with conclusion.

## 2   Design of the Proposed Methodology

The proposed system consists of PV panel, switched mode boost DC-DC converter, multilevel cascaded H-Bridge (CHB) inverter, HBA controller and COA controller. This approach is used to get maximum power and also enhance the efficiency and accuracy tracking. Figure 1 illustrates the block diagram of the overall PV system. The PV panel is integrated with MPPT block for getting maximum power. This latter is converted from DC to AC with the help of the inverter to supply AC loads [15].

The maximum power is obtained by using HBA based MPPT algorithm. Then, the stability is provided by the COA controller.

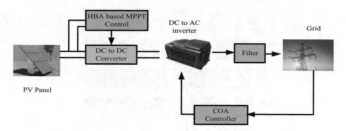

**Fig. 1.** Block diagram of the proposed PV system.

## 2.1 HBA Based Algorithm for Maximum Power Generation

The power generated from PV panel should be maximized using HBA which is based on the hunting behavior of honey badger animal. It finds the prey location by smelling and digging or follows the honey guide bird to directly locate beehive in which it can dig fifty holes of radius forty kilometer in a day. In this algorithm, there are two phases, digging phase and honey phase. Each step of this algorithm is explained in [16]. This flowchart is shown in Fig. 2, it is based on exploration and exploitation phases, and the population candidate solution is given in Eq. (1)

$$\begin{bmatrix} x_{11} & x_{12} & \dots & x_{1D} \\ x_{21} & x_{22} & \dots & x_{2D} \\ \dots & \dots & \dots & \dots \\ x_{n1} & x_{n2} & \dots & x_{nD} \end{bmatrix} \tag{1}$$

Honey badger is in the position of i which is expressed in Eq. (2),

$$x_i = x_i^1 + x_i^2 + \dots + x_i^D \tag{2}$$

In initialization phase n is the number of honey badgers and badger's position is given by Eq. (3)

$$x_i = lb_i + r_1 \times (ub_i - lb_i) \tag{3}$$

$r_1$ is a random number from 0 to 1, $lb_i$ and $ub_i$ are the lower and upper values, respectively, of search domain.

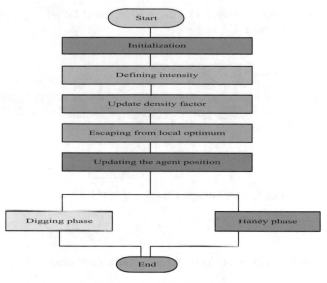

**Fig. 2.** Flow chart of HBA algorithm.

Intensity is defined as the prey concentration strength at distance between prey and ith honey badger position. Here $I_i$ is the smell intensity and if it increases, the speed will also increase, which is expressed as inverse square law in Eqs. (4–6)

$$I_i = r_2 \times \left( \frac{S}{4\pi d_i^2} \right) \tag{4}$$

$$S = (x_i - x_{i+1})^2 \tag{5}$$

$$d_i = x_{prey} - x_i \tag{6}$$

S is the concentration strength of prey. The density factor ($\alpha$) makes smooth transition by controlling time-varying randomization from exploration to exploitation. For decreasing the iteration with time, update $\alpha$ which is given in Eq. (7).

$$\alpha = C \times \exp\left( \frac{-t}{t_{max}} \right) \tag{7}$$

where, $t_{max}$ is the maximum iteration, C is a constant, $C > 1$ (default $= 2$).

The search direction is changed by flag F which alters search direction, as shown in Fig. 3, to escape from local optima by scanning of the search space [17].

As mentioned in the above explanation, the update process ($x_{new}$) is divided as digging phase and honey phase. In honey phase, the movement of honey badger is in cardioid shape so the cardioidal motion is given in Eq. (8),

$$x_{new} = x_{prey} + F \times \beta \times I \times x_{prey} + F \times r_3 \times \alpha \times d_i \times |\cos(2\pi r_4) \times [1 - \cos(2\pi r_5)]| \tag{8}$$

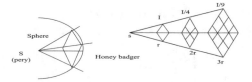

**Fig. 3.** Search direction of honey badger.

where, $x_{prey}$ is the best fit prey position, $\beta$ is greater than or equal to 1 (default = 6), the distance between prey and $i^{th}$ honey badger is denoted as $d_i$ and $r_3$, $r_4$, and $r_5$ are three different random numbers between 0 and 1. Value of F is expressed in Eq. (9)

$$F = \begin{cases} 1 & if \ r_6 \leq 0.5 \\ -1 & else \end{cases} \tag{9}$$

where, $r_6$ is a random number from 0 to 1. Moreover, in digging phase, any disturbance F received by badger is used to find better prey location. The below Eq. (10) expresses the honey phase to reach the beehive.

$$x_{new} = x_{prey} + F \times r_7 \times \alpha \times d_i \tag{10}$$

The search is varied from time ($\alpha$) and disturbance F at this stage.

## 2.2 COA Based Stability Enhancement Optimization

In the proposed method, COA optimization is adopted for maintaining the efficiency of the system by considering the placement of PV panel. Its core idea is based on canis latrans species that locate in America. The balance between exploitation and exploration process is developed for optimization. The algorithm is explained by the flowchart in Fig. 4.

COA is developed based on the social behavior and hunting strategy of coyotes [17]. The latter are distinguished by cooperative functionalities as they head towards the prey in the close chain. The location of the prey is identified by the smelling sense, and they attack in groups. The social behavior of the coyote is defined in Eq. (11)

$$SOC_c^{p,t} = \vec{x} = (x_1, x_2, x_3, ...x_D) \tag{11}$$

The coyote social condition is denoted as $fit_c^{p,t} \in \Re$.
The current social condition of coyotes behavior is assessed by Eq. (12):

$$fit_c^{p,t} = f\left(SOC_c^{p,t}\right) \tag{12}$$

Initially the coyotes are in random packs. After, they will move and change their packs. These packs are interchanged between coyotes increasing the consideration with their interaction, as expressed by alpha, given in Eq. (13).

$$alpha = \left\{ SOC_c^{p,t} \Big|_{\substack{arg \\ c=\{1,2,...N_c\}}} \min f(SOC_c^{p,t}) \right\} \tag{13}$$

Cultural tendency is defined as the existing information shared by coyotes when all coyotes are arranged and exchange in social culture which is given in Eqs. (13, 14),

$$
cult_j^{p,t} = \begin{cases} O^{p,t}_{\left(\frac{N_c+1}{2}\right)} & ; \quad N_c \ is \ odd \\ \dfrac{O^{p,t}_{\frac{N_c}{2}j} + O^{p,t}_{\left(\frac{N_c}{2}+1\right)j}}{2} ; & otherwise \end{cases}
\tag{14}
$$

where $O^{p,t}$ is social condition ranking of all coyotes at instant t and the cultural tendency of considered pack is determines coyotes mean in the social condition from a specific pack. In this $age_c^{p,t}$ is the birth of new coyotes,

$$
pup_j^{p,t} = \begin{cases} SOC_{r1,j}^{p,t} & ; rnd \ \alpha \ P_s \quad or \ j = j_1 \\ \dfrac{SOC_{r2,j}^{p,t}}{2} & ; rnd \ \alpha \ P_s + P_a \ \ or \ j = j_2 \\ R_j & ; otherwise \end{cases}
\tag{15}
$$

where, $SOC_{r1,j}^{p,t}$ is the Social condition of coyote $r_1$, $SOC_{r2,j}^{p,t}$ is the Social condition of coyote $r_2, j_1, j_2$ are the dimensions of the optimization problem, Pa, Ps are the probability of association and probability of scatter respectively, R j is the number in the range of variable bounds.

The cultural variety of coyotes is calculated by Pa and Ps from pack which is given below,

$$
P_s = \frac{1}{D}, P_a = \frac{1 - P_S}{D}
$$

For simulation we have to consider two parameters. the fitness function w is the bad function and pack effect δ in which the coyotes quantity is denoted Q and the function is mentioned in Eqs. (16, 17)

$$
\delta = alpha^{p,t} - SOC_{cr2}^{p,t}
\tag{16}
$$

$$
\delta = cult^{p,t} - SOC_{cr2}^{p,t}
\tag{17}
$$

The alpha and pack effect are important while updating coyotes which is shown in Eq. (18),

$$
SOC_c^{p,t,new} = SOC_c^{p,t,old} + r1.\delta1 + r2.\delta2
\tag{18}
$$

Update of social condition of coyotes is given by Eq. (19),

$$
SOC_c^{p,t+1} = \begin{cases} SOC_c^{p,t,new} fit_c^{p,t} \alpha fit_c^{p,t} \\ SOC_c^{p,t} otherwise \end{cases}
\tag{19}
$$

Thus the output is maintained efficiently and accurately by using the above mentioned optimization technique. By using the simulated output of this system, the effectiveness of maximum power generation progress is verified.

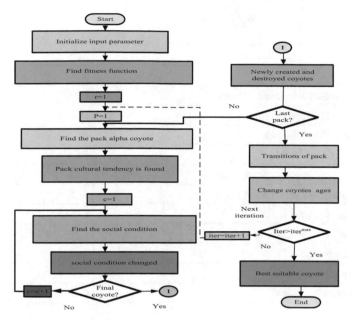

**Fig. 4.** Flow chart of COA algorithm.

# 3 Results and Discussion

The proposed method is simulated under Matlab/Simulink environment and the key results are shown. In this section, HBA technique is used for the MPPT controller to track the maximum power during a minimum time and COA provides high efficiency and stability as compared to others recent and efficient methods (FPA, GSA and PSO). The simulation circuit of the proposed system consists of PV panel, HBA and COA controllers, boost DC-DC converter, CHB inverter. The PV panels with different irradiance levels are the input of the proposed system. In the proposed system the tracking efficiency is improved by using proposed technique in MPPT controller for tracking maximum power during minimum time. The key results show the PV panel output power, load and converter output voltage. The simulation parameters are given in Table 1. For both the normal and abnormal state boundaries like voltage, current and power are utilized to evaluate. For the testing process, the time needed for simulation is 0–5 s.

The characteristics of PV pannel are represented in Fig. 5, which consists of I-V and P-V curves. The generation of maximum power is improved with the help of the boost DC-DC converter on the basis of HBA algorithm. The generated power comparison analyses the proposed technique with others recent techniques are illustrated in Fig. 6. The induced power and the tracking efficiencies are described in Table 2. As well as the induced power is transmitted to the grid by means of the VSI and three phase filter. The comparison analysis proves that the proposed hybrid technique provides a very high tracking speed (less than 1.5 s), very high tracking efficiency (more than 99.5 %) with low distortion wave forms, and very low oscillations around the steady state (less than 2 W). From the above analysis, it can be concluded that the proposed method offers better results to manage power among generation side and grid side.

**Table 1.** Simulation parameters

| Parameters | Values |
| --- | --- |
| PV | |
| PV panel generated Maximum Power (W) | 9 kW |
| Cells per module (Ncell) | 96 |
| Open circuit voltage Voc (V) | 64.2 |
| Short-circuit current Isc (A) | 5.96 |
| Voltage at maximum power point Vmp (V) | 54.7 |
| Current at maximum power point Imp (A) | 5.58 |
| Light-generated current IL (A) | 6.0092 |
| Diode saturation current I0 (A) | 6.30e-12 |
| Diode ideality factor | 0.94504 |
| Shunt resistance Rsh (ohms) | 269 |
| Series resistance Rs (ohms) | 0.37 |
| Three phase supply | |
| Nominal phase-to-phase voltage Vn (Vrms) | 260 |
| Nominal frequency fn (Hz) | 60 |

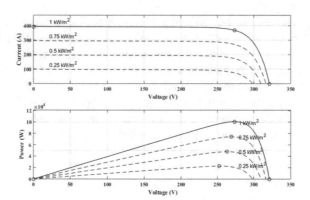

**Fig. 5.** IV and PV characteristics of PV panel.

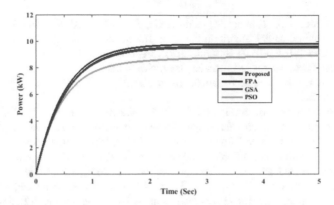

**Fig. 6.** Comparison analysis for different partial shading conditions.

**Table 2.** Key simulation results for tracking time and efficiency.

| Optimization Technique | Tracking time (s) | | | Tracking Efficiency (%) |
|---|---|---|---|---|
| | Irradiance = 1000 (W/m$^2$) Pmpp = 2490.75 W | Irradiance = 900(W/m$^2$) Pmpp = 2235.4 W | Irradiance = 700 (W/m$^2$) Pmpp = 1728.23 W | |
| PSO | 8.20 | 8.01 | 7.92 | 85.73 |
| FPA | 4.26 | 3.77 | 4.15 | 95.86 |
| GSA | 3.65 | 3.15 | 2.87 | 99.38 |
| Proposed | 1.25 | 1.53 | 1.55 | 99.49 |

## 4 Conclusion

In this paper, the proposed hybrid HBA-COA technique was presented to enhance the performance of the MPPT of a grid-tied PV system under PCS and climate changes. The proposed method offers a very high tracking speed (less than 1.5 s), very high tracking efficiency (more than 99.5%) with low distortion waveforms, and very low oscillations around the steady state (less than 2 W). The proposed technique was demonstrated under the environment of Matlab/Simulink, was compared to other recent and effective techniques and proved its superiority over these techniques in terms of tracking speed, efficiency, stability, distortions in the waveforms of the PV response. In future work, MPPT with the hybrid HBA-COA technique will be evaluated on an experimental hardware platform using a PV emulator.

## References

1. Jiang, J., et al.: On application of a new hybrid maximum power point tracking (MPPT) based photovoltaic system to the closed plant factory. Appl. Energy **124**, 309–324 (2014)

2. Sher, H., Murtaza, A., Noman, A., Addoweesh, K., Al-Haddad, K., Chiaberge, M.: A new sensor less hybrid MPPT algorithm based on fractional short-circuit current measurement and P&O MPPT. IEEE Trans. Sustain. Energy 6(4), 1426–1434 (2015)
3. Reza Tousi, S.M., Moradi, M., Saadat Basir, N., Nemati, M.: A function-based maximum power point tracking method for photovoltaic systems. IEEE Trans. Power Electron. 31(3), 2120–2128 (2016)
4. Alisson Alencar Freitas, A., Lessa Tofoli, F., Mineiro Sa Junior, E., Daher, S., Luiz Marcelo Antunes, F.: High-voltage gain DC–DC boost converter with coupled inductors for photovoltaic systems. IET Power Electronics 8(10), 1885–1892 (2015)
5. Sharma, D.K., Purohit, G.: Differential power algorithm based maximum power point tracking for a standalone solar PV system. Energy Southern Afr. 26(2), 103 (2015)
6. Tang, S., Sun, Y., Chen, Y., Zhao, Y., Yang, Y., Szeto, W.: An enhanced MPPT method combining fractional-order and fuzzy logic control. IEEE J. Photovolt. 7(2), 640–650 (2017)
7. Chen, Y., Lai, Z., Liang, R.: A novel auto-scaling variable step-size MPPT method for a PV system. Sol. Energy 102, 247–256 (2014). https://doi.org/10.1016/j.solener.2014.01.026
8. Liu, J., Tian, R., Nie, J.: Effect of impeller solidity on the generating performance for solar power generation. J. Electron. Sci. Technol. 19, 100132 (2021)
9. Prabaharan, N., Palanisamy, K.: Analysis and integration of multilevel inverter configuration with boost converters in a photovoltaic system. Energy Convers. Manage. 128, 327–342 (2016)
10. Moghassemi, A., Ebrahimi, S., Padmanaban, S., Mitolo, M., Holm-Nielsen, J.B.: Two fast metaheuristic-based MPPT techniques for partially shaded photovoltaic system. Int. J. Electric. Power Energy Syst. 137, 107567 (2022)
11. Celikel, R., Yilmaz, M., Gundogdu, A.: A voltage scanning-based MPPT method for PV power systems under complex partial shading conditions. Renewab. Energy 184, 361–373 (2022)
12. Raj, A., Raj Arya, S., Gupta, J.: Solar PV array-based DC–DC converter with MPPT for low power applications. Renew. Energy Focus 34, 109–119 (2020)
13. Attou, A., Massoum, A., Saidi, M.: Photovoltaic power control using MPPT and boost converter. Balkan J. Electric. Comput. Eng. 2, 1 (2014)
14. Singh, Y., Pal, N.: Reinforcement learning with fuzzified reward approach for MPPT control of PV systems. Sustain. Energy Technol. Assessm. 48, 101665 (2021)
15. Wang, F., et al.: Optimal design of solar-assisted steam and power system under uncertainty. J. Clean. Prod. 336, 130294 (2022)
16. Hashim, F.A., Houssein, E.H., Hussain, K., Mabrouk, M.S., Al-Atabany, W.: Honey Badger Algorithm: new metaheuristic algorithm for solving optimization problems. Math. Comput. Simul. 192, 84–110 (2022)
17. Janamala, V., Sreenivasulu Reddy, D.: Coyote optimization algorithm for optimal allocation of interline –photovoltaic battery storage system in islanded electrical distribution network considering EV load penetration. J. Energy Storage 41, 102981 (2021)

# Frequency Characterization of Water Solutions Used in Electrolysis for Production of H₂

Selma Zenati[(✉)], Hocine Moulai, Abderrahmane Ziani, and Ryad Haridi

University of Science and Technology Houari Boumedienne, FEI, LCDEP, Algiers, Algeria
zenatiselma8@gmail.com, s.zenati@usthb.com

**ABSTRACT.** The paper discusses about characterization of solutions obtained by two methods, the first one is by extraction of samples randomly from water, and the second one is by using solutions obtained after electric discharges applied in water environments. An experimental comparative study was led with methodology in this paper. Water was used with two types: sea water and tap water for measuring impedances and phases in a frequency bandwidth increasing from 20 Hz to 10 MHz. The impedance and the phase were measured with a device which analyzes the impedance (LCR meter), variations of impedances and phases depending on frequency range. It is enlightened that the impedance graph for water before and after electrolysis may be influenced by factors such as electrolyte type and concentration, duration and intensity of electrolysis, and specific experimental conditions. Comparing the impedance graphs before and after electrolysis allows direct comparison of changes induced by the electrolysis process.

**Keywords:** Electrical discharges · Variation of impedance · Green hydrogen · Water solutions

## 1 Introduction

World energy demand has increased considerably for several decades due to population increase, economic growth and technologies implemented [1]. Hydrogen ($H_2$) is the most abundant element in the universe. However, hydrogen is not an energy source, but an energy carrier, produced by a chemical reaction from a primary resource [2]. It is known to be an excellent energy carrier, likely to play an important role in the near-future energy system [3]. A hydrogen atom consists of one proton and one electron, making it the lightest element in with unique properties like high energy density (120 MJ/kg) and lower volumetric energy density (8 MJ/L). However, it is not readily accessible directly on land, it is available in combination chemical forms from water, fossil fuels, and biomass [4].

Water does not naturally decompose into hydrogen and oxygen. It must necessary to provide energy for that. Electrolysis of water is therefore a "forced" electrochemical reaction carried out in a specific device called an electrolyser. Such devices are supplied with electric current (using a generator) and liquid water, as well as hydrogen and gaseous oxygen. This means using electricity to transform one compound (water) into another ($H_2$ and $O_2$) [5].

Green hydrogen can be produced by using electricity to electrolyze water, splitting it into hydrogen and oxygen. Powering without emitting greenhouse gases; all the energy used to run the process comes from renewable sources [6].

Hydrogen has a high mass energy density and is lightweight. Energy can be transported through a simple electrochemical conversion. Across geographic regions through pipelines or forms liquid fuels such as ammonia on cargo ships [6].

Water electrolysis technology can be classified as follows: Applied electrolyte separating the two half-reactions at the anode (oxygen evolution reaction) and cathode (hydrogen evolution reaction) of the electrolytic cell [7]. PEM electrolysis has been known for over sixty years and was developed by General Electric, where a solid sulfonated polystyrene membrane is used as the electrolyte.

This technology is also named polymer electrolyte membrane electrolysis. The proton exchange membrane acts as both the gas separator and the electrolyte. In the electrolysis process, sea water and tap water are injected into the cell without any electrolytic additives. There are a variety of advantages of PEM electrolysis, such as high current density, greater energy efficiency, low gas permeability, wider operating temperatures (20–80°C) and easy handling and maintenance [8].

The purpose of this work is to study and characterize the electrochemical properties of water in order to study the production of hydrogen by electrolysis. So, to study the characterization of water for this case, we used four samples of water, two of them consist of tap water and the two others are sea water: two are investigated before electrical discharges occurrence and two others after discharges. The characteristics of these samples will allow us to estimate the percentage of the product obtained and make a comparison between them. Obtaining all these characteristics enabled to make a comparison before and after discharges application to enlighten their effect on the water characteristics and the chemical compounds obtained.

## 2   Experimental Setup

### 2.1   Materials Used

**1) Impedance Analyzer (LCR Meter):** LCR meters are measuring instruments that measure a physical property known as impedance. Impedance, which is expressed using the quantifier Z, indicates resistance to the flow of an AC current. It can be calculated from the current I flowing to the measurement target and the voltage V across the target's terminals. The appartus is equipped with 3 channels, with 23000 memory points and ensuring the transfer of data thanks to the USB port. The measurement of the discharge current will be visualized through the channel 1, and the voltage on the channel 2, The cathode-ray allows measurements of voltages, frequencies and phase, which allow to deduce the current and impedance. In general the accuracy of the measurements is low (at best 5%). For voltage measurements the input impedance is high to avoid degradation of the oscilloscope. Since the time base is calibrated with an accuracy of a few %, it is possible to determine the frequency of the signal. The oscilloscope has calibration signals, which are standard amplitude and frequency signals, which are used for oscilloscope calibration. These calibration signals are square signals with 1 kHz or 1 MHz frequencies and 0.1 V or 1 V amplitude and are available and marked with "CAL."

In this study, we used the model GwINSTEK LCR-8110G precision LCR Meter DC 20–10 MHz (Fig. 1) that can measure many electrical and physical parameters of any type of load or insulating material (resistance, impedance, phase, capacitance, inductance, permittivity, tan $\delta$, etc.). Performance of this materials: - 20Hz ~ 10MHz wide test frequency (LCR-8110G), 6 digit measurement resolution, 10mV ~ 2V measurement drive level (DC/20Hz ~ 3MHz), -0.1% basic measurement accuracy.

**Fig. 1.** The impedance analyzer: model GwINSTEK LCR-8110G.

**2) The Test Cell:** The test cell is in the form of a glass container where the different electrodes and insulations are placed It is a closed isolation. It is used to apply breakdowns of the distilled water and compare it with the three hydrogen peroxide samples with different percentages (3%, 30%, 35.28%). So a millimetric paper is inserted behind the testing cell to measure the distance between the electrodes. The distance used is L = 1.5mm. Figure 2 shows the achieved test cell.

**Fig. 2.** The test cell.

**3) Software LCR 8110G:** It consists of an open source software made available on the internet by the manufacturer. It is used to collect the measurements made by the impedance analyzer directly on PC under the form of DATA files.

## 2.2  Precautions to Be Taken Prior to Starting the System

- Adjust the gap between the two electrodes to 1 cm.
- Rinse the vessel and electrodes with the electrolyte to be analyzed.
- Fill the vessel with the electrolyte to avoid air bubbles until the electrolyte level exceeds the top of the electrodes

   The study of samples of water is done in 4 parts:

- Influence of frequency increase on impedance and phase.
- Influence of the electrodes surface.
- Influence of temperature increase.
- Effect of thermal ageing.

The study was made to investigate the behavior of the impedance and phase increasing each time a frequency range up to 10MHz using the materials described in the previous part in a hemispheric electrode configuration. This analysis treated the tap water without any additives introduced into the plate electrolyser. Figure 3 shows the software window used for the measurements by displaying the changes in the impedance and phase of the sample.

## 2.3  Experimental Work Methodology

Electrical discharges were generated in a sphere-sphere electrode system for testing distilled and mineral water. To measure the discharge current in both samples, measurements are made by the voltage drop over of a non-inductive resistor. A cubic insulating glass cell has been used. It is pierced on the sides to introduce a separate point-sphere electrode system of 4mm gap, the tip having a radius of curvature of 50 μm and the sphere having a diameter of 2.5cm. The results are displayed on the PC as DATA files and the experimental configuration with these components is shown in Fig. 3.

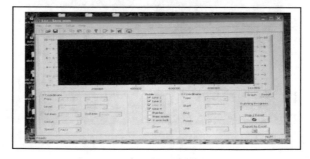

**Fig. 3.**  Software GW INSTEK.

# 3    Results and Discussions

## 3.1    Measurement of Impedances Before Discharge

The results of the impedancemetry are measured between 20 Hz and 10 MHz of all water samples used in the electrolysis process. The frequency ranges are divided into five to better appreciate the behavior in each of the ranges. In addition to impedance measurements, we will investigate phase variations between current and applied voltage. The variations of impedance and phase are measured and compared for different frequencies under a distance of 1 cm between the electrodes. Two bandwidths for each case have been chosen to represent them and for interpretation. The first one (20Hz to 1 kHz) and the second one (1 MHz to 10 MHz).

### 1) Impedance in Tap Water Before Discharge:
Tap water is characterized with LCR meter, and measurements of impedance and phase are performed in the two ranges of frequency. Figures 4 and 5 present the variations of impedance and phase angle for both considered frequency ranges.

Before applying discharges in samples of water, one remarks in the four graphs (Figs. 4 and 5) the impedance is decreasing in almost exponentially trend with frequency, except for the bandwidth 20 Hz - 1 kHz where increasing pulses are recorded at 118.8 kHz and the impedance magnitude passes from $3.88 \times 10^7$ Ω to almost $3.15 \times 10^9$ Ω. Except this singular increasing pulse, the impedance of the tap water in bandwidth 1MHz to 10 MHz decreases from $2.79 \times 10^4$ Ω at 20 Hz to $2.66 \times 10^3$ Ω, and the phase increases with high difference (Fig. 5). It is in general negative and decreases as the frequency increases indicating a more capacitive nature of the water as the frequency increases. However, in the first range, at the frequency of 336 kHz, the last positive point is recoded at $+ 15.46$ Ω. After this point, the impedance became negative: -116 Ω at 346 kHz (Fig. 4). Many resonance pulses appear, making to the water an inductive behavior. This phenomenon affects the results in a high decrease of the water impedance in this frequency range.

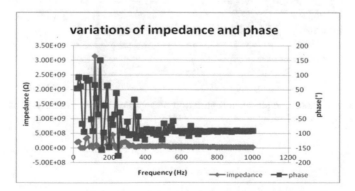

**Fig. 4.**  Impedance variation of tap water before electrolysis between 20Hz and 1kHz, distance between electrodes: 1 cm.

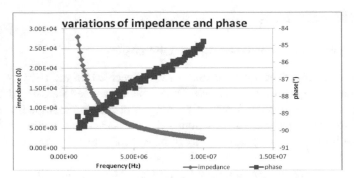

**Fig. 5.** Variation of impedance and phase shift of tap water before electrolysis between 1MHz to 10MHz, distance 1 cm.

## 2) Impedance of Sea Water Before Discharge:

In the case of sea water in the bandwidth 20 Hz -1 kHz, increasing pulses are recorded at 69.49 kHz and the impedance magnitude value is $6.12 \times 10^8$ Ω. Except these singular increasing pulses, the impedance of the sea water in bandwidth 1MHz to 10MHz has a maximum value of $1.14 \times 10^6$ Ω at the frequency $9.36 \times 10^6$ MHz, and the phase increases with high difference. The phase is in general negative and decreases as the frequency increases indicating a more capacitive nature of the water as the frequency increases. However, in the first range, the maximum frequency is recorded at $+ 178.81$ Ω (Fig. 6), and in the second range, three positive pulses are recorded ($+174.44$ Ω, $+ 160.79$ Ω and $156.04$ Ω). They are presented on Fig. 7.

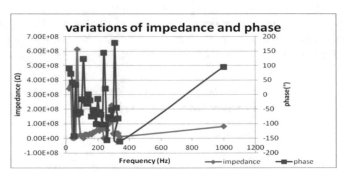

**Fig. 6.** The impedance and phase shift variation of seawater before electrolysis between 20 Hz and 1kHz, distance 1 cm.

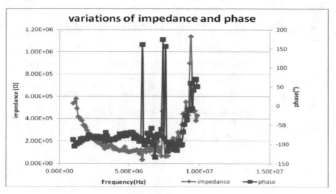

**Fig. 7.** Impedance and phase shift variations of seawater before electrolysis between 1MHz and 10MHz, distance 1 cm

## 3.2  Electrical Discharges in Sea and Tap Water

### 1) Impedance of Tap Water After Discharges:

The same conditions as the previous tests have been retained. The cell is filled with water and the impedance and phase variations are measured. The results are shown on Figs. 8 and 9.

After applying a series of discharges, a noticeable change is detected. In tap water For the bandwidth 20 to 100 Hz (Fig. 8), an increasing pulses are detected at 37.77 and 58.79 Hz with so little magnitude, The phase between the resistive and reactive components is in general negative and decreases as the frequency increases indicating a more capacitive nature of the water as the frequency increases. However, in the frequency range of 1 MHz to 10MHz the impedance is completely negative (Fig. 9).

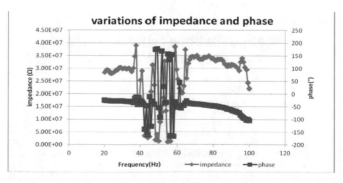

**Fig. 8.** The impedance and phase shift variation of tap water after electrolysis between 20 Hz and 100 Hz, distance 1 cm

### 2) Impedance of Sea Water After Discharges:

In sea water for the range of 20 to 100 Hz, one remarks two events: first we have an anti resonance at 25.65 Hz with the impedance value of 47.58 $\Omega$, and for the phase at

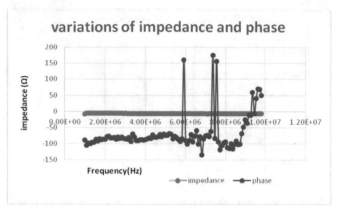

**Fig. 9.** The impedance and phase shift variation of tap water after electrolysis between 1MHz and 10MHz, distance 1 cm.

65.25 Hz that has the maximum value of + 82.05° (Fig. 10). And in the bandwidth of 1MHz to 10 MHz, one remarks that when the impedance is decreasing the phase is increasing (Fig. 11). This can be due to the compensation between the inductive and capacitive components of the sea water samples. It is well known that water conductivity is increased when additives are mixed to it [9, 10], then enabling to increase the current flow and hence hydrogen production.

The mechanisms of high voltage discharges in water are very complex and not yet well understood. Air bubbles that are initially present in the liquid or that form because of local heating are involved and accelerate the phenomenon. If the electric field is sufficiently intense, an avalanche of electrons becomes the starting point for the propagation of electrical discharges for higher applied between the electrodes.

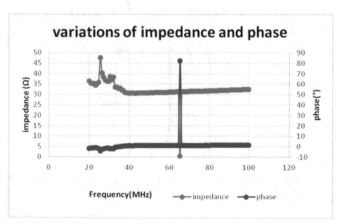

**Fig. 10.** The impedance and phase shift variation of seawater after electrolysis between 20 Hz and 100 Hz, distance 1 cm.

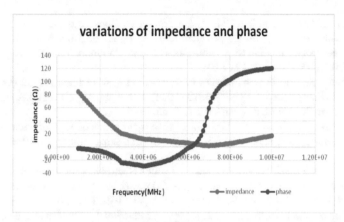

**Fig. 11.** Variation d'impédance et de déphasage de l'eau de mer après l'électrolyse entre 1MHz et 10MHz, distance 1 cm

## 4 Conclusion

Hydrogen production is a key area of interest in the quest for sustainable and clean energy solutions. Several methods of hydrogen production exist, such as steam methane reforming, electrolysis, and water splitting. Among these, water electrolysis stands out as a promising approach due to its potential for utilizing renewable energy sources and producing hydrogen without greenhouse gas emissions.

Characterization of water using impedance spectroscopy is a versatile and invaluable technique in the fields of science and engineering. By analyzing the electrical response of water at different frequencies, researchers can gain a deeper understanding of its structural, chemical, and dynamic properties. Impedance spectroscopy enables the identification of various ions, contaminants, and solutes present in water, aiding in water quality assessment and environmental monitoring.

Electrical discharges in water produce hydrogen, oxygen and other compounds. Its quantity and percentage vary and depend on the following parameters: The applied voltage, the distance between the electrodes, the type of electrodes and the conductivity of water.

The impedance of both tap water and sea water decreases as the frequency increases. It is the lowest at 10 MHz in coparison with the DC voltage that is currently used in electrolysers. A chopped voltage source could be then a most promising alimentation to increase the efficiency of hydrogen production.

## References

1. Qureshy, A.M.M.I., Dincer, I.: Multi-component modeling and simulation of a new photo-electrochemical reactor design for clean hydrogen production. Energy **224**, 120196 (2021). https://doi.org/10.1016/j.energy.2021.120196
2. Richstein, J.C.: Project-based carbon contracts: a way to finance innovative low-carbon investments. SSRN Electron. J. (2017).https://doi.org/10.2139/ssrn.3109302

3. Caravaca, A., Garcia-Lorefice, W.E., Gil, S., de Lucas-Consuegra, A., Vernoux, P.: Towards a sustainable technology for H2 production: direct lignin electrolysis in a continuous-flow polymer electrolyte membrane reactor. Electrochem. Commun. **100**, 43–47 (2019). https://doi.org/10.1016/j.elecom.2019.01.016

4. Shiva Kumar, S., Lim, H.: An overview of water electrolysis technologies for green hydrogen production. Energy Rep. **8**, 13793–13813 (2022). https://doi.org/10.1016/j.egyr.2022.10.127

5. Millet, P.: par électrolyse de l'eau sur membrane acide

6. Oliveira, A.M., Beswick, R.R., Yan, Y.: A green hydrogen economy for a renewable energy society. Curr. Opin. Chem. Eng. **33**, 100701 (2021). https://doi.org/10.1016/j.coche.2021.100701

7. Buttler, A., Spliethoff, H.: Current status of water electrolysis for energy storage, grid balancing and sector coupling via power-to-gas and power-to-liquids: a review. Renew. Sustain. Energy Rev. **82**, 2440–2454 (2018). https://doi.org/10.1016/j.rser.2017.09.003

8. Shiva Kumar, S., Himabindu, V.: Hydrogen production by PEM water electrolysis – a review. Mater. Sci. Energy Technol. **2**(3), 442–454 (2019). https://doi.org/10.1016/j.mset.2019.03.002

9. Yueyue, D., Ying, Z., Zheng, X., Dou, B., Cui, G.: Correlating electrochemical biochar oxidation with electrolytes during biochar-assisted water electrolysis for hydrogen production. Fuel **339**, 126957 (2023). https://doi.org/10.1016/j.fuel.2022.126957

10. Xu, Y., Wang, C., Huang, Y., Fu, J.: Recent advances in electrocatalysts for neutral and large-current-density water electrolysis. Nano Energy **80**, 105545 (2021)

# Robotics and Electrical Vehicle

# The Traction Chain of an Electric Vehicle

Touhami Nawal[1]($\boxtimes$), Mansouri Smail[1], Ouled Ali Omar[2], and Benoudjafer Cherif[3]

[1] LEESI laboratory, Ahmed Draia University, Adrar, Algeria
touhami.nawal@univ-adrar.edu.dz
[2] LDDI laboratory, Ahmed Draia University, Adrar, Algeria
[3] SGRE laboratory, Tahri Mohamed University, Bechar, Algeria

**Abstract.** Due to the oil crisis and environmental limitations, numerous car manufacturers are compelled to invest in significant research initiatives for the advancement of electric and hybrid vehicles. Consequently, discovering clean and efficient propulsion solutions becomes crucial in preserving our mobility experience. Hence, effectively controlling and managing energy within electric vehicles presents a substantial challenge.

In this framework, the focus of this study primarily revolves around the propulsion system of an electric vehicle, aiming to enhance energy management efficiency and facilitate the seamless integration of energy sources within the vehicle [1].

**Keywords:** Electric vehicle (EV) · the traction chain of an EV · batteries · electric motors

## 1 Introduction

Air pollution is currently a highly sensitive issue, with greenhouse gas-induced global warming being one of its significant consequences. Among greenhouse gases, carbon dioxide (CO2) is recognized as the most hazardous.

Vehicles, responsible for 17% of global greenhouse gas emissions, rank as the third-largest source of emissions after energy generation and industrial activities. Their impact on the environment is noticeable both on a global and local scale. Consequently, many countries are urged to commit to limiting their greenhouse gas emissions [2].

To achieve cleaner and more sustainable transportation in densely populated urban areas worldwide, the adoption of electric vehicles emerges as a potential solution [3].

Electric vehicles utilize specific electric machines for propulsion and rely on batteries as an energy source. These machines include direct current machines, synchronous machines (particularly those with permanent magnets), variable reluctance machines, and cage asynchronous machines [4].

Furthermore, electric vehicles offer easier maintenance and greater reliability compared to conventional internal combustion engine vehicles.

Electric motors generally have a significantly longer lifespan than their internal combustion counterparts [5]. Additionally, the electric motor's remarkable advantage lies in

M. Hatti (Ed.): IC-AIRES 2023, LNNS 984, pp. 195–201, 2024.
https://doi.org/10.1007/978-3-031-60629-8_20

its ability to swiftly and precisely generate the required torque under all driving conditions. This precise torque control is particularly advantageous in wheeled vehicles, where the interaction between the tires and the road surface heavily influences the vehicle's traction force [5].

The objective of this study is to delineate the various components comprising the traction chain of an electric vehicle.

## 2 Definition of an Electric Vehicle (EV)

A vehicle is a collection of interconnected components designed to facilitate both the movement of the chassis and the comfort of the passengers it carries. It consists of a chassis and a ground connection system that includes tires, wheels, axles, and suspensions. The steering system's dynamics are not considered, and the angle of the steering wheel directly affects the position of the wheels [6].

An electric vehicle (EV) is a type of vehicle that relies solely on electric energy to generate propulsion. Unlike traditional fuel-powered vehicles, the power is transmitted to the wheels through one or more electric motors, depending on the chosen transmission solution.

Due to advancements in power electronics, such as energy management systems, significant scientific and technological progress has been made in this field. As a result, numerous ideas and new designs are being explored to further develop electric propulsion systems. The architecture of an electric propulsion system is relatively simple, consisting of an electric actuator, a transmission device, and wheels [7].

## 3 The Traction Chain of an Ev

The primary element of an electric vehicle is the electric traction chain.

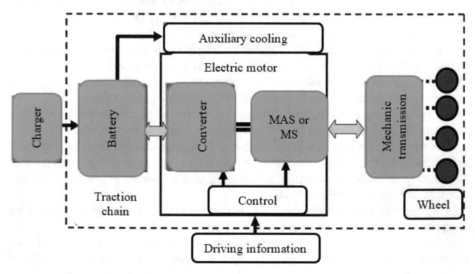

**Fig. 1.** Functional diagram of the traction chain of an electric vehicle [5].

## 3.1  Battery

Batteries, also referred to as electrochemical batteries, are devices that use electrochemical reactions to convert electrical energy into potential chemical energy when they are being charged, and then convert the stored chemical energy back into electrical energy during discharge.

Various types of batteries exist, such as lead-acid, nickel-metal hydride, and lithium-ion batteries, among others. However, lithium-ion batteries have emerged as the preferred choice for electric vehicle (EV) applications due to their specific energy capacity and relatively high power output. The properties of batteries commonly used in electric vehicles are summarized in Table (Table 1) [1].

**Table 1.** Properties of batteries used in electric vehicles [1].

| Battery technology | Specific energy (Wh/Kg) | Specific power (W/Kg) | Number of cycles at 80% discharge | Cost (euro/KWh) |
|---|---|---|---|---|
| **Pb-Acide** | 40–50 | 140–250 | 800–1500 | 100–190 |
| **Ni-Mh** | 60–80 | 500–1400 | 500–2000 | 400–2000 |
| **Li-Ion** | 70–130 | 600–3000 | 800–1500 | 700–2000 |

## 3.2  Power Converters

Depending on the use of direct current or alternating current machines, the energy converters will have to be different, namely:

- **AC/DC CONVERTERS**

In electric vehicles, the AC/DC converters are employed to convert the incoming alternating current (AC) electrical energy from either the general distribution network or an onboard alternator, which is connected to a heat engine. The purpose is to transform it into direct current (DC) electrical energy that can be stored in an electrochemical accumulator battery and/or a high-capacity battery [5].

- **DC/DC CONVERTERS**

A DC/DC converter is an electrical device that allows for the conversion of a substantially constant direct current voltage from a voltage source into controlled and adjustable voltages and currents, which can differ from the input values. These converted values are tailored to meet the specific power requirements of various components, such as motors and batteries, in an electric vehicle.

In an electric vehicle, DC/DC converters serve two vital purposes:

a. They play a crucial role in supplying power to direct current motors used for propulsion.
b. They are necessary to adapt the voltage from the main battery to the voltage requirements of electronic auxiliary components like sensors and regulators.

By employing a chopper, it becomes possible to maintain the desired motor current while allowing for smooth and gradual adjustment without significant loss of motor voltage. Additionally, it enables the adjustment of torque and speed, thereby controlling the vehicle's traction and electric braking capabilities [5].

• *DC/AC INVERTERS*

Electric vehicles equipped with an alternating current (AC) motor require the use of an inverter as a conversion device between the energy source and the traction motor(s). The primary function of the inverter is to convert direct current (DC) electrical energy into alternating current (AC) electrical energy. Additionally, it enables the control of motor torque and vehicle speed adjustment.

There are multiple methods available for DC-AC conversion, but in the context of road vehicles and industrial efficiency, a specific inverter structure has gained preference. This structure incorporates six bidirectional switches, each consisting of an insulated-gate bipolar transistor (IGBT) and a diode mounted in antiparallel configuration. These switches are controlled using a Pulse Width Modulation (PWM) technique. This configuration allows for the combination of a voltage source (such as a battery) with a current source type receiver (such as an asynchronous motor, wound synchronous motor, permanent magnet motor, or variable reluctance motor).

The PWM control method offers two significant advantages:

a) It pushes the harmonics of the output voltage towards higher frequencies, making it easier to filter out those frequencies from the voltage.
b) It is employed to adjust the fundamental frequency of the output voltage [5].

### 3.3   Electric Motors

Electric motors are essential components in electric vehicle (EV) technologies because they transform electrical energy from the battery into mechanical energy, facilitating the movement of the vehicle. Additionally, they function as generators during regenerative braking, allowing the return of energy to the power source. Over the past few decades, various kinds of electric motors, such as AC and DC motors, have been created to cater to the requirements of the automotive industry. Among these options, permanent magnet motors (MSAP) are widely preferred [8].

• *Motorizations for the traction chain of an EV*

*a. Separately Excited DC Motor*
DC motors with independent field excitation are regulated using two power choppers. One chopper controls the motor armature, while the other regulates the excitation or

inductor. This control scheme is extensively employed in automotive electric propulsion systems and is particularly prevalent in the LEROY Sommer motor, which powers the majority of electric vehicles currently in use. The LEROY Sommer motor is based on models originally developed for industrial traction. By utilizing a relatively uncompli-cated and cost-effective thyristor system, the motor's performance can be optimized by adjusting the desired torque/speed characteristics through the manipulation of armature and excitation current levels [5].

### b. Permanent Magnet DC Motor

The industrial sector is witnessing a growing trend towards the use of permanent magnet DC motors. This technology enables the development of lighter motors with superior efficiency compared to their predecessors. However, the composition of these magnets involves the use of rare earth elements, which are costly. Additionally, these magnets are highly susceptible to temperature increases [5].

### c. Asynchronous Motor

There are two main types of asynchronous machines: slip-ring rotor asynchronous machines and squirrel-cage asynchronous machines, with the latter being the more com-monly used variant. The squirrel-cage asynchronous machine has gained popularity due to its high level of robustness, reliability, and cost-effectiveness, making it a widely adopted choice.

Controlling these machines necessitates the utilization of an electronic inverter. The primary function of the inverter is to convert the direct current supplied by the batteries into three-phase alternating current, allowing precise control of the motor's operation during acceleration and cruising phases. Moreover, the inverter converts the alternating current generated by the motor during deceleration and braking into direct current to recharge the batteries and enhance motor braking.

To ensure proper regulation, the inverter adjusts the frequency of the alternating current supplied to the motor based on the driver's demand, typically through the position of the accelerator pedal. This regulation function is expected to be carried out in the near future by a vector control system.

The introduction of electronics has significantly increased the rotational speed capa-bilities of these motors, allowing them to operate at speeds ranging from 3,000 to 15,000 revolutions per minute [5].

### d. Permanent Magnet Synchronous Motor

Permanent magnet synchronous motors offer exceptional power-to-weight ratios and efficiency. However, their high cost remains a major drawback. While ferrite magnets can be utilized as an alternative, they do not exhibit outstanding performance. On the other hand, rare earth magnets such as Samarium-Cobalt or Iron-Neodyne-boron are more promising options.

One limitation of this motor is the inability to adjust the excitation. The magnetic field of the permanent magnets undergoes slight variations over time and temperature changes but not significantly.

To achieve high speeds, it becomes necessary to increase the stator current in order to demagnetize the motor. Unfortunately, this leads to higher stator joule losses as a consequence [5].

### e. Wound-rotor Synchronous Motor

This particular engine offers a highly compelling alternative. In comparison to the permanent magnet synchronous motor, it possesses an extra degree of freedom: the ability to adjust the excitation flux. This additional feature provides a broad spectrum of control algorithms, including synchronous compensator, unity power factor, maximum torque, and loss minimization [5].

### f. Variable Reluctance Motor

This engine is characterized by its affordability. However, the primary challenge lies in its design. At high speeds, this engine generates a highly pulsating torque, which leads to mechanical vibration issues and produces higher acoustic noise compared to its competitors. These effects are direct consequences of its operating principle [5].

- **The advantages and disadvantages of the different electric motors for powering the traction chain of an Electric Vehicle**

The advantages and disadvantages of various electric motors used to power the traction system of a vehicle are outlined in Table (2) [9, 10].

**Table 2.** Advantages and Disadvantages of the various electric motors for powering the traction chain.

|  | Advantages | Disadvantages |
|---|---|---|
| **DC motor with separate excitation** | - Control by a single power DC/DC converter on the armature and a low power DC/DC converter for controlling the excitation current | - High engine price<br>- The manufacturing process is difficult to automate<br>- Relatively low power density |
| **Permanent magnet DC motor** | - Control with a single DC/DC converter<br>- braking relatively simple to implement<br>- excellent efficiency | - high price |
| **Asynchronous motor** | - Easily industrializable, therefore low cost<br>- High specific power<br>- acceptable efficiency<br>- robust motor | - Relatively expensive electronics<br>- high supply voltage to facilitate the manufacture of the motor |
| **Permanent magnet synchronous motor** | - High specific power<br>- high efficiency<br>- relatively easy dynamic braking | - High price<br>- high priced electronics |
| **Coil rotor synchronous motor** | - simple and less expensive food | - Uncommon technology<br>- fragility of the brush ring system |
| **Variable reluctance motor** | - High mass torque | - Poor performance<br>- High price of power electronics<br>- Motorswithhighspecific performances have a small air gap<br>and are relatively fragile |

# 4  Conclusion

In this study, we provided a definition of electric vehicles and discussed the different components that constitute the powertrain of an electric vehicle, highlighting the advantages and disadvantages of each component. Electric vehicles, known for their eco-friendliness and cleanliness, are expected to gain increasing prominence in the market in the near future, eventually replacing conventional gasoline-powered cars, which are highly polluting and not sustainable in the long term. However, it is important to note that the energy production required for operating and manufacturing electric vehicles contributes to global warming. Therefore, the ecological impact of electric cars depends on the source of energy used, with renewable energy sources like solar or hydropower being more environmentally friendly options. Currently, there are various electric vehicle architectures available, each offering different performances and functionalities [11].

# References

1. Bouguenna, I.F.: Commande Robuste d'une Chaine de Traction d'un Véhicule Electrique Multi sources. Université Djillali Liabes Sidi Bel Abbes (2020)
2. L'observatoire Cetelem.: Le système de voiture électrique. Observatoire auto (2019)
3. Kumar, K.N., Tseng, K.J.: Impact of demand response management on chargeability of electric vehicles. Energy 111, 190–196 (2016). https://doi.org/10.1016/j.energy.2016.05.120
4. Benbouya, B., Ouabbas, D.: Etude des éléments de la chaîne de traction d'un véhicule électrique à base d'une machine asynchrone double étoile. Univérsité A.Mira Bejaia (2018–2019)
5. Miloudi, S., Derradj, A., Bouhlal, A.: Simulation Numérique D'une Chaine De Traction D'un Véhicule Electrique. Université Mohamed Boudiaf M'sila 2011/2012
6. Maakaroun, S.: Modélisation et simulation dynamique d'un véhicule urbain innovant en utilisant le formalisme de la robotique. PhD theses (2011)
7. Benariba, H.: Contribution à la commande d'un véhicule électrique. PhD theses (2018)
8. Labdouni, N., Latreche, S.: Répartition d'Energie d'un Véhicule Electrique par la Technique de Réseaux de Neurones. Université Ibn Khaldoun Tiaret (2021/2022)
9. Guy, G., Guy, C.: Actionneurs Electriques, Principes Modèles Commande. Eyrolles (2000)
10. Multon, B., Hirsinger, L.: Problème De La Motorisation D'un Véhicule Electrique. Ecole normale supérieure de Cachan, Mars 1996
11. Abdelmalek, B., Hamza, F.: Etude et simulation des éléments de chargeurs intégrés pour véhicule électrique. Université Abdelhamid Ibn Badis Mostaganem (2018/2019)

# A Comparative Study of Speed Control Strategies for In-Wheel Direct Current Machines in Electric Vehicles: PI, Fuzzy Logic, and Artificial Neural Networks

Ali Hamil[✉], Fethia Hamidia, and Amel Abbadi

Lrea Laboratory, Yahia Feres University Medea, Medea, Algeria
Alihamil15@gmail.com

**Abstract.** This paper proposes a comparative study of three speed controllers for DC motor speed in electric vehicles (EVs). The controllers studied are the Proportional Integral Controller (PI), the Fuzzy Logic Controller (FLC), and the Artificial Neural Networks (ANNs). The primary objective of this study is to evaluate the performance, efficiency, and adaptability of these control techniques for speed control of DC motors. Rigorous simulations using MATLAB/Simulink are used to evaluate the accuracy, response time, and overall control effectiveness of each approach. The PI controller, widely used for its simplicity and robustness, is used as the basis for comparison. The FLC shows advantages in effectively controlling the speed of DC motors. The study also explores the potential of ANNs, using their learning and adaptive capabilities, to control DC motors.

**Keywords:** DC-Motor · Electric Vehicles (EVs) · Proportional Integral control · Fuzzy Logic control · Artificial Neural networks

## 1 Introduction

Electric vehicles (EVs) have gained significant attention as a sustainable and eco-friendly transportation solution. Among the various EV configurations, those equipped with four in-wheel motors have emerged as a promising technology, offering advantages such as improved gear ratios and reduced vehicle weight. Achieving precise speed control in these in-wheel motors is crucial for optimizing vehicle performance, energy efficiency, and overall driving experience.

This paper presents a comprehensive comparative study of different speed control methods for in-wheel direct current (DC) motors in electric vehicles. The focus is primarily on three control approaches: Proportional Integral (PI) control [1–5], fuzzy logic control (FLC) [6–9], and artificial neural networks (ANNs) [10, 11]. These methods have been extensively investigated in various applications, but their specific suitability and effectiveness in the context of in-wheel motor control warrant further exploration and evaluation. The primary objective of this study is to assess and compare the performance,

M. Hatti (Ed.): IC-AIRES 2023, LNNS 984, pp. 202–213, 2024.
https://doi.org/10.1007/978-3-031-60629-8_21

efficiency, and adaptability of these control techniques in achieving precise speed regulation in in-wheel DC motors. Rigorous simulations using MATLAB/Simulink are conducted to evaluate the control accuracy, response time, robustness, and energy efficiency of each method. Through a comprehensive analysis, this study aims to provide valuable insights into the strengths, limitations, and practical implications of these control approaches. The PI control approach, widely recognized for its simplicity and robustness, serves as a fundamental benchmark for comparison. The FLC method is examined for its performance in achieving precise speed control in in-wheel motors. Furthermore, the study explores the potential of ANNs, leveraging their learning and adaptive capabilities, to optimize speed control in this specific context.

The paper is organized as follows: Sect. 2 details three methods for speed control of in-wheel drive electric vehicles, namely the Proportional Integral (PI), fuzzy logic, and artificial neural network (ANN) controllers. This section provides an in-depth examination of the principles, characteristics, and advantages of each method in achieving precise speed regulation in the context of in-wheel drive electric vehicles. Moving on, Sect. 3 illustrates the simulation of the studied methods using Matlab Simulink and discusses the results obtained. The simulation process enables a comprehensive analysis of the performance, accuracy, and efficiency of each control method. The findings are then discussed to provide insights into their effectiveness and potential applications. Lastly, Sect. 4 presents the conclusion of the paper, summarizing the key findings derived from the comparative analysis. Additionally, this section offers valuable perspectives and recommendations for future research directions in the field of speed control for in-wheel drive electric vehicles.

## 2 Methods

In the realm of electric vehicles (EVs), the control of in-wheel direct current (DC) machines is essential for achieving precise and efficient operation. Control methods play a vital role in regulating the speed and performance of these machines, ensuring optimal vehicle dynamics, energy efficiency, and overall driving experience. Three EV control methods for in-wheel DC machines will be discussed in this section namely PI, Fuzzy and artificial neural network control.

These control strategies main goal is to manage the in-wheel DC machines speed, which has a direct impact on the vehicle's stability, acceleration, and deceleration. The ability to accurately regulate the speed makes it possible to improve the performance of the vehicle in a number of ways, including power distribution, torque management, and regenerative braking. To achieve the desired speed, each control approach uses a distinct set of principles and algorithms. The PI control technique continually adjust the control signal depending on the difference between the desired and actual speed using a mix of proportional and integral terms. Fuzzy logic control, on the other hand, employs linguistic rules and membership functions to capture and process imprecise and uncertain information, making it suitable for handling complex and non-linear systems. Artificial Neural networks, utilize interconnected nodes and layers to learn and model the relationship between input and output variables, enabling adaptive and self-learning control.

In this section, we will thoroughly examine each method for speed control in direct current machines integrated within the wheels of electric vehicles.

## 2.1  Proportional-Integral (PI) Controller

To regulate and stabilize systems, the PI control approach [1–4] combines the proportional term (P) with the integral term (I) (see Fig. 1). The proportional term produces an output directly proportional to the current error, facilitating a fast dynamic response. Accuracy and stability of the system are improved by the integral term's accumulation and correction of long-term aberrations and enduring mistakes. This controller achieves a balance between long-term error correction and rapid dynamic response.

We can mathematically represent it as:

$$u(t) = Kp \times e(t) + Ki \times \int_0^t e(\tau)d\tau \quad (1)$$

where $u(t)$ is the output, $Kp$ is the proportional gain, $Ki$ is the integral constant, $e(t)$ is the error, and $\int_0^t e(\tau)d\tau$ is the integral of the accumulated error. Optimal performance depends on proper tuning of the controller parameters, such as $Kp$ and $Ki$.

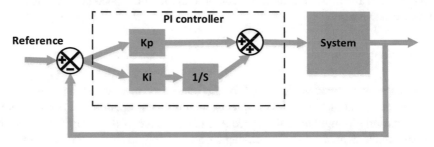

**Fig. 1.** A typical PI control structure

A simulation study (refer to Fig. 2) was done to evaluate the performance of the PI control method and compare it to the others. The simulation provides a controlled analysis of speed regulation in in-wheel direct current machines of electric vehicles. Aspects such as speed tracking accuracy, stability, and response time may be examined by modifying the control settings and analyzing the system's behavior. The simulation model consists of the in-wheel DC machine, the chopper circuit, the PWM generator, the PI controller [5], and the speed reference signal.

## 2.2  Fuzzy Logic Controller

Fuzzy Logic Control (FLC) [6, 7] is a control approach that handles complicated and unpredictable circumstances by employing language rules and fuzzy logic. It is particularly suitable for nonlinear systems and imprecise input data.

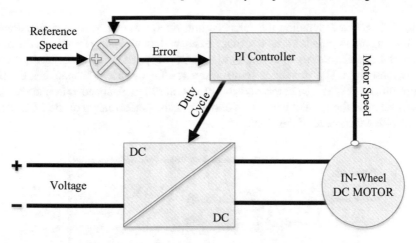

**Fig. 2.** Speed control structure using PI controller

FLC uses linguistic words and fuzzy sets to build control rules, allowing the system to make decisions depending on the degree of membership of linguistic variables. The fuzzification converts crisp inputs into fuzzy linguistic variables, capturing system imprecision. The inference engine applies fuzzy rules to these variables, determining appropriate control actions. Finally, the defuzzification converts the fuzzy control outputs into crisp values for system utilization.

To enhance the understanding of the Fuzzy Logic Control (FLC) system, we can incorporate a visual representation, such as an illustrative diagram (see Fig. 3), which effectively illustrates the components and information flow within the system.

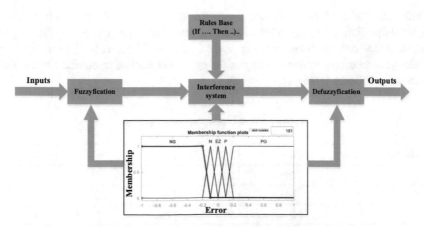

**Fig. 3.** A typical Fuzzy control structure

Similarly a simulation study was conducted (Fig. 4) to assess the effectiveness of the Fuzzy Logic Control (FLC) [8, 9] method for speed regulation in in-wheel direct current machines of electric vehicles. The simulation model incorporates the key components of the FLC system, including the fuzzification, inference engine, and defuzzification. By adjusting the linguistic variables, fuzzy rules, and membership functions, the performance of the FLC system can be analyzed in terms of speed tracking accuracy, stability, and response time. The simulation setup comprises the in-wheel DC machine, the chopper circuit, the PWM generator, and the FLC controller. A speed reference signal is used to evaluate the system's behavior and assess the performance of the FLC method in controlling the speed of the machine.

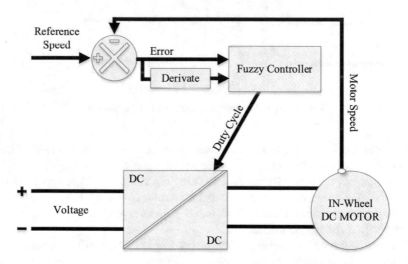

**Fig. 4.** A Speed control structure using Fuzzy controller

The rule-based inference system forms the core of the FLC approach. It consists of a set of predefined fuzzy rules that define the relationship between the linguistic input variables and the corresponding control actions. These rules capture the expert knowledge or heuristic understanding of the system behavior. In our case, the rules are illustrated as follows (Table 1):

**Table 1.** Rule-based inference system

| Error\Error Derivate | BP | P | EZ | N | BN |
|---|---|---|---|---|---|
| BP | BP | BP | BP | BP | N |
| P | BP | P | P | EZ | N |
| EZ | P | P | EZ | N | N |
| N | P | EZ | N | N | BN |
| BN | EZ | BN | BN | BN | BN |

## 2.3 Artificial Neural Network Controller

Another control mechanism used for speed regulation in in-wheel direct current machines of electric cars is the Artificial Neural Network [10, 11] Controller. Unlike traditional control approaches, Artificial Neural Networks employ a data-driven approach that allows the system to learn and adapt from training examples. Artificial Neural networks consist of linked artificial neurons organized in layers that analyze input data and generate output signals. The network learns by altering the weights and biases of the neurons to minimize the error between the desired and actual outputs. The topology of the artificial neural network, including the number of layers, the number of neurons in each layer, and the activation functions utilized, has a significant impact on the performance of the controller.

Artificial Neural Network Controllers learn from training examples through a process known as training. This involves feeding the network with input-output pairs and iteratively adjusting the weights and biases of the neurons to minimize the error between the desired and actual speed outputs. In our case, we used the data we got from the fuzzy controller to train our artificial neural network controller. To provide a comprehensive visual representation of the neural network architecture used in the study, a figure is presented below. The figure highlights the number of layers and the number of neurons in each layer, offering a clear and concise overview of the network configuration Fig. 5.

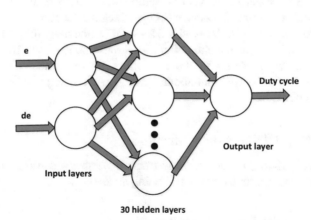

**Fig. 5.** Neural Network Architecture for Speed Control in In-Wheel Motor Drive Electric Vehicles

An illustrative simulation diagram showcasing the implementation of the artificial Neural Network Controller for speed regulation in in-wheel direct current machines of electric vehicles is presented in (Fig. 6).

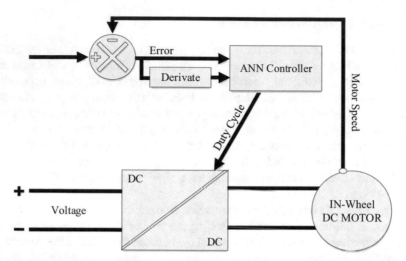

**Fig. 6.** Speed control structure using Artificial Neural Network controller

## 3   Simulation and Results

In this section, we present the simulation setup and discuss the results obtained from implementing different control methods for speed regulation in in-wheel direct current machines of electric vehicles. We used a 240-V DC-machine with a rotation speed of 1750 rounds per minute for the simulation. This machine is driven by a chopper that is regulated by a duty cycle set by the controllers. We examined two scenarios: one with a step speed of 1500 rpm and another with a random setpoint involving various accelerations and brakings.

### 3.1   Step Speed of 1500 rpm (Blue Signal)

The performance criteria, including speed-tracking accuracy, settling time, overshoot, and rising time, are represented in the following table:

### 3.2   Random Setpoint Speed (Green Signal)

The control results of the in-wheel motors speed obtained by applying: PIC, FLC and ANNC are shown in Figs. 7, 8, 9, 10, 11 and 12 and respectively. To compare the different algorithms the performance of each model is quantatively evaluated using 4 criteria namely: Overshoot, Settling time, Rising time and Accuracy (see Tables 2 and 3).

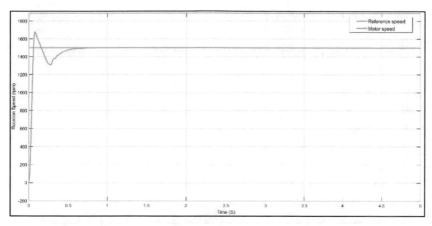

**Fig. 7.** Illustration of the PI controller response to 1500-rpm reference

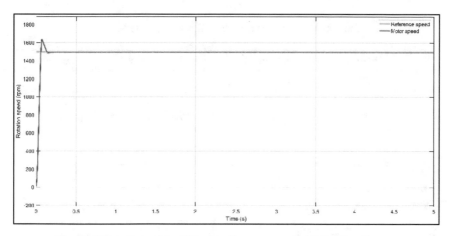

**Fig. 8.** Illustration of the FL controller response to 1500-rpm reference

We assessed each control method's capacity to correctly regulate the speed of the DC machine. The PI control approach displayed quick dynamic response and accurate speed tracking. However, it exhibited slight speed oscillations when subjected to sudden load changes. The Fuzzy Logic control demonstrated resilience in dealing with unexpected changes in the reference speed while maintaining stable operation. The Neural Network approach displayed its adaptability and learning capability, achieving precise speed regulation. According to table (2, 3) the FLC and ANNC model provide superior results compared to the results obtained in PI model.

While each control method had its advantages, there were limitations to consider. The PI control methods reliance on precise parameter tuning could be challenging in complex systems. The Fuzzy Logic methods rule-based approach required extensive rule development and might introduce subjective decisions. The Neural Network methods training process could be time-consuming and demand a significant amount of data.

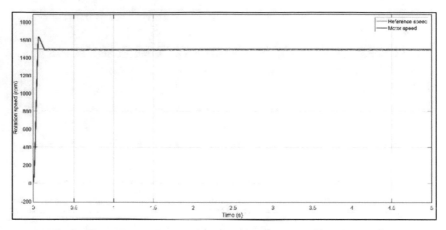

**Fig. 9.** Illustration of the ANN controller response to 1500-rpm reference

**Table 2.** Performance criteria to the step speed response

|       | Overshoot | Settling time | Rising time | Accuracy |
|-------|-----------|---------------|-------------|----------|
| PI    | 11.9%     | 0.7 s         | 0.06 s      | 99.95%   |
| FL    | 9.26%     | 0.14 s        | 0.058 s     | 99.39%   |
| ANN   | 9.26%     | 0.14 s        | 0.058 s     | 99.39%   |

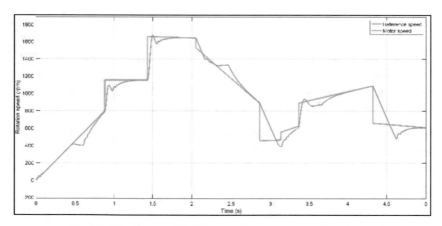

**Fig. 10.** Illustration of the PI controller response to the reference

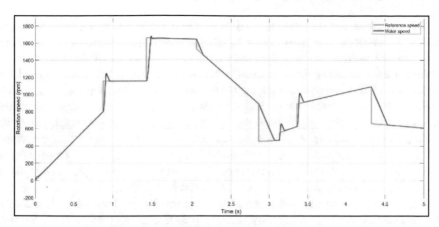

**Fig. 11.** Illustration of the FL controller response to the reference

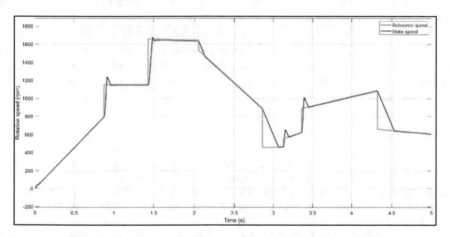

**Fig. 12.** Illustration of ANN controller response to the reference

**Table 3.** Accuracy to the random setpoint speed response

|          | PI      | FL      | ANN     |
|----------|---------|---------|---------|
| Accuracy | 98.39%  | 99.47%  | 99.55%  |

## 4  Conclusion

In summary, this research delved into a comprehensive analysis of three distinct speed control algorithms designed for in-wheel electric vehicle motors. The findings from extensive simulations shed light on the substantial advantages offered by both the fuzzy logic (FL) and artificial neural network (ANN) controllers over the traditional PI controller.

The outcomes clearly demonstrated that, when subjected to FL and ANN controllers, the in-wheel motor exhibited significantly enhanced performance characteristics. Notably, these controllers showcased quicker reaction times and markedly greater control precision, making them superior choices for regulating the motor's speed. The significance of this improvement extends beyond mere technical prowess; it directly translates into tangible benefits for electric vehicles. One of the most significant implications of this improved performance is the direct impact it has on vehicle safety and handling stability. Vehicles equipped with FL and ANN controllers are better equipped to respond swiftly and accurately to dynamic driving conditions, enhancing overall road safety. Moreover, the enhanced control precision contributes to smoother and more stable handling, ultimately providing a more comfortable and secure driving experience for passengers. Looking ahead, further research could involve implementing FL and ANN controllers in real electric vehicles. By utilizing tools like LABVIEW and FPGA boards, researchers can assess the feasibility of deploying these controllers in practical settings. This could bridge the gap between theory and application, potentially leading to safer, more efficient, and environmentally friendly electric vehicles.

In conclusion, the study highlights the potential of utilizing advanced control algorithms such as fuzzy logic and artificial neural networks in electric vehicle technology. These advancements have the capacity to go beyond academic interest and may catalyze a transformation in the electric vehicle industry, ushering in an era marked by improved safety, performance, and sustainability.

# References

1. Gasbaoui, B., Nasri, A., Abdelkhalek, O.: An efficiency PI speed controller for future electric vehicle in several topology. Procedia Technol. **22**, 501–508 (2016)
2. Al-Hazam, H.A.: Prediction of corrosion inhibitor efficiency of some aromatic hydrazides and Schiff bases compounds by using artificial neural network. J. Sci. Res. **2**(1), 108–113 (2010)
3. Semenov, A.S., Semenova, M.N., Bebikhov, Y.V.: Mathematical simulation of a PI speed controller for a separately excited DC motor. In: 2021 International Ural Conference on Electrical Power Engineering (UralCon), pp. 298–302. IEEE (2021)
4. Balamurugan, S., Umarani, A.: Study of discrete PID controller for DC motor speed control using MATLAB. In: 2020 International Conference on Computing and Information Technology (ICCIT-1441), pp. 1–6. IEEE (2020)
5. Khan, H.S., Kadri, M.B.: DC motor speed control by embedded PI controller with hardware-in-loop simulation. In: 2013 3rd IEEE International Conference on Computer, Control and Communication (IC4), pp. 1–4. IEEE (2013).
6. Xingyan, S.X.: BLDC motor speed servo system based on novel P-fuzzy self-adaptive PID control. In: IEEE International Conference on Information, Networking and Automation (pp. 186–190) (2010).
7. Arulmozhiyal, R., Kandiban, R.: Design of fuzzy PID controller for brushless DC motor. In 2012 International Conference on Computer Communication and Informatics, pp. 1–7. IEEE (2012).
8. Shchur, I., Turkovskyi, V. Comparative study of brushless DC motor drives with different configurations of modular multilevel cascaded converters. In: 2020 IEEE 15th International Conference on Advanced Trends in Radioelectronics, Telecommunications and Computer Engineering (TCSET) (pp. 447–451). IEEE (2020)

9. Poornesh, K., Mahalakshmi, R., Reddy, G.: Speed control of BLDC motor using fuzzy logic algorithm for low cost electric vehicle. In: 2022 International Conference on Innovations in Science and Technology for Sustainable Development (ICISTSD), pp. 313–318. IEEE (2022)
10. Sharma, M.: DC motor speed control using artificial neural network. Int. J. Eng. Sci. Math. **7**(9), 1–14 (2018)
11. Hamoodi, S.A., Sheet, I.I., Mohammed, R.A.: A Comparison between PID controller and ANN controller for speed control of DC Motor. In 2019 2nd International Conference on Electrical, Communication, Computer, Power and Control Engineering (ICECCPCE), pp. 221–224. IEEE, February 2019
12. Almatheel, Y.A., Abdelrahman, A.: Speed control of DC motor using fuzzy logic controller. In: 2017 International Conference on Communication, Control, Computing and Electronics Engineering (ICCCCEE), pp. 1–8. IEEE (2017)

# Development of an Integral Fractional Order Super Twisting Sliding Mode Controller for Electric Vehicle

M. K. B. Boumegouas[1](✉), K. Kouzi[1], and E. Ilten[2]

[1] Laboratory of Materials, Energy Systems, Renewable Energies and Energy Management, Amar Telidji University of Laghouat, P. O. B. 37G, 03000 Laghouat, Algeria
mk.boumegouas@lagh-univ.dz
[2] Electrical-Electronics Engineering, Balıkesir University, Balıkesir 10145, Turkey

**Abstract.** EV speed controller has been a worldwide research field due to its high importance in the traction chain of the electric vehicle. Much advanced controller has been carried out, the most common and chosen is the sliding mode controller, which this known for its high performance and robustness, and ease of implementation. However, chattering phenomena are its limitations. For that, a speed controller design based on fractional order super twisting sliding mode controller is proposed in this paper for controlling the EV speed using dual star PMSM as an electric drive powered by li-ion battery cylindric NMC21700 cells. This controller is used for increasing the performance of the traction chain of the EV. The simulation results obtained confirm the improvement using fractional integral order with STSMC in the EV in terms of robustness and dynamic performance while the battery storage system provides the energy requirements during all the phases successfully.

**Keywords:** Electric vehicle (EV) · Fractional order integral (FOI) · Super twisting sliding mode controller (STSM) · Fractional order integral super twisting sliding mode controller (FOISTSMC) · Dual star permanent magnet synchronous motor (DS-PMSM) · Li-ion NCM 21700

## 1 Introduction

An electric vehicle (EV) is a very complex system that consists of many subsystems. Each subsystem is playing different rules and important part in the design of the traction chain of the EV. While the battery system supplies the energy needed by the electric drive, this last is responsible for providing the speed, torque, and power requirements of the EVs [1]. This confirms the electric car traction chain is an intricate system but interconnected and integrated. The battery system is the key to providing the energy and power necessary to the traction chain during all the phases of the EV, many types of battery for EV has been investigated. However, lithium-ion cells gain a high reputation in EV applications for their high performance ensured by it and providing high energy, power, and life cycle compared to other types. Among lithium-ion batteries, we are interested in cylindrical type with NMC 21700 cells [2, 3]. Recently, this type is owning highly recommended for use in storage systems such as in Tesla electric cars.

© The Author(s), under exclusive license to Springer Nature Switzerland AG 2024
M. Hatti (Ed.): IC-AIRES 2023, LNNS 984, pp. 214–220, 2024.
https://doi.org/10.1007/978-3-031-60629-8_22

Furthermore, the dual star PMSM has been widely used in the EVs for its high advantage offers compared to traditional PMSM in terms of reducing ripples, increase power, and fault tolerance which rising the safety of the traction chain of the EV. The PMSMs family is worked basically with vector control using classical regulators such as PI, PID, etc. It offers acceptable results and performance. However, the sensibility to parameter variations, and lowest robustness to external perturbations. Thus, an advanced controller is necessary for the improvement of the controller system performance for the electric vehicle by increasing the robustness and insensitivity to parameter variations.

Hence, many advanced controllers have been suggested in literature such as adaptive control [4], Fuzzy logic controller [5], and sliding mode control [6, 7]. The most used one is sliding mode due to its robustness against perturbations and parameter variations, fast response, and ease of implementation. But the major Inconvenient is the chattering phenomena produced by the saturation function. Solving this issue has been a challenge to the researchers in last decade and suggesting many methods to reduce the chattering phenomena such as smooth function, and high-order sliding mode controller.

The super twisting sliding mode controller is known as the solution for the SMC problem by reducing the chattering phenomena by increasing the order of the sliding mode controller. To improve this solution even more in state error and increase its robustness and stability, a fractional order based on integrator has been added to the STSM algorithm control [8, 9]. Meanwhile, this structure is proposed to improve the dynamic performance of the traction chain of the electric vehicle.

This paper is organized as a mathematical model of the dual star PMSM is presented in Sect. 2, Sect. 3 is introduced the suggested controller, Sect. 4 denotes the battery system while the simulation results are illustrated in Sect. 5, ending with a conclusion appearing in section 6.

## 2  System Model

### 2.1  Dual Star PMSM Mathematical Model

Dual star PMSM in the dq frame mathematical model can be expressed as [10]:

$$
\begin{aligned}
\frac{d\,i_{d1}}{dt} &= \frac{1}{L_{d1}}(u_{d1} - Ri_{d1} + L_{q1}\omega_e i_{q1}) \\
\frac{d\,i_{q1}}{dt} &= \frac{1}{L_{q1}}(u_{q1} - Ri_{q1} - L_{d1}\omega_e i_{d1} - \omega_e \Phi_f) \\
\frac{d\,i_{d2}}{dt} &= \frac{1}{L_{d2}}(u_{d2} - Ri_{d2} + L_{q2}\omega_e i_{q2}) \\
\frac{d\,i_{q2}}{dt} &= \frac{1}{L_{q2}}(u_{q2} - Ri_{q2} - L_{d2}\omega_e i_{d2} - \omega_e \Phi_f) \\
\omega_e &= \frac{P}{2}\omega_r
\end{aligned}
\tag{1}
$$

While, the mechanical equation written as:

$$
T_e = J\frac{d\omega_r}{dt} + B\omega_r + T_L
\tag{2}
$$

## 2.2 Proposed Controller

The proposed controller consists of adding the fractional order integrator to the super twisting sliding mode controller, firstly, an ordinary STSM is written as [1, 7]:

$$u = -K_p|y|^r sign(y) - \int -K_i sign(y) \tag{3}$$

where; $K_p$, $K_i$ are the STSMC gains, $r$ is a positive constant, $y$ is the sliding surface which represents the error between the reference value and actual value.

The integrator fractional order equation can be expressed as follows [8]:

$$_0I_t^\lambda x(t) = {_0}D_t^{-\lambda}x(hM) = \frac{1}{h^{-\lambda}} \sum_{j=0}^{M} w_j^{(-\lambda)} x(hM - jh) \tag{4}$$

Thus, the proposed controller is given as:

$$u = -K_p|y|^r sign(y) - \int -K_i sign(y) + I^\lambda \tag{5}$$

where, $I^\lambda$ is presented in Eq. 4

## 3  EV Battery System

During Acceleration phase the battery supply the power and energy to the traction chain with a continue voltage called Bus continue (Vdc). This voltage is the last part of the storage system while the battery storage system (BSS) consists Battery pack, converter (DC/DC), and controller for these two parts for charging and discharging. Hence, during acceleration the battery discharge the power toward the traction chain, and while the deceleration phases the traction chain charges the battery which is called charging mode. This phenomenon is shown and presented in the figure that illustrates the BSS integration in the system and the two equations that represent the behavior of the battery charge/discharge [11] (Fig. 1).

$$f_1(i_t, i^*, i) = E_0 - K\frac{Q}{Q - i_t} \times i^* - \frac{Q}{Q - i_t} \times i_t + A.\exp(-B \times i_t) \tag{6}$$

$$f_2(i_t, i^*, i) = E_0 - K\frac{Q}{i_t + 0.1 \times Q} \times i^* - \frac{Q}{Q - i_t} \times i_t + A.\exp(-B \times i_t) \tag{7}$$

While the dynamic model of the EV is detailed and explained in [12, 13, 14].

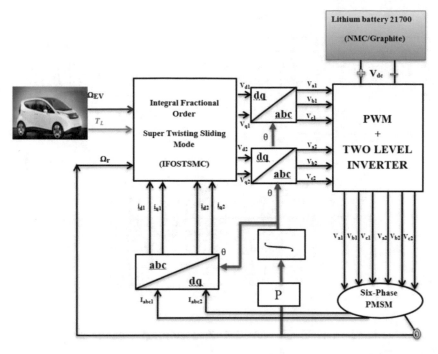

**Fig. 1.** Global schema of the system for the suggested traction chain of the EV.

## 4  Simulation Results

This part is presented the simulations results of the proposed controller using IFOSTSMC for EV application. Table 1 a, b and c represent the parameter of dual star PMSM, and NMC 21700 cell respectively.

**Table 1.** Parameters of the systems; a) Dual star PMSM, b) Battery cell specifications.

| Parameter six-phase PMSM | Quantity |
| --- | --- |
| Rs | 1.9 Ω |
| Ld | 0.000835 H |
| Lq | 0.000835 H |
| P | 4 |
| J | 0.015 |
| B | 0.0954 |
| φf | 0.353 Wb |
| Vdc | 800 V |

a)

| Cell specification | Voltage | Capacity | Energy | Weight |
| --- | --- | --- | --- | --- |
| Value | 3.6 V | 4.9 Ah | 255Wh/Kg | 69g |

b)

b) Battery cell specifications.

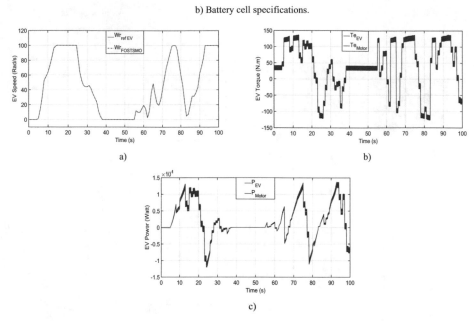

**Fig. 2.** EV performance results using IFOSTSMC. a) EV speed results, b) Torque of the EV, c) Power of the EV.

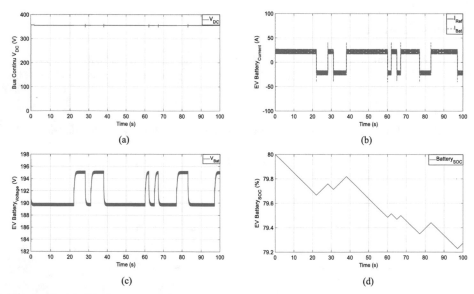

**Fig. 3.** Battery Storage System performance results, a) Bus continue results, b) Battery current results, c) Battery voltage results, d) State of charge (SOC) results.

Based on the results obtained, it is noticed that the speed controller by using IFOS-TSMC follows greatly the speed of the EV in the acceleration and deceleration phase with no overshoot, fast tracking, and no state error during all the phases. Meanwhile, the torque developed by the motor shows very well performance during the acceleration and deceleration by reducing the ripple, following the Torque of the EV. As for the power performance, it is obvious that the power demanded by the EV has been ensured by the dual-star PMSM during the phases of the driving cycle. With almost no errors or losses. These main advantages improve the traction chain of the EV performances both static and dynamic performances which influence directly the comfortable of the EV.

Since the system has been fed and powered successfully for the traction chain of the electric vehicle using lithium cylindrical NMC21700 cells as a battery storage system. It confirms the feature of the suggested cells for the BSS for the EV. it can be said that the bus continuing performance during the driving cycle remains at the desired constant value (355 V) at each change of the mode neither charge nor discharge mode, which is required for the traction chain of the EV as shown in Fig. 3a. As for the battery current, the result illustrates in Fig. 3b. it displays the tracking of the reference current controlled by the battery currents at each mode of charge and discharge with an accepted ripple. Moreover, the battery voltage results are presented in Fig. 3c, demonstrate the charge and discharge of the battery storage system during the phases as required by the EV. Furthermore, the battery state of charge (SOC) performs in Fig. 3d, successfully confirm the discharge and the charge mode of the EV during all the phases necessary. Hence, the results obtained in Fig. 2 and 3 respectively have clearly concluded the merits of both controller and the BSS chosen in improving the traction chain and the BSS of the electric vehicle.

## 5 Conclusion

A speed controller for electric vehicle application has been carried out based on fractional order integral super twisting sliding mode controller associated with dual star permanent magnet synchronous motor. A lithium-ion cylindrical NMC 21700 cells type ensured the supply the energy and power to the traction chain of the EV. The NMC21700 cells offered a fascinating option for the battery storage system for EVs by supplying the energy and power required by the EV during all the phases successfully. it demonstrated an attractive performance for controlling EV speed and improved the performance and robustness of the traction chain of the EV.

## References

1. Boumegouas, M.K.B., Kouzi, K.: A New synergetic scheme control of electric vehicle propelled by six-phase permanent magnet synchronous motor. Adv. Electr. Electron. Eng. **20**, 1–14 (2022)
2. Noël, J.-P., Guilbault, R., Perreault, C., Zaghib, K.: Numerical optimization of lithium-ion cell performance using metaheuristic algorithms. J. Energy Storage **71**, 108235 (2023)
3. Ohneseit, S., Finster, P., Floras, C., et al.: Thermal and Mechanical safety assessment of type 21700 lithium-ion batteries with NMC, NCA and LFP cathodes-investigation of cell abuse by means of accelerating rate calorimetry (ARC). Batteries **9**, 237 (2023)

4. Ren, Y., Wang, R., Rind, S.J., et al.: Speed sensorless nonlinear adaptive control of induction motor using combined speed and perturbation observer. Control. Eng. Pract. **123**, 105166 (2022)
5. Kouzi, K., Naït-Saïd, M.S.: Adaptive fuzzy logic speed-sensorless control improvement of induction motor for standstill and low speed operations. COMPEL- Int. J. Comput. Math Electr. Electron. Eng. **26**, 22–35 (2007)
6. Song, Q., Jia, C.: Robust speed controller design for permanent magnet synchronous motor drives based on sliding mode control. Energy Proc. **88**, 867–873 (2016)
7. Benariba, H., Boumédiène, A.: Super twisting sliding mode control of an electric vehicle. In: 2015 3rd International Conference on Control, Engineering & Information Technology (CEIT), pp 1–6. IEEE (2015)
8. Ilten, E., Demirtas, M.: Fractional order super-twisting sliding mode observer for sensorless control of induction motor. COMPEL- Int. J. Comput. Math. Electr. Electron. Eng. **38**, 878–892 (2019)
9. Ilten, E.: Conformable fractional order controller design and optimization for sensorless control of induction motor. COMPEL-Int. J. Comput. Math. Electr. Electron. Eng. (2022)
10. Boumegouas, M.K.B., Kouzi, K., Birame, M.: A robust decoupled control of electric vehicle using Type-2 fuzzy logic controller. In: International Conference on Artificial Intelligence in Renewable Energetic Systems, pp 419–426. Springer (2022)
11. Tremblay, O., Dessaint, L.-A.: Experimental validation of a battery dynamic model for EV applications. World Electr. Veh. J. **3**, 289–298 (2009)
12. Boumegouas, M.K.B., Kouzi, K., Birame, M.: Robust synergetic control of electric vehicle equipped with an improved load torque observer. Int. J. Emerg. Electr. Power. Syst. (2023)
13. Boumegouas, M.K.B., Kouzi, K.: Design of a robust nonlinear controller for electric vehicles driven by a six-phase permanent-magnet synchronous motor. Elektroteh Vestn **90**, 23–31 (2023)
14. Boumegouas, M.K.B., Kouzi, K.: Novel synergetic control of electric vehicle propelled by six phases permanent magnet synchronous motor. In: Hatti, M. (ed.) IC-AIRES 2021. LNNS, vol. 361, pp. 633–642. Springer, Cham (2022). https://doi.org/10.1007/978-3-030-92038-8_63

# Bidirectional Charging Impact Analysis of Electric Vehicle Battery Cycle Aging Evaluation in Real World, Under Electric Utility Grid Operation

Fouzia Brihmat[1,2(✉)]

[1] Research Center in Industrial Technologies – CRTI, Cheraga, 16014 Algiers, Algeria
f.brihmat@crti.dz, fouziabrimat726@gmail.com
[2] Communication and Photovoltaic Conversion Device Laboratory, Department of Electronics,
The National Polytechnic School El Harrach, 16200 Algiers, Algeria

**Abstract.** This scientific article examines the integration of electric vehicles (EVs) into electricity grids using Vehicle-to-Grid (V2G) technology. The study compares the economic implications of two EV charging approaches: firm EV charging and managed EV charging. Interestingly, the research finds that the firm EV load is associated with a higher Net Present Cost (NPC) compared to the managed charging load, highlighting the economic benefits of the latter approach.

The facility being studied consumes 26,498 kWh of electricity daily and has a peak demand of 2,022 kW. To meet its electricity needs, the proposed system incorporates several generation sources, notably a Canadian Solar PV system with a nominal capacity of 4,000 kW and a Tesla storage system with a nominal capacity of 10,079 kWh. These sources produce an annual total of 6,918,968 kWh.

Additionally, the study evaluates two tariff plans, with a focus on the 'EV' tariff. The analysis reveals that adopting the 'EV' tariff can result in significant cost savings for the hospital. However, it emphasizes that the hospital must individually meter its EV charging station to qualify for this advantageous tariff.

Furthermore, the research addresses the issue of EV battery degradation, particularly concerning its implications for V2G technology integration. Despite widespread concerns, empirical evidence in various grid service contexts is limited. To address this gap, the paper introduces a comprehensive method to quantify EV battery degradation. A unique aspect of this approach is its multiyear perspective, enabling a robust comparison of scenarios involving driving alone and those incorporating multiple vehicle-grid services. This innovative methodology provides valuable insights into the complex relationship between EV battery degradation and the evolving energy landscape.

**Keywords:** Firm EV · Managed EV · Tesla storage · Multi-year planning · Degradation · PV

## 1 Introduction

The growing adoption of electric vehicles (EVs) has spurred advancements in battery technology, aiming to enhance durability and cost-efficiency. This progress has led to new possibilities like connecting batteries directly to the grid at substations, renewable

© The Author(s), under exclusive license to Springer Nature Switzerland AG 2024
M. Hatti (Ed.): IC-AIRES 2023, LNNS 984, pp. 221–230, 2024.
https://doi.org/10.1007/978-3-031-60629-8_23

energy facilities, or large consumption centers. To make effective use of these distributed storage resources, integration into control centers is essential. This integration plays a crucial role in managing the power grid, considering that EVs spend about 95% of their lifetime unused, and the average EV typically uses less than 80% of its battery [1, 2] capacity for daily commuting. However, integrating renewable energy into EVs poses technical and socio-economic challenges driven by environmental concerns, fluctuating oil prices, and government policies. Photovoltaic and wind energy have increased independent and decentralized production, disrupting traditional network models. The research problem at hand is focused on ensuring stable grid operation when incorporating renewables and EVs. This involves optimizing system design, resource control, and real-time management to reduce consumption peaks and costs. Potential solutions may require innovative electrical architectures like microgrids, and optimization algorithms will play a central role in minimizing energy bills for multi-source systems. Furthermore, the intermittent nature of EV charging, influenced by factors like weather and driving conditions, adds complexity. Energy Storage Systems (ESS) are essential but challenging to manage. EV usage also impacts load profiles, causing consumption peaks [3]. Efforts to manage power at the network level have limitations as a long-term solution. The concept of "vehicle-to-grid" (V2G) emerges, allowing EV batteries to serve as mobile storage capacity. This approach could provide an additional energy storage resource, provided it is technologically and economically feasible [4]. It represents a departure from traditional mass energy storage methods.

## 2 Methodology

This study employs an innovative and comprehensive methodology that combines various elements related to managed electric vehicle (EV) charging, daily charging patterns, and individual charging session behaviors. It operates within a multiyear optimization framework, recognizing the dynamic landscape of electric mobility and grid management. The methodology includes the following key components [5, 6]:

1. *Managed EV Charging Context*: The study acknowledges the importance of managed EV charging for effectively utilizing charging infrastructure and balancing grid electricity demand. Managed charging utilizes intelligent control systems to optimize when and how EVs charge, enhancing grid efficiency while minimizing costs.
2. *Daily Profile Analysis*: A pivotal aspect involves the analysis of daily charging profiles. This entails examining how multiple EVs collectively charge over a day, considering variables like start times, duration, and charging rates. The goal is to understand typical daily charging trends and demand fluctuations.
3. *Single Session Profile Examination*: The study dives into individual charging session profiles, scrutinizing how individual vehicles charge during specific sessions. This analysis considers factors such as start times, charging speeds, interruptions, and overall energy consumption. The aim is to gain insights into the preferences and behaviors of EV users during single charging events.
4. *Multiyear Optimization Concept*: A unique feature of this methodology is its consideration of multiyear factors and projections. It recognizes that electric mobility and grid management are long-term endeavors, allowing for the development of strategies

and policies that adapt to evolving trends. This approach accommodates changes in EV adoption rates and grid infrastructure development over time.

By integrating these components, the study aims to offer a comprehensive understanding of managed EV charging practices. It provides insights into both short-term and long-term implications for grid stability, cost-efficiency, and the user experience. This holistic approach is essential for guiding policies and strategies that promote sustainable electric mobility and resilient grid management in the future.

## 2.1 The Adopted System

In this scientific article, we introduce an integrated system that plays a pivotal role within a firm electric vehicle (EV) and managed EV charging infrastructure situated in Algeria. This integrated system is comprised of two vital components: a Canadian Solar PV system and a Tesla storage system. Together, these components are engineered to address the daily electricity consumption of 24 MWh for the primary load. Importantly, this integrated system is also connected to the grid, making it an integral part of the local electrical network. This innovative system offers an efficient and sustainable approach to meet the energy demands of the EV charging infrastructure while contributing to the broader grid ecosystem. The system architecture is illustrated on Fig. 1.

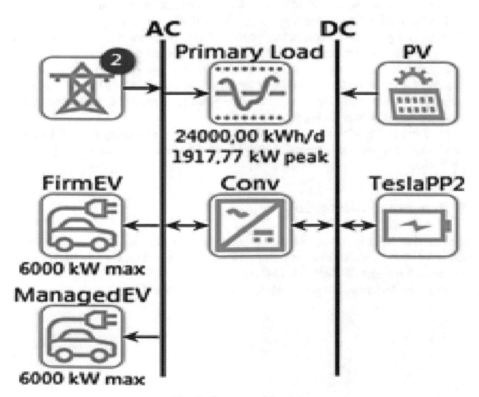

**Fig. 1.** System architecture.

## 2.2 Importance of Multi-year Approaches

Multiyear studies are essential for enhancing our understanding of various phenomena, providing a comprehensive perspective that informs decision-making and policy development. They empower stakeholders across industries to make informed decisions, optimize performance, ensure economic sustainability, and enhance the resilience and sustainability of microgrid systems. The knowledge gained from these studies contributes to a more reliable, cost-effective, and environmentally conscious energy future impact [7].

Given that economic and environmental changes often occur gradually over time, multiyear studies help researchers adapt to evolving circumstances and plan resource utilization more effectively, maximizing their impact. Additionally, for complex phenomena, extended observations are necessary to uncover intricate cause-and-effect relationships and understand how different factors interact over time [8]. Consistency in trends observed over multiple years strengthens the credibility and validity of research findings.

## 2.3 The Used Storage Technology

Tesla's battery technology is a key contributor to its EV success, with various sizes and configurations available to suit different vehicle models and performance needs.

The lithium-ion battery pack, placed in the vehicle's floor for optimal weight distribution, is a prominent Tesla battery product. Customers can choose from different battery options with varying capacities to match their desired range.

Tesla's battery technology extends beyond EVs, encompassing energy storage solutions like the Powerwall and Powerpack for homes and businesses. These systems utilize similar battery technology to store electricity, aiding in managing peak demand and providing backup power during outages [7, 9].

In summary, Tesla's battery innovations are pivotal in advancing electric mobility and clean energy solutions, impacting not only electric vehicles but also energy storage applications.

# 3 Results and Discussion

The Canadian Solar PV system has a nominal capacity of 4 000 kW. The annual production is 6 918 968 kWh/yr. The Tesla storage system's nominal capacity is 10 079 kWh, the annual throughput is 3 459 084 kWh/yr.

Their characteristics are given respectively on the Tables 1 and 2.

**Table 1.** Solar PV system characteristics.

| Rated Capacity | 4 000 kW | | Total Production | 6 918 968 kW |
|---|---|---|---|---|
| Capital Cost | 9,00 $M | | Maintenance Cost | 40 000 $/yr |
| Specific Yield | 1 730 kWh/kW | | LCOE | 0,0884 $/kWh |
| PV Penetration | 79,0 % | | | |

**Table 2.** Tesla power pack characteristics.

| Rated Capacity | 10 079 kWh | | Expected Life | 20,0 yr |
|---|---|---|---|---|
| Annual Throughput | 3 459 084 kWh/yr | | Capital Costs | 6,00 $M |
| Maintenance Cost | 48 000 $/yr | | Losses | 432 111 kWh/yr |
| Autonomy | 10,1 hr | | | |

## 3.1 Electric Vehicles

On the Figs. 2 and 3 are illustrated two examples the daily profile and the single session profile and their characteristics respectively of the studied system operation, on two different days and seasons.

In this paper, battery lifetime estimation of an electric vehicle (EV) using different driving styles on arterial roads integrating recharging scenarios in the neighborhood of the vehicle-to-grid integration is studied. Aggressive driving and recharging behavior significantly reduce battery life [10]. It is to notice that daily utility services impose extra degradation on the battery.

- The daily profile, for January 1

  Number of sessions: 1
  Energy served: 17850kWh
  Average energy per session (kWh): 1050,0
  Total charging time: 119

- The for July 1

  Selected session: ManagedEV 17:42#3589
  Session start: 23-June 30 17:42
  Vehicle type: EV ambulance
  Session duration: 7,0 h
  Total energy (kWh): 1050,0
  Charger name: ManagedEV
  Session type: Managed

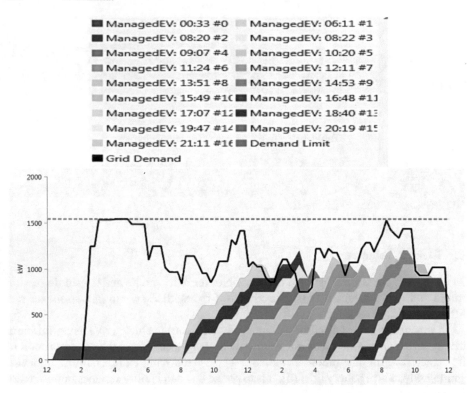

**Fig. 2.** Managed EV daily profile.

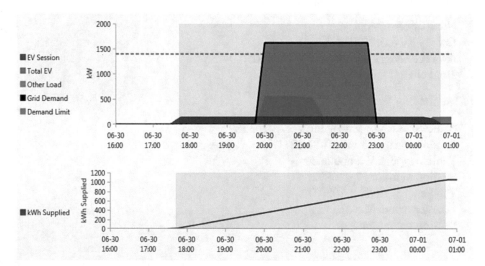

**Fig. 3.** Managed EV single session profile profile.

## 3.2  Production

On the Fig. 4 is presented the monthly average electricity generation of PV and utility.

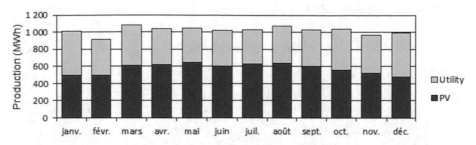

**Fig. 4.** Monthly average electricity generation.

## 3.3  Renewable Contribution

On the Fig. 5, we observe a clear degradation in the PV characteristics. "PV max production over project lifetime, capacity factor, daily mean output and solar penetration".

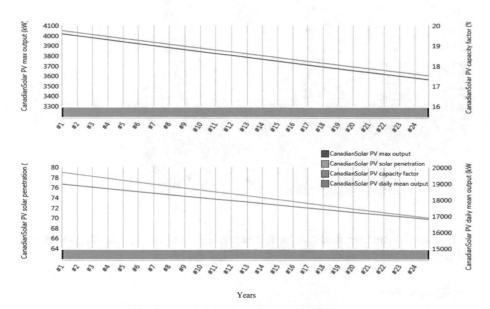

**Fig. 5.** PV characteristics over project lifetime.

## 3.4 Utility Bill Details

For a project lifetime of 25 years, an expected inflation rate of 2,0%, a nominal discount rate of 6,0% and a real interest rate of 3,9%, utility bill details are given for the whole project lifetime, on the Table 3.

**Table 3.** Utility Bill Details.

| Year | 1 | 2 | 3 | 4 | 5 | 6 | 7 | 8 | 9 | 10 |
|---|---|---|---|---|---|---|---|---|---|---|
| Bonus Depreciation | 0,00 $ | 0,00 $ | 0,00 $ | 0,00 $ | 0,00 $ | 0,00 $ | 0,00 $ | 0,00 $ | 0,00 $ | 0,00 $ |
| CanadianSolar PV | (40 000 $) | (40 000 $) | (40 000 $) | (40 000 $) | (40 000 $) | (40 000 $) | (40 000 $) | (40 000 $) | (40 000 $) | (40 000 $) |
| General - Time of Use, Demand Metered, Rate E (NEM 2.0) | (175 747 $) | (189 449 $) | (192 225 $) | (203 679 $) | (209 433 $) | (219 127 $) | (238 162 $) | (246 252 $) | (256 247 $) | (252 409 $) |
| SGIP- Large Storage- Step 2- Claiming ITC | 0,00 $ | 0,00 $ | 0,00 $ | 0,00 $ | 0,00 $ | 0,00 $ | 0,00 $ | 0,00 $ | 0,00 $ | 0,00 $ |
| Tesla Powerpack 2 | (48 000 $) | (48 000 $) | (48 000 $) | (48 000 $) | (48 000 $) | (48 000 $) | (48 000 $) | (48 000 $) | (48 000 $) | (48 000 $) |

| Year | 11 | 12 | 13 | 14 | 15 | 16 | 17 | 18 | 19 | 20 |
|---|---|---|---|---|---|---|---|---|---|---|
| Bonus Depreciation | 0,00 $ | 0,00 $ | 0,00 $ | 0,00 $ | 0,00 $ | 0,00 $ | 0,00 $ | 0,00 $ | 0,00 $ | 0,00 $ |
| CanadianSolar PV | (40 000 $) | (40 000 $) | (40 000 $) | (40 000 $) | (40 000 $) | (40 000 $) | (40 000 $) | (40 000 $) | (40 000 $) | (40 000 $) |
| General - Time of Use, Demand Metered, Rate E (NEM 2.0) | (267 926 $) | (285 940 $) | (307 557 $) | (313 245 $) | (328 092 $) | (337 026 $) | (345 481 $) | (375 954 $) | (391 822 $) | (408 201 $) |
| SGIP- Large Storage- Step 2- Claiming ITC | 0,00 $ | 0,00 $ | 0,00 $ | 0,00 $ | 0,00 $ | 0,00 $ | 0,00 $ | 0,00 $ | 0,00 $ | 0,00 $ |
| Tesla Powerpack 2 | (48 000 $) | (48 000 $) | (48 000 $) | (48 000 $) | (48 000 $) | (48 000 $) | (48 000 $) | (48 000 $) | (48 000 $) | (48 000 $) |

| Year | 21 | 22 | 23 | 24 | 25 |
|---|---|---|---|---|---|
| Bonus Depreciation | 0,00 $ | 0,00 $ | 0,00 $ | 0,00 $ | 0,00 $ |
| CanadianSolar PV | (40 000 $) | (40 000 $) | (40 000 $) | (40 000 $) | (40 000 $) |
| General - Time of Use, Demand Metered, Rate E (NEM 2.0) | (420 092 $) | (428 432 $) | (450 545 $) | (482 913 $) | (499 834 $) |
| SGIP- Large Storage- Step 2- Claiming ITC | 0,00 $ | 0,00 $ | 0,00 $ | 0,00 $ | 0,00 $ |
| Tesla Powerpack 2 | (48 000 $) | (48 000 $) | (48 000 $) | (48 000 $) | (48 000 $) |

It's to notice, through the Fig. 6 the increasing of the utility total consumption charge, and it's total fixed and demand charge over the system lifetime.

## 3.5 Storage Pack

Figure 7 illustrates the storage depletion and throughput over time during to charge and discharge cycles.

We can notice a degradation in both characteristics, due to the fact that high temperatures speed up reactions and can cause thermal issues in Li-ion batteries, while very low temperatures raise internal resistance and decrease capacity.

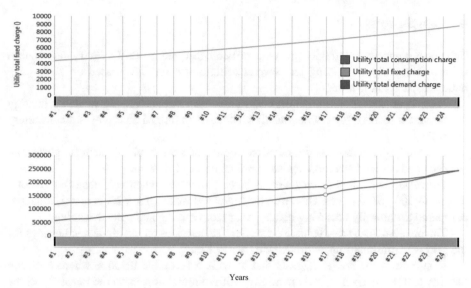

**Fig. 6.** Performance evolution over life time of the utility total consumption charge, and its total fixed and demand charge.

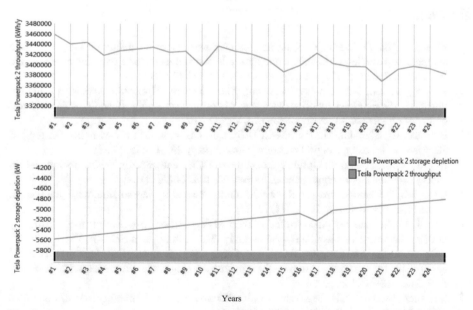

**Fig. 7.** Batteries experience performance degradation over time during to charge and discharge cycles.

# 4  Conclusions

The paper proposes a methodology to quantify electric vehicle (EV) battery degradation from driving only vs. driving and several vehicle-grid services, based on Tesla pack battery capacity fade model.

This paper aims to minimize the cost of electric vehicle charging while accounting for estimated costs of battery degradation, because optimized charging extends battery lifetime.

The study compares the degradation impact of grid services, such as peak load shaving, and net load shaping, against baseline cases of driving.

The results show that battery wear is indeed increased when vehicles offer vehicle-to-grid (V2G) grid services. However, the increased wear from V2G is inconsequential compared to naturally occurring battery wear from driving and calendar aging.

The results suggest that vehicles can offer grid services on the highest value days for the grid with minimal impact on vehicle battery life.

In summary, the study suggests that there is a favorable balance where EVs can provide grid services on high-demand days with minimal impact on the longevity of the vehicle's battery.

# References

1. Zheng, Y., Niu, S., Shang, Y., et al.: Integrating plug-in electric vehicles into power grids: a comprehensive review on power interaction mode, scheduling methodology and mathematical foundation. Renew. Sustain. Energy Rev. **112**, 424–439 (2019)
2. Ray, S., Kasturi, K., Patnaik, S., Nayak, M.R.: Review of electric vehicles integration impacts in distribution networks: placement, charging/discharging strategies, objectives and optimisation models. J. Energy Storage **72**, Part D (2023)
3. Bjørndal, E., Bjørndal, M., Bøe, E.K., Dalton, J., Guajardo, M.: Smart home charging of electric vehicles using a digital platform. Smart Energy **12**, 100118 (2023)
4. Han, T., Yan, Y., Safar, B.: Optimal integration of CCHP with electric Vehicle parking lots in energy hub. Sustain. Energy Technol. Assess. **58**, 103324 (2023)
5. Edge, J.S., O'Kane, S., Prosser, R., et al.: Lithium ion battery degradation: what you need to know. Phys. Chem. Chem. Phys. (14) (2021)
6. Gao, T., Lu, W.: Mechanism and effect of thermal degradation on electrolyte ionic diffusivity in Li-ion batteries: a molecular dynamics study. Electrochim. Acta **323**(10), 134791 (2019)
7. Sobrinho, D.M., Almada, J.B., Tofoli, F.L., Leão, R.P.S., Sampaio, R.F.: Distributed control based on the consensus algorithm for the efficient charging of electric vehicles. Electr. Power Syst. Res. **218**, 10923.1 (2023)
8. Miskolczi, M., Földes, D., Munkácsy, A., Jászberényi, M.: Urban mobility scenarios until the 2030s. Sustain. Cities Soc. **72**, 103029 (2021)
9. Zmud, J., Ecola, L., Phleps, P., Feige, I.: The Future of Mobility: Scenarios for the United States in 2030, 09 November 2013
10. Kiviluoto, K., et al.: Towards sustainable mobility – transformative scenarios for 2034. Transport. Res. Interdisc. Perspect. **16**, 100690 (2022)

# Hyperstability Criterion in Model Reference Adaptive Control: Enhancing Induction Motor Efficiency for Autonomous Electric Vehicles

A. Guendouz[1](✉) and A. Bouhenna[2]

[1] Department of Automation and Control Systems, Yahia Fares Medea University, Algiers, Algeria
guendas1524@gmail.com
[2] Department of Electrical Engineering, National Polytechnic School of Oran – Maurice Audi-MA, Oran, Algeria
abderrahmane.bouhenna@enp-oran.dz

**Abstract.** The combination of Electric Autonomous Vehicles (EAVs) and renewable energy aligns with sustainability goals. However, to achieve these goals, it's necessary to implement a robust control system for the key component that drives the vehicle, specifically, the electric motor. This paper explores the control of Induction Motor pursuing the Required Performance Standards for Autonomous Vehicles. The primary objective is to build a robust control system to achieve this, the first approach is the Indirect Rotor Field Oriented Control (IFOC), the Latter is implemented along with a PI controller, Lastly Model Reference Adaptive control by Hyperstability Criterion is introduced. A detailed analysis is carried out using MATLAB Simulink, which shows the efficiency of adaptive control not only in speed tracking, but also in parameter variation and disturbance. The comparative analysis between PI and Hyperstability control method validates the theory asserting the robustness and insensitivity of adaptive control to parameter uncertainty and disturbances. Furthermore, it convincingly demonstrates that the latter outperforms traditional classic controllers.

**Keywords:** Autonomous vehicles · Electric motors · Model Reference Adaptive Control · Hyperstability Criterion · indirect field-oriented control

## 1 Introduction

Electric Autonomous Vehicles (EAVs), "the revolutionary breakthrough in the realm of transportation", offering the potential to deliver significant environmental advantages. As the world grapples with the challenges of climate change, EAVs have emerged as a promising strategy for reducing carbon emissions and pollution. Operating on electric power, which can be generated from renewable sources, thereby reducing reliance on fossil fuels. This shift has the potential to significantly decrease greenhouse gas emissions, particularly carbon dioxide, which is primarily responsible for global warming [1].

© The Author(s), under exclusive license to Springer Nature Switzerland AG 2024
M. Hatti (Ed.): IC-AIRES 2023, LNNS 984, pp. 231–242, 2024.
https://doi.org/10.1007/978-3-031-60629-8_24

The operation of an autonomous vehicle requires actuation of major controls: acceleration (throttle actuation), direction (steering actuation), and stopping (brake actuation).In this regard, Different types of electric motors are used in EAVs including DC motor, brushless DC motor (BLDC), induction motor (IM), permanent magnet synchronous motor (PMSM), and switched reluctance motor (SRM) [3]. Considering the Tesla Model S, In its rear-drive configuration, the propulsion system consists of a single induction motor, an inverter, a fixed gearbox, and a differential, as illustrated in (Fig. 1) [2]. The focus of this study is on the heart of the system: the induction motor, the favored one for its reliability, higher maximum speed, robustness, and, most importantly, its affordability [4]. Despite its advantages, this motor possesses a notably high level of complexity, making it challenging to control. However, with the Indirect Rotor Field Control (IRFOC) which performs decoupling between the torque and the flux of the machine, makes it easy to introduce control strategies. The objective is to conduct a comparative study between the adaptive control approach and a conventional PI control. Model Reference Adaptive Control (MRAC) by Hyperstability is introduced for its several advantages compared to other control methods, such as insensitivity to parameter uncertainty and disturbance. The structure of the remaining sections of the paper is as follows: Sect. 2 provides the model of Induction Motor, in Sect. 3 the motor's control systems are introduced, Simulation results are elaborated in Sect. 4. Lastly, Sect. 5 contains the conclusions drawn from the study.

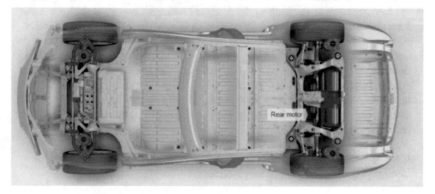

**Fig. 1.** Tesla model S rear drive. Taken from [2]

## 2   Induction Motor Modelling

The chosen state variables are the stator currents $i_{ds}$, $i_{qs}$, rotor fluxes $\phi_{dr}$, $\phi_{qr}$ and speed $\Omega$, considering Mechanical angular frequency: $\omega = \dot{\theta} = p\Omega = \omega_s - \omega_r$. The resulting model of the IM derived from [4] is as follows:

$$\frac{di_{ds}}{dt} = -\gamma i_{ds} + \omega_s i_{qs} + Kp\Omega\phi_{qr} + \frac{K}{T_r}\phi_{dr} + \frac{1}{\sigma L_s}V_{ds} \tag{1}$$

$$\frac{di_{qs}}{dt} = -\gamma i_{qs} - \omega_s i_{ds} - Kp\Omega\phi_{dr} + \frac{K}{T_r}\phi_{qr} + \frac{1}{\sigma L_s}V_{qs} \tag{2}$$

$$\frac{d\phi_{dr}}{dt} = -\frac{1}{T_r}\phi_{dr} + \frac{M}{T_r}i_{ds} - (\omega_s - p\Omega)\phi_{qr} \tag{3}$$

$$\frac{d\phi_{qr}}{dt} = -\frac{1}{T_r}\phi_{qr} + \frac{M}{T_r}i_{qs} - (\omega_s - p\Omega)\phi_{dr} \tag{4}$$

$$J\frac{d\Omega}{dt} = C_{em} - f\Omega - C_{res} \tag{5}$$

where:
$\gamma = \frac{R_s}{\sigma L_s} + \frac{R_r M^2}{\sigma L_s L_r^2}, T_r = \frac{L_r}{R_r}, K = \frac{M}{\sigma L_s L_r}$ and $\sigma = 1 - \frac{M^2}{L_s L_r}$ (the Blondel dispersion coefficient).

And:
$R_s$: the stator resistance, $R_r$: the rotor resistance, $L_s$: the stator inductance, $L_r$: the rotor inductance, $M$: the mutual inductance, $C_{em}$: The electromagnetic torque, $C_{res}$: Load torque, $f$: coefficient of friction, $p$: number of poles and $J$ is the moment of inertia.

## 3 Induction Motor Control System

### 3.1 Indirect Rotor Flux-Oriented Vector Control (IRFOC)

The IRFOC is a powerful tool that was designed by Blaschke in 1972 for alternating current machines to independently control both torque and flux.

Its principle is based on an analogy with the expression of the DC motor torque [5, 6], knowing that the torque in the $(d, q)$ reference frame is given by:

$$C_{em} = P\frac{M}{L_r}\left(\phi_{dr}i_{qs} - \phi_{qr}i_{ds}\right) \tag{6}$$

Orienting the d-axis in the direction of the rotor flux $\phi_{qr}$ (Fig. 2) in order to nullify the product $\phi_{qr}i_{ds}$, allows for:

$$C_{em} = P\frac{M}{L_r}\left(\phi_{dr}i_{qs}\right) \tag{7}$$

Hence, the previously obtained model becomes [5, 6]:

$$V_{ds} = R_s i_{ds} - \sigma L_s \omega_s i_{qs} + \sigma L_s \frac{di_{ds}}{dt} + \frac{M}{L_r}\frac{d\phi_r}{dt} \tag{8}$$

$$V_{qs} = R_s i_{qs} + \sigma L_s \omega_s i_{ds} + \frac{M}{L_r}\omega_s\phi_r + \sigma L_s \frac{di_{qs}}{dt} \tag{9}$$

$$0 = \frac{R_r}{L_r}\phi_r - \frac{R_r M}{L_r}i_{ds} + \frac{d\phi_r}{dt} \tag{10}$$

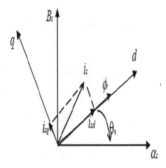

**Fig. 2.** The principle of rotor flux orientation.

$$0 = -\frac{R_r M}{L_r} i_{qs} + \omega_r \phi_r \tag{11}$$

$$J\frac{d\Omega}{dt} = \frac{pM}{L_r}\phi_r i_{qs} - f_m\Omega - C_{res} \tag{12}$$

By applying the Laplace transform,

$$V_{ds} = (R_s + \sigma L_s s)I_{ds} - \sigma L_s \omega_s I_{qs} + \frac{M}{L_r}s\phi_r \tag{13}$$

$$V_{qs} = (R_s + \sigma L_s s)I_{qs} + \sigma L_s \omega_s I_{ds} + \frac{M}{L_r}\omega_s\phi_r \tag{14}$$

$$\phi_r = \frac{M}{T_r s + 1}I_{ds} \tag{15}$$

$$\omega_r = \frac{M}{T_r\phi_r}I_{qs} \tag{16}$$

$$\Omega(s) = \frac{1}{Js+f_m}(C_{em} - C_{res}) \tag{17}$$

For speed control, a PI controller have been used, where the control loop simulated in MATLAB is given below (Fig. 3 and Fig. 4). The parameters $K_p, K_i$ were chosen by performing identification using the second-order characteristic equation, where $\xi = 1$ and $\omega_n = 19\ rad/s$.

**Speed Tracking Test:** In this test, a reference speed of 150 rad/s have been applied, associated with a Load torque of 10 N.m, during $\Delta t = 0.6$ s. The graphical outputs from the simulation are given in (Fig. 5 and Fig. 6).

The response of speed converges to the desired value of 150 rad/s after an estimated time of t = 0.4 s, indicating good tracking. It's also noteworthy that disturbance rejection is observed after the application of a load torque at time instants $t_1 = 0.8$ s and $t_2 = 1.4$ s.

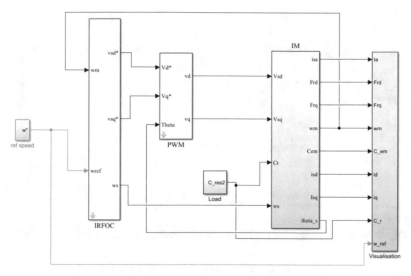

**Fig. 3.** Simulation diagram of the indirect vector control of the IM

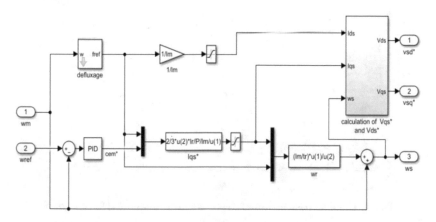

**Fig. 4.** Block diagram of indirect vector control with a PI controller.

## 3.2 Adaptive Control

Adaptive control is a robust control strategy that enables the controller to dynamically adjust its parameters in real-time, responding to variations of a process. Its primary objective is to achieve or sustain a desired level of performance even when the parameters of the process under control exhibit time-varying behavior. Adaptive control systems are three: pre-programmed gains, model references, self-adjusting controllers [7].

Model Reference Adaptive control (MRAC) is introduced with three methods: the gradient method, Lyapunov method and the Popov's hyperstability [8]. The method used for this study is the Hyperstability Criterion, this concept was developed by Popov for the stability analysis of dynamic systems [10], based on the principle of energy conservation in systems. The control law generated for this method is the sum of two commands: one

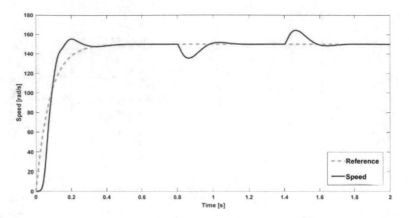

**Fig. 5.** Graphical output for the Reference (desired speed) and Speed (actual speed).

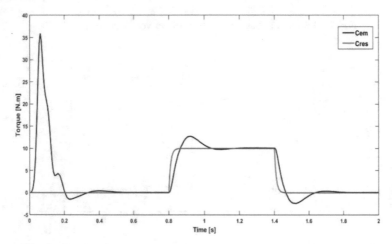

**Fig. 6.** Graphical output for the electromagnetic torque and Load torque.

linear $u_1(t)$ and the other nonlinear $u_2(t)$, given in Eqs. (18), (19), (20) along with the equations associated to the process of this control (derived from [9]).

$$u(t) = u_1(t) + u_2(t) \tag{18}$$

$$u_1(t) = -K_p x(t) + K_u r(t) + K_e e(t) \tag{19}$$

$$u_2(t) = K_x x(t) + K_w r(t) + K_\theta e(t) \tag{20}$$

$K_e$, $K_u$, $K_p$ are Linear adaptation gains, $K_x$, $K_w$, $K_\theta$ are Nonlinear adaptation gains, $x(t)$ is the output signal, $r(t)$ is the reference signal and $e(t)$ is the error between the output signal and the output of the reference model.

State space models of a dynamic system and a reference model are illustrated by Eqs. (21), (22), (23), (24)

$$\dot{x}(t) = Ax(t) + BU(t) + f_{NL}(x, t) \tag{21}$$

$$y(t) = Cx(t) \tag{22}$$

$$\dot{x}_m(t) = A_m x_m(t) + Br(t) \tag{23}$$

$$y_m(t) = C_m x_m(t) \tag{24}$$

$$e(t) = x_m(t) - x(t) \tag{25}$$

The Linear gains are obtained by using the condition: $\lim\limits_{t \to +\infty} e(t) = 0$,

$$K_e = (B^T B)^{-1} B^T (A_m - A_e) \tag{26}$$

$$K_p = (B^T B)^{-1} B^T (A - A_m) \tag{27}$$

$$K_u = (B^T B)^{-1} B^T B_m \tag{28}$$

The Nonlinear gains are obtained by LANDAU criterion [9, 10]:

$$K_x = \int_0^t M \cdot v \cdot x^T \cdot S dt + L \cdot v \cdot x^T \cdot S \tag{29}$$

$$K_w = \int_0^t G \cdot v \cdot r^T \cdot T dt + F \cdot v \cdot r^T \cdot T \tag{30}$$

$$K_\theta = \int_0^t Q \cdot v \cdot e^T \cdot U dt + P \cdot v \cdot e^T \cdot U \tag{31}$$

According to Landau's theory, $v = De$ is the output vector of the linear block. The matrices $S, T\ M, G, Q$, and $U$ are positively defined, and the matrices $F, L$ and $P$ are semi-positively defined.

This control theory applied on the Induction Motor is given by the equations below:

$$u_1(t) = -K_p \Omega + K_u \Omega^* + K_e e \tag{32}$$

$$u_2(t) = K_x \Omega + K_w \Omega^* + K_\theta e \tag{33}$$

The mechanical equation of the machine is given by:

$$J \frac{d\Omega}{dt} = C_{em} - f\Omega - C_{res} \tag{34}$$

The chosen reference model is as follows:

$$\tau \frac{d\Omega_m}{dt} + \Omega_m = k\Omega^* \tag{35}$$

The error between the reference model and the motor's model is given below.

$$e = \Omega_m - \Omega \tag{36}$$

By applying identification, the following parameters are obtained: $A_m = -\frac{1}{\tau}; B_m = \frac{k}{\tau}; A = -\frac{f}{J}; B = \frac{1}{J}$.

Hence, $K_u = \frac{Jk}{\tau}, K_p = \frac{J}{\tau} - f, K_e = J(1 - \frac{1}{\tau})$.

The Nonlinear gains are computed in Eqs. (37), (38), (39) respectively.

$$K_x = \int_0^t M \cdot v \cdot \Omega dt + L \cdot v \cdot \Omega \tag{37}$$

$$K_w = \int_0^t G \cdot v \cdot \Omega^* dt + F \cdot v \cdot \Omega^* \tag{38}$$

$$K_\theta = \int_0^t Q \cdot v \cdot e dt + P \cdot v \cdot e \tag{39}$$

where: $v = De, S = T = U = 1, F = L = P = k_1, G = M = Q = k_2$.

## 4   Simulation Results

The work done previously is captured in the Simulink diagram given in (Fig. 7), $\tau = 7.2 * 10^{-4}s, k_1 = 0.001, k_2 = 0.0001$.

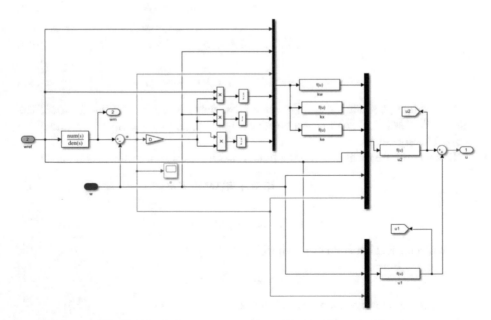

**Fig. 7.** Block diagram of Adaptive control by Hyperstability.

After conducting the initial test (Speed tracking test) using the Adaptive control, the output signals (speed and torque) were plotted on the same initial graph. Which allows for a direct comparison between the PI controller and MRAC. The graphical outputs for this simulation are illustrated in (Fig. 8 and Fig. 9).

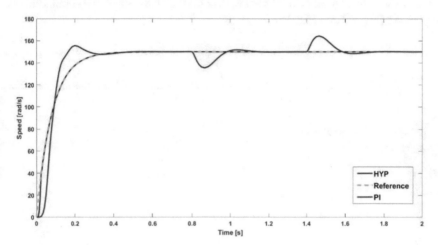

**Fig. 8.** Simulation results after applying the tracking test (HYP: the speed using MRAC, PI: the speed using PI controller).

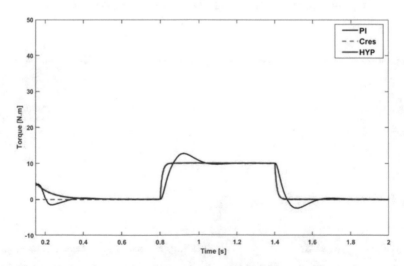

**Fig. 9.** Simulation results after applying the tracking test, (HYP: the electromagnetic torque using MRAC, PI: electromagnetic torque using PI controller).

The resulting speed, when implementing adaptive control using Hyperstability, perfectly follows the reference model trajectory along with a remarquable disturbance rejection, unlike the PI response, which exhibits significant overshoot. Furthermore, the torque response using Hyperstability did not exhibit any overshoot, unlike the torque response in the PI control.

**Robustness Test:** the rotor resistance plays a crucial role in controlling the motor's performance which is why this test consists of varying the rotor resistance $R_r$ by 200% to observe how the control systems behave towards the disturbance, the simulation results are given in (Fig. 10 and Fig. 11).

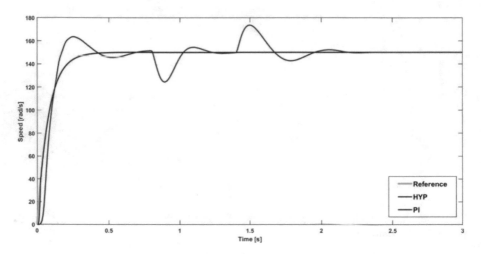

**Fig. 10.** Speed simulation result after the rotor resistance variation.

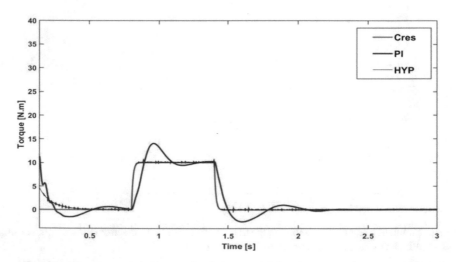

**Fig. 11.** Electromagnetic Torque simulation result after the rotor resistance variation.

It is very easy to observe that the speed response of the Hyperstability control is superimposed on the reference speed, whereas the response of the classical control exhibits noticeable distortions. Similarly, torque results shown in (Fig. 10) confirm this observation.

The error range is calculated for each speed signal output, see (Table 1).

Following the various tests conducted, it is easy to conclude that the adaptive control has effectively fulfilled its role as a regulator, of ensuring the attainment of good performance in the presence of parametric variations and disturbances.

**Table 1.** Speed Error results of the $R_r$ variation test.

| $R_r$ variation of 200% | PI controller | MRAC by Hyperstability |
|---|---|---|
| Error (rad/s) | 5 | 0.002 |
| Error (%) | 3.33 | 0.001 |
| Rise Time(s) | 0.7 | 0.3 |

## 5 Conclusion and Further Work

This paper has introduced a modeling and simulation approach for the control of induction motors, aiming to explore the capabilities of both traditional control and adaptive control in the context of electric motors and, consequently, electric autonomous vehicles (EAVs). After the development of the motor's model, the Indirect Rotor Flux-Oriented Vector Control (IRFOC) was initially implemented and simulated with a classical PI controller to control speed.

Furthermore, Model reference adaptive control (MRAC) by Hyperstability was used for the speed control. Both the PI and the MRAC by Hyperstability control strategies achieved a satisfactory speed tracking performance. However, it is reasonable to assert that the latter exhibited superior performance compared to the PI control.

External environmental factors such as extreme temperatures, poor ventilation, and wind resistance can affect the resistance of a motor. This, in turn, can impact the performance of an Electric Autonomous Vehicle (EAV) that uses this motor, which is why the resistance variation test was brought to this study. The Hyperstability method achieved a better performance: a faster speed tracking, a less speed error and steady state error, it outperformed PI control.

Adaptive control has demonstrated promising results in alleviating the negative impact on the Electric Autonomous Vehicle (EAV). Nevertheless, there is a substantial amount of additional work to be done. The next steps would involve developing the vehicle model within the simulation and including further details into the control performance and implementing a real scenario model.

## References

1. Hannappel, R.: The impact of global warming on the automotive industry. In: AIP Conference Proceedings, vol. 1871, no. 1. AIP Publishing, August 2017

2. Un-Noor, F., Padmanaban, S., Mihet-Popa, L., Mollah, M.N., Hossain, E.: A comprehensive study of key electric vehicle (EV) components, technologies, challenges, impacts, and future direction of development. Energies **10**(8), 1217 (2017)

3. Torres-Romero, L.A., González-Jiménez, L.E., Ruiz-Cruz, R.: Doctoral Program in Engineering Sciences at ITESO (2019)

4. Aktas, M., Awaili, K., Ehsani, M., Arisoy, A.: Direct torque control versus indirect field-oriented control of induction motors for electric vehicle applications. Eng. Sci. Technol. Int. J. **23**(5), 1134–1143 (2020)

5. Talla, J., Leu, V.Q., Šmídl, V., Peroutka, Z.: Adaptive speed control of induction motor drive with inaccurate model. IEEE Trans. Industr. Electron. **65**(11), 8532–8542 (2018)

6. Iqbal, A., Reddy, B.P., Rahman, S., Meraj, M.: Modelling and indirect field-oriented control for pole phase modulation induction motor drives. IET Power Electron. **16**(2), 268–280 (2023)

7. Nguyen, N.T.: Model-reference adaptive control. In: Model-Reference Adaptive Control. Advanced Textbooks in Control and Signal Processing. Springer, Cham (2018). https://doi.org/10.1007/978-3-319-56393-0

8. Zhang, D., Wei, B.: A review on model reference adaptive control of robotic manipulators. Annu. Rev. Control. **43**, 188–198 (2017)

9. Landau, I.: A hyperstability criterion for model reference adaptive control systems. IEEE Trans. Autom. Control **14**(5), 552–555 (1969)

10. Landau, I.: A generalization of the hyperstability conditions for model reference adaptive systems. IEEE Trans. Autom. Control **17**(2), 246–247 (1972)

# Pathfinding of a Mobile Robot by the Algorithm A Star (A*)

Newres Bouam[(✉)]

Faculty of Technology, Department of Electrical Engineering, Yahia Fares University of Medea, Medea, Algeria
bouam05_ab3@yahoo.fr

**Abstract.** In many industrial fields, mobile robots are widely used these days. Research on the mobile robot's path planning is one of the most important aspects of improvements on the mobile robot field. A mobile robot's path planning involves finding a collision-free trajectory, through the robot's environment with obstacles, from a specified starting location to a desired destination while meeting certain optimization criteria. The objective of this work consists in the development of a computer program on Matlab software, which allows the determination of the navigation optimal trajectory in a two-dimensional environment at minimal time. The method is based on the A (*) star algorithm, which is widely used in practice according to its advantages and the results obtained, compared to other algorithms. The simulation results obtained are in good agreement with those of the literature and show the ability of this algorithm to solve the given problem.

**Keywords:** A star (A*) algorithm · Navigation · Pathfindig

## 1 Introduction

Mobile robots [1] are getting more important in daily life as a result of their increasing role in making human's life easier. According to this point of view, it is obvious that Mobile Robots will be an indispensable part of future life [2]. To make Mobile Robots able to perform an assigned task, they have to know their locations [3], how to navigate [4] within their environment [5] and how to comprehend what the world around them looks like in order to choose the optimal path [6] from the start to the end point in order to reach the target. The path selection depends on the selected criteria to evaluate the path (ex: travel time, number of visited nodes, etc.). The optimal route [7] can be obtained by a variety of methods like the graph theory [8], probabilistic or heuristic [9] based optimization methods [10]. At every point in the grid map representing the working area of the robot, the selected algorithm assigns a direction for the robot out of the possible four directions: Right, Left, Up, and Down.

M. Hatti (Ed.): IC-AIRES 2023, LNNS 984, pp. 243–251, 2024.
https://doi.org/10.1007/978-3-031-60629-8_25

The A* algorithm is a technique that has been used for a long time in the research community. Its efficiency, simplicity and modularity are often strong advantages over other algorithms. Due to its ubiquity and widespread use, A* has become a common option for researchers trying to solve problems.

The objective of this work consists in the development of a program on Matlab which allows the determination of the optimal trajectory of a navigation in a two-dimensional environment at a minimum time. The method is based on the A (*) star algorithm, which is widely used in practice according to its advantages and the results obtained, compared to other algorithms. The simulation results obtained are in good agreement with those of the literature and show the ability of this algorithm to solve the given problem.

## 2   Path Planning by the A* Algorithm

### 2.1   Procedure Algorithm

In accordance with the international standard, the communication should be written in English. When it comes to exploring a graph, especially in the field of trajectory planning, it is quite simple to use the distance from the arrival node as a heuristic. The heuristic $h(n)$ is an estimate of the minimum cost to reach the arrival node from the current node $n$. By knowing the cost $g(n)$ between the starting node and the current node n, we therefore obtain an estimate of the total cost of the solution, noted $f(n)$. The exploration strategy of the A* algorithm then consists of choosing the next node with the lowest estimated total cost, $f(n)$.

$$f(n) = g(n) + h(n) \tag{1}$$

or

$$h(n) = \text{distance}(n, \text{nfinal}) \tag{2}$$

and

$$g(n) = \sum_{d\acute{e}part}^{n} cout(noeud_i, noeud_j) \tag{3}$$

The A* algorithm is optimal under certain conditions, which we will describe in this paragraph. In the case of graph exploration, the optimality condition of A* is that the heuristic must be consistent (Russel, 1995) [11]. A heuristic $h(n)$ is consistent if, for any node $n_1$ and for any neighbor $n_2$ of $n_1$, the estimated cost $h(n_1)$ of the path between $n_1$ and the arrival node is not greater than the sum of the cost of the step between $n_1$ and $n_2$ and the estimated cost $h(n_2)$ of the path between $n_2$ and the arrival node.

$$h(n_1) \leq co\hat{u}t(n_1, n_2) + h(n_2) \tag{4}$$

This condition can be satisfied by using, for example, the Cartesian distance between node $n_1$ and the arrival node as a heuristic (Fig. 1):

$$h(n) = dist(n, n_{final}) = \sqrt{(x - x_{final})^2 + (y - y_{final})^2} \tag{5}$$

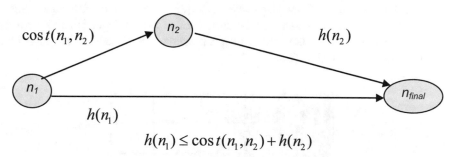

$$h(n_1) \leq \cos t(n_1, n_2) + h(n_2)$$

**Fig. 1.** Exploration consistency A\*.

In this case, the triangle inequality requires that:

$$dist(n_1, n_{final}) \leq dist(n_1, n_2) + dist(n_2, n_{final}) \tag{6}$$

and:

$$\cos t(n_1, n_2) \geq dist(n_1, n_2) \tag{7}$$

It also implies that when we first reach the final node, we have already reached our goal with the optimal path.

## 2.2 Description of the Algorithm A\* [11–13]

We start with the starting node, which becomes the current node.
We examine all neighboring nodes of the current node.
If a neighboring node is an obstacle, we ignore it.
If a neighboring node is already in the closed list, we ignore it.
If a neighboring node is already in the open list, we update the open list if the corresponding node has a lower quality (and we don't forget to update its parent).
Otherwise, we add the neighboring node to the open list with the current node as parent.
We are looking for the best node in the open list. If the open list is empty, it means there is no solution, and we end the algorithm.

We move this node into the closed list and delete it from the open list.
We repeat the process with this node as the current node until the current node is the destination node.

Here is a representation of an iteration of the algorithm. Our goal is to move from the orange point to the blue point. The neighboring nodes of the orange node are represented by the boxes marked in green, and they are added to the open list. For each neighboring node, we calculate the costs G and H to reach the orange node and to reach the destination. In this example, I used the distance as the crow flies as the heuristic. The box with the lowest total cost is the green box below, so it is moved to the closed list. The algorithm will now iterate from this box (Fig. 2).

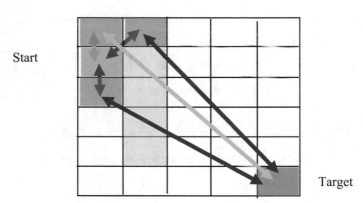

**Fig. 2.** Illustration of an iteration of the A* algorithm.

## 3   Flowchart of the Simulation

The flowchart is presented at the Fig. 3.

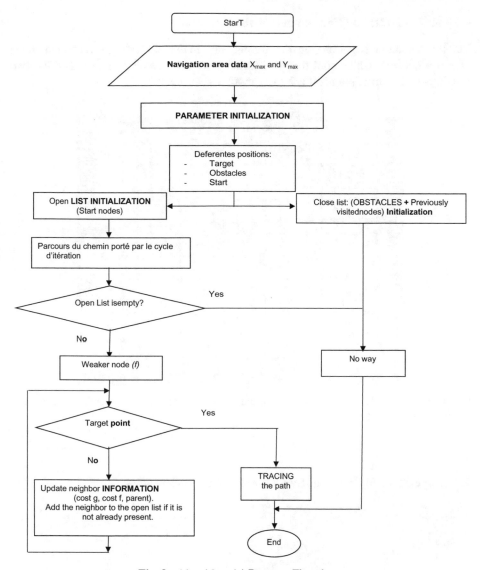

**Fig. 3.** Algorithm A* Program Flowchart.

## 4   Simulation Results

In this part, we will apply the A* planning algorithm for a navigation of a trajectory in several environments and we will present the optimal path traveled to ensure the smoothing of the path.

The A* algorithm, the starting position, the goal position, the position of the obstacles as problem data and the planned path defined by a set of points as results. The simulation was done by the Matlab R2014a software (8.3.0.532), [14].

248      N. Bouam

## 4.1  *EXPERIMENT 1:* Environment without Obstacles

The Fig. 4 shows navigation from the starting point to the end point, the path in the form of a straight line, which confirms the shorter route it is a straight line linked the two starting and finishing points, it is a route more optimized.

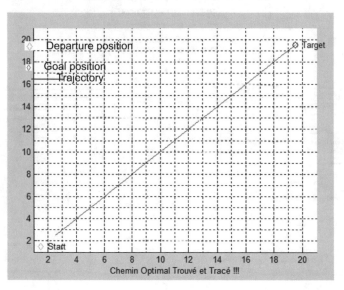

**Fig. 4.** Trajectory planned by A* in an environment without obstacles. - Execution time: 0.025 s - path length: 26.1725 um

## 4.2  Experience 2: Environment Contains Walls as Obstacles

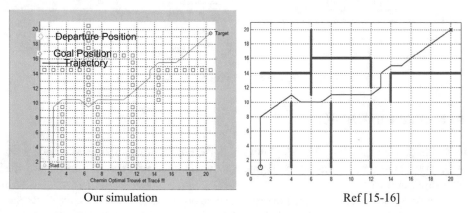

Our simulation                                          Ref [15-16]

**Fig. 5.** Path planned by A* in an environment containing walls as obstacles.

In this experiment we used an environment in 2D form with several obstacles proposed as walls. The results obtained, Fig. 6 are in agreement with those of reference [15], which confirms the proper functioning of our program (Fig. 5).

### 4.3 Experience 3: Moving Robot in a Labyrinth

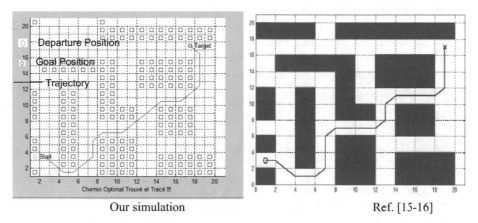

Our simulation                                    Ref. [15-16]

**Fig. 6.** Path planned by A\* in a maze.

During this test, we used an environment in 2D form "a labyrinth" has several obstacles proposed as walls. The results obtained, Fig. 6 are in agreement with those of reference [15, 16], which confirms the proper functioning of our program.

## 5 Conclusion

Robotics can be defined as a field bringing together techniques and studies aimed at designing mechanical, computer or hybrid systems capable of replacing the motor, sensory and intellectual functions of humans.

In particular, cooperative mobile robotics is a very active area of research and development for industry. The difficulty of working on this field lies in its multidisciplinary aspect on the one hand and distributed on the other. Indeed, multi-robot systems require very advanced artificial intelligence techniques.

The objective of this work of a study of Use of the algorithm A\* for the planning and the optimization of trajectory of a mobile robot consists in the continuation of trajectories of a mobile robot in a two-dimensional plan comprises of obstacles.

During this study, the technique of the A\* algorithm was detailed and adapted to the establishment of a navigation program of a two-dimensional environment without and with obstacles using the Matlab R2014a software.

Although researchers have done a lot of research on the A\* algorithm, which is confirmed that the performance of the A\* algorithm is mainly reflected in the path planning speed and the robustness of the planned path.

As far as the real robot is concerned, the principle remains the same, except for the transmission of data via the card.

Based on the results obtained, we can draw the conclusion that the initial goal of this work is achieved, because these results are acceptable and are in good agreement with those of the literature. However, many actions can follow this work in order to improve the behavior of robots through the following points:

Assignment of the simulation program to a single robot using an Arduino card in an environment with several moving obstacles,

Assignment of the simulation program to several robots using Arduino cards in an environment with several obstacles (static study and dynamic study).

When planning, it is important to take into account the real dimensions of the robot as well as its mechanical constraints (such as non-holonomy, etc.).

# References

1. Mester, G.: Applications of mobile robots. In: Conference: 7th International Conference on Food Science At: Szeged, Hungary (2006)
2. Teleweck, P.E., Chandrasekaran, B.: Path planning algorithms and their use in robotic navigation systems. J. Phys. Conf. Ser. **1207**(1), 12018 (2019)
3. Pop, C.M., Mogan, G.L., Neagu, M.: Localization and path planning for an autonomous mobile robot equipped with sonar sensor. Appl. Mech. Mater. **772**, 494–499 (2015)
4. Hoang, T.T., Hiep, D.T., Duong, P.M., Van, N.T.T., Duong, B.G., Vinh, T.Q.: Proposal of algorithms for navigation and obstacles avoidance of autonomous mobile robot. In: 2013 IEEE 8th Conference on Industrial Electronics and Applications (ICIEA), pp. 1308–1313 (2013)
5. Stentz, A.: Optimal and efficient path planning for partially known environments. In: Hebert, M.H., Thorpe, C., Stentz, A. (eds.) Intelligent Unmanned Ground Vehicles, vol. 388, pp. 203–220. Springer, Boston (1997). https://doi.org/10.1007/978-1-4615-6325-9_11
6. Zhou, I.-H., Lin, H.-Y.: A self-localization and path planning technique for mobile robot navigation. In: 2011 9th World Congress on Intelligent Control and Automation, pp. 694–699 (2011)
7. Jan, G.E., Chang, K.-Y., Parberry, I.: A new maze routing approach for path planning of a mobile robot. In: Proceedings 2003 IEEE/ASME International Conference on Advanced Intelligent Mechatronics (AIM 2003), vol. 1, pp. 552–557 (2003)
8. Katsuki, R., Tasaki, T., Watanabe, T.: Graph search based local path planning with adaptive node sampling. In: 2018 IEEE Intelligent Vehicles Symposium, vol. IV, pp. 2084–2089 (2018)
9. Reeves, C.R.: Heuristic search methods: a review. Oper. Res. Pap., 122–149 (1996)
10. Floudas, C.A., Pardalos, P.M.: Encyclopedia of Optimization. Springer, New York (2008). https://doi.org/10.1007/978-0-387-74759-0
11. Kedad-Sidhoum, S.: Plus courts chemins : Algorithme A*, cours d'algorithmique avancée, Laboratoire d'Informatique de Paris 6 (LIP6) (2005–2006)
12. Li, Y., et al.: Mobile Robot Path Planning Algorithm Based on Improved A* Algorithm. IEEE Access (2022)
13. Sudhakara, P., Ganapathy, V.: Path planning of a mobile robot using amended A-Star algorithm. IJCTA **9**(37), 489–502 (2016)
14. Logiciel Matlab: R2014a (8.3.0.532) (2014)

15. Tazebinte, H., Lameche, K.: Planification et optimisation de trajectoire d'un robot mobile par les algorithmes évolutionnaires. Mémoire de master en Automatique, Ecole Nationale Polytechnique, El-harrach, Alger (2012)
16. Garcia, D.: Planification de trajectoire dans un atlas de cartes. Mémoire de maîtrise des sciences appliquées (automatique et systèmes), Ecole Polytechnique de Montréal, CANADA (2008)

# Task Scheduling in a Multiprocessor System

Ouided Hioual[✉], Nour El Houda Zeghoud, and Elhamza Degaichia

Computer Science Department, Abbes Laghrour University, 4000 Khenchela, Algeria
ouided.hioual@gmail.com

**Abstract.** The scheduling problem involves allocating resources to a set of tasks while respecting a set of constraints. In the case of a multiprocessor system, the stakes are even higher as it aims to maximize resource utilization while minimizing task execution time. To optimize resource utilization, this study proposes a task scheduling approach in a homogeneous multiprocessor system based on partition scheduling heuristics combined with the Gantt task representation technique. The approach aims to optimize task scheduling by reducing execution time and minimizing resource conflicts (processors). Although our approach may not guarantee the optimal solution, it can provide satisfactory results in many practical cases.

**Keywords:** Multiprocessors · Scheduling · Gantt Chart · Scheduling Heuristics

## 1 Introduction

In a computer system, multiple tasks or processes may be waiting for execution. These tasks can be programs, applications, or services that require resources such as the CPU, memory, or input/output devices to perform their work. Therefore, it is necessary to have a scheduler that intervenes to decide which task should be executed and when it should be executed. Scheduling is a fundamental concept in the field of computer systems, particularly in operating systems. It deals with how tasks or processes are planned and executed by the central processor of a computer system (A. Jean et al, 2006). The objective of scheduling is to optimize the utilization of available resources, improve the system's efficiency, and meet time constraints. The choice of the execution order of tasks can have a significant impact on the overall performance of the system. A good scheduling algorithm can reduce waiting times, minimize response times, and maximize resource utilization, thereby enhancing system efficiency (A. Nadine, 2014). On the other hand, poor scheduling can lead to delays, bottlenecks, and inefficient resource usage. Scheduling can be influenced by various factors such as task priorities, time constraints, resource availability, task interdependencies, and more. It's important to note that the task scheduling problem can be NP-hard, meaning it's challenging to find an optimal solution in a reasonable amount of time for large instances of the problem. Therefore, many approaches focus on finding satisfactory or near-optimal solutions rather than perfect ones. In multiprocessor systems, partition scheduling, also known as multiprocessor scheduling, is a variant of the task scheduling problem that involves allocating tasks to specific processors. Tasks are exclusively assigned to each processor, reducing

M. Hatti (Ed.): IC-AIRES 2023, LNNS 984, pp. 252–264, 2024.
https://doi.org/10.1007/978-3-031-60629-8_26

the costs associated with preemption and eliminating migration costs. Furthermore, this simplifies the multiprocessor scheduling problem by breaking it down into a set of separate single-processor scheduling problems to solve. In summary, partition scheduling provides efficient task allocation and simplifies scheduling management across multiple processors (B. Khaoula, 2020). In order to optimize resource utilization, in paper, we propose a task scheduling approach in a homogeneous multiprocessor system based on partition scheduling heuristics, combined with the Gantt chart task representation technique. Our approach aims to optimize task scheduling by reducing execution time and minimizing resource conflicts (processors). While our approach may not guarantee the optimal solution, it can provide satisfactory results in many practical cases (N. Thanh Dat, 2020).

## 2   Related Work

In this section, we present some related work related to partition scheduling. These works illustrate advancements in the field of partition scheduling, considering various constraints and modern hardware architectures.

The authors of (B. Björn et al., 2014) presented an efficient partition scheduling algorithm for periodic tasks on multi-core platforms. The algorithm takes into account preemption constraints, task migrations, and deadlocks.

In (L. Nan, et al., 2019, the authors proposed a partition scheduling algorithm for heterogeneous real-time systems with resource augmentation. The algorithm optimizes overall performance by dynamically adjusting the resources allocated to each partition.

In (H. Gang, et al., 2020), the authors studied the impact of preemption costs on partition scheduling. They propose methods to minimize preemption costs and improve CPU utilization in partitioned systems.

In (S.S. Krishna, et al., 2021), the authors presented a hierarchical partitioned scheduling algorithm for mixed-criticality systems. The algorithm ensures isolation between high-criticality and low-criticality tasks while maximizing CPU utilization.

In (E. Arvind, et al., 2022), the authors proposed a partitioned scheduling algorithm with dynamic priority for mixed-criticality real-time systems. The algorithm dynamically adjusts task priorities to ensure deadline compliance while efficiently utilizing resources.

## 3   Our Proposal

Partition scheduling, also known as multiprocessor scheduling, is a variant of the task scheduling problem that allows for the exclusive assignment of tasks to individual processors, reducing the costs associated with preemption and eliminating migration costs. Each task must be executed on a single processor, and processors operate independently and in parallel. This presents significant advantages. Additionally, it simplifies the multiprocessor scheduling problem by breaking it down into a set of separate single-processor scheduling problems to solve. In summary, partition scheduling provides efficient task allocation and simplifies scheduling management across multiple processors.

In this section, we propose a task scheduling approach in a homogeneous multiprocessor system based on partition scheduling heuristics, combined with the Gantt chart task representation technique.

Our approach aims to optimize task scheduling by reducing execution time and minimizing resource conflicts (processors). While our approach may not guarantee the optimal solution, it can provide satisfactory results in many practical cases.

### 3.1 Overview of the Proposed Approach

The proposed architecture includes two main components (see Fig. 1).

- Processing Component: This component is responsible for task scheduling using partition scheduling heuristics. It determines how tasks are allocated to processors, taking into consideration various factors such as task characteristics, resource availability, and scheduling policies.
- Reporting Component: In this part, the results generated by the processing component are presented in the form of a graphical report. The Gantt chart is utilized as a visualization tool to display task schedules and their execution times.

These two components work together to optimize task scheduling in a homogeneous multiprocessor system, aiming to reduce execution time and minimize resource conflicts.

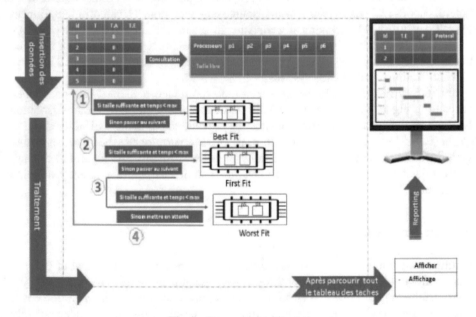

**Fig. 1.** Proposed Architecture

## 3.2   Assumptions and Constraints

For the proper functioning of our approach, we have established the following assumptions:

- Execution Time Limit: In order to prevent overheating and processor failures, we have set a time execution limit (four hours) for each pair of processors (Px, Py). This assumption helps maintain processor performance and avoid delays in task execution.
- Memory Capacity: We also ensure that the memory capacity of each machine is sufficient to accommodate the size of the task. This assumption can prevent problems caused by memory limitations, which could result in task slowdowns or even complete task failure.
- We have proposed conducting preliminary tests before assigning tasks to each processor. This step can help ensure that the resources (processors, memory) are available to execute the task, which can reduce waiting times and improve the overall efficiency of the system.
- We assume that tasks are independent, meaning two tasks can execute simultaneously.
- We assume that each processor has its own memory.

Overall, these assumptions can contribute to improving processor utilization and reducing waiting times for task execution.

## 3.3   Addressed Aspects

In our work, we have addressed three essential aspects to ensure the proper functioning of a homogeneous multiprocessor system. These three aspects are as follows:

- Execution Time: The objective of this aspect is to reduce task execution time by assigning them to fast processors with as low a probability of failure as possible. This objective can be achieved by using scheduling heuristics in a precisely defined order.
- Available Memory Space: The aim of this aspect is to prevent issues arising from limited memory when executing a specific task. Insufficient memory can significantly reduce performance or even lead to the complete interruption of task execution.
- Prevention of Processor Failures: This aspect is crucial in scheduling problems. It involves controlling execution time to prevent processor overheating and failures. There are several techniques to prevent these two issues, such as periodic cooling, regular cleaning, and temperature monitoring. We have opted for the third technique, which involves monitoring processor heat levels to ensure that the total consecutive execution time does not exceed four hours. If the temperature exceeds critical thresholds, we must reduce the workload.
- Waiting Time: This aspect aims to reduce task waiting time. This objective can be achieved by using scheduling heuristics in a precisely defined order.

## 3.4   Functioning of the Proposed Approach

In this section, we explain how our approach operates and achieves the objectives established at the beginning of our work. As mentioned earlier, we utilize partition scheduling heuristics along with the Gantt chart for reporting (generating reports). The user needs to input the available memory sizes for the 6 processors. The first two are designated for BestFit, the next two for FirstFit, and the last two for WorstFit. Next, the user can input the tasks, specifying their size and respective estimated execution time.

Afterward, the user can click on "allocation." Once the processors and tasks are created, we can proceed to the next step.

Initially, all tasks are allocated to the first two processors using BestFit to attempt to allocate all tasks. If all tasks are allocated to the first two processors and if both conditions are met (i.e., the task size is less than or equal to the machine size, and the total execution time for each pair of processors does not exceed 240 min), there is no need to proceed to the next processors.

If the first two processors (BestFit) are unable to allocate all tasks, the remaining ones will be assigned to the next processors using FirstFit. If the latter also fail to allocate all tasks, the remainder will be assigned to the last two processors using WorstFit. It is important to check the two conditions mentioned earlier before assigning tasks to processors.

In the end, all tasks that were successfully allocated to processors are recorded, and those that could not be allocated will restart from the beginning once all previous tasks have been completed. This process will be repeated until there are no more tasks in the waiting list.

In the user interface, tasks that have found an allocation with the corresponding processor, along with their ranking and execution time, will be displayed on the screen.

Finally, the user can click on "show Gantt chart" to visualize the Gantt chart on the screen.

### Processing Section

We have developed an architecture that utilizes scheduling heuristics through partitioning, specifically the Best Fit, First Fit, and Worst Fit protocols, with the aim of improving task scheduling on a homogeneous multiprocessor system. It is worth noting that we have excluded Next Fit as it is similar to First Fit (C. Maxime et al, 2015) (F. Frédéric et al, 2011).

The heuristics operate as follows, making both processor selection and protocol choice simultaneously:

- Best Fit (BF): The task is placed in the partition with the smallest size that is sufficient to accommodate the task. This heuristic can reduce fragmentation but is slower than the first heuristic. Assigns a task to the processors (P1, P2) with the least unused capacity that can schedule it.
- First Fit (FF): The task is placed in the first partition that can accommodate it. This heuristic is fast but can lead to significant fragmentation. Assigns a task to the first processor that can schedule it, starting from the first processor (P3, P4).

– Worst Fit (WF): The task is placed in the largest available partition. This heuristic can result in significant fragmentation but is faster than the second heuristic. Assigns a task to the processor (P5, P6) with the most unused capacity that can schedule it.

These heuristics are used to find a solution for each subset of tasks using traditional single-processor scheduling. Each heuristic has its own advantages and disadvantages, and the choice depends on the system's characteristics and scheduling objectives (D. François, 2010) (N.H. Mohammed, 2020).

**Reporting Section**

To visualize the scheduling of tasks on the allocated processors, along with the protocols and execution times associated with each task, we have used the Gantt chart. This chart allows us to represent all tasks in an orderly manner.

Gantt charts are the oldest of the three planning techniques mentioned above. They are named after their creator, Henry Laurence Gantt, an American engineer and consultant who developed them in 1917. They are also known as bar charts, scheduling graphs, or workload charts. The purpose is to visualize the use of resources over time in order to optimize their allocation (P.Esquirol et al., 2001).

The principle of this type of chart is to represent different tasks as rows within a table, and units of time as columns (expressed in months, weeks, days, etc.) (B.R, M. Dibon, 1996).

# 4   Case Study

We performed this work on a machine with an Intel(R) Core(TM) i7-3632QM CPU @ 2.20GHz processor, equipped with 6GB of memory, running on a 64-bit Windows 10 operating system.

As part of this project, we used several computer tools. Specifically, we used Python as the main programming language, PyCharm as the integrated development environment (IDE) for Python project development. Additionally, we chose Anaconda, a specialized distribution for data-oriented projects, which allows the creation of dedicated virtual environments for Python-based projects.

## 4.1   Application Overview

In this section, we provide a detailed presentation of the proposed application, which implements the First-Fit, Best-Fit, Next-Fit, and Worst-Fit heuristics.

**Startup Interface:** Figure 2 showcases the graphical startup interface of our application. In this interface, two icons are distinguishable. They are labeled "Open Windows" and "Exit." Clicking on the "Open Windows" icon directs us to the next window. The "Exit" icon allows you to exit the application.

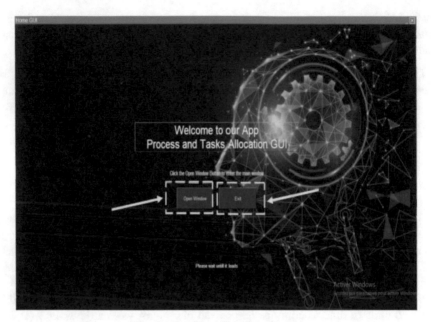

**Fig. 2.** Graphical User Interface 1.

**First Experiment**

Upon clicking the "Open Windows" icon in the previous graphical interface, the main interface of our application is displayed (see Fig. 3). To demonstrate the functionality of our application, we enter processors from 1 to 6 in the window located on the far left and display their corresponding information, such as their number (Process number) and their respective size (Process Size).

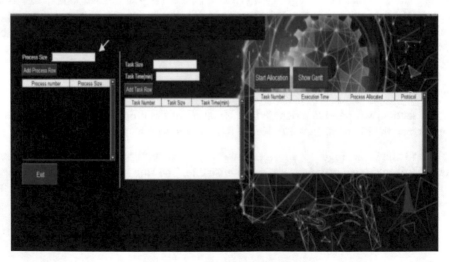

**Fig. 3.** Graphical User Interface 2

Figure 4 below illustrates the steps involved in this stage. First, it is necessary to determine the size of the machine's memory. This operation involves entering the desired size in the 'Process Size' field and confirming the addition of each processor (from P1 to P6) by clicking the 'Add Process Row' button. Processors may have the same size, or they may be heterogeneous with different sizes.

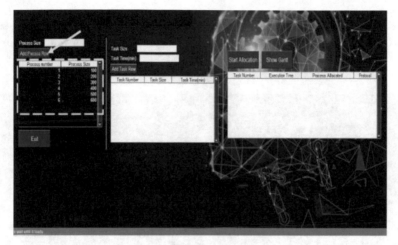

**Fig. 4.** Step 1

After completing the first step and selecting the sizes for six processors of the machine, we will proceed to the second step.

Figure 5 illustrates the process of introducing a set of tasks, each having a size (Task Size) and an execution time (Task Times). This procedure involves entering the desired size for each task in the "Task Size" field and providing its corresponding execution time. Then, to confirm the addition of all tasks, we click on the "Add Task Row" button. We are free to choose the sizes and execution times for each task.

Note: "Time is measured in units per minute."

**Fig. 5.** Step 2

Figure 6 explains the two previous steps and continues by describing the step required to assign tasks and allocate processors based on their respective protocols and priority. To perform this step, simply click the "Start Allocation" button to begin scheduling tasks based on their size relative to that of the machine, the maximum execution time required for them, and to prevent processor overheating that could lead to a machine breakdown. Additionally, it's important to consider the priority of the protocols, as illustrated in the following figure.

Through this window, we can observe information related to task execution, such as their order, the protocol used to complete them, the processor assigned to each task, and the duration required to accomplish them.

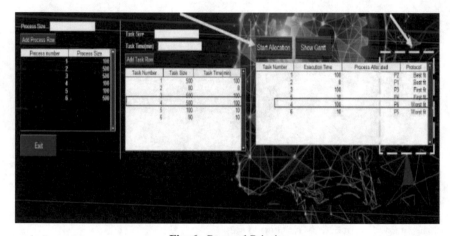

**Fig. 6.** Protocol Priority

To visualize the sequence of tasks and their processor allocation, it is necessary to click on the "Show Gantt" button. The figure below presents a Gantt chart that illustrates the expected execution duration for each task as well as the corresponding processor allocation as indicated in the window.

**Second Experiment**

To demonstrate that our application functions correctly, we conducted a second experiment with 7 tasks, as illustrated in Fig. 8. Thus, the results of this second experiment demonstrate that the protocol effectively distributes tasks correctly, as evidenced by the following assignments:

- The task allocation results for tasks 1, 2, and 7 have shown that the Best Fit protocol is consistently able to choose the appropriate space before allocating processors.
- Observations from tasks 3 and 6 have confirmed that the First Fit protocol indeed chooses the first available space to allocate processors.
- As long as the size of task 4 is greater than that of P5, it is executed on P6 because it is available and fulfills all the required conditions.
- Finally, task 5 revealed how the protocols are prioritized.

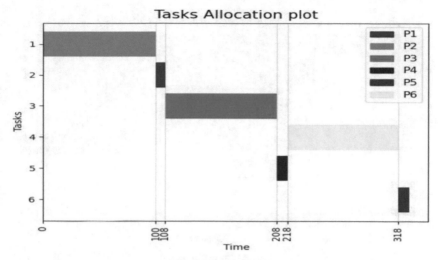

**Fig. 7.** Gantt Chart 1

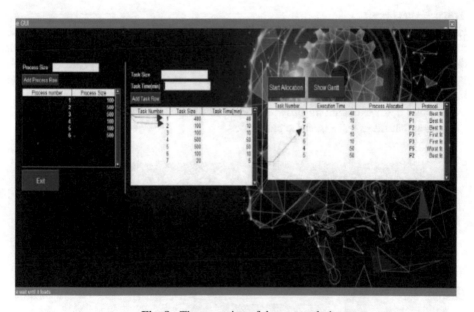

**Fig. 8.** The operation of the protocols 1

The Gantt chart shown in Fig. 9 provides a visual way to schedule the various tasks to be performed. This graphical representation offers a clear and structured overview of the activities to be carried out.

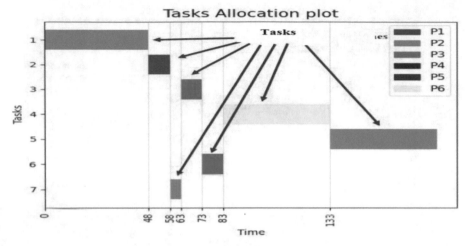

**Fig. 9.** Gantt Chart 2

**Third Experiment**

Figure 10 illustrates that if the task size is sufficiently large and if the execution time exceeds or equals the maximum working capacity of the machine, represented by a duration of 240 min or 4 h, then the task cannot be executed, and processor allocation will never take place. This scenario is reflected by task 2, which demonstrates the correlation between the time required to complete a task and the state of the machine. Furthermore, tasks 1 and 6 are executed on processor P1, and their execution time is less than 240, indicating that these tasks must be performed.

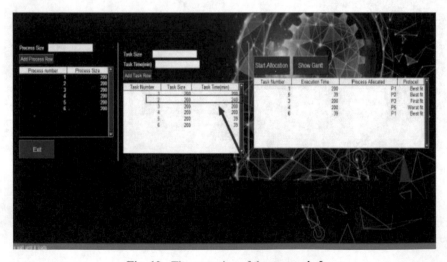

**Fig. 10.** The operation of the protocols 2

The Gantt chart shown in Fig. 11 allows for the visual organization of the various tasks to be performed. This graphical representation provides a clear and structured overview of the activities to be carried out.

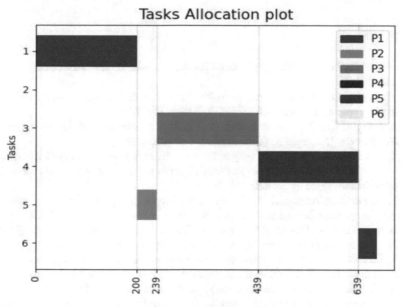

**Fig. 11.** Gantt Chart 3

## 5 Conclusion

Scheduling is a crucial process in task or process management within a computer system. In a multiprocessor system, there are various scheduling approaches, one of which is partition scheduling. It allows for the distribution of tasks among multiple processors independently, optimizing resource utilization and enhancing overall system performance.

To achieve efficient partition scheduling, several criteria must be considered. It is important to take into account the size and complexity of tasks as well as communication constraints between partitions. A well-balanced distribution of tasks among partitions is essential to ensure equitable resource utilization.

The main objective of this paper is how to perform partitioning scheduling in a homogeneous multiprocessor system where each processor has its own memory using different heuristics, namely: BEST FIT, FIRST FIT, and and GANTT chart visualization. The goal is to minimize execution time and ensure better processor allocation while taking into consideration the probability of failure.

The results obtained demonstrate the effectiveness of our approach for task scheduling in a multiprocessor system, as they provide favorable indicators.

The subject being very broad, there is still much to be done to improve this work. Therefore, potential avenues for further exploration include adding additional constraints, integrating other scheduling techniques and methods, as well as any other ideas deemed useful, feasible, and beneficial.

# References

Jean, A., Yves, C., Yves, D., Fabre, J.-C., Jean-Claude, L., David, P.: Tolérance aux fautes. Journal, pp. 240–270 (2006)

Nadine, A.: Partitionnement temps réel multiprocesseur sous contraintes de qualité de service et d'énergie. Thèse de doctorat, l'Université de Nantes (2014)

Khaoula, B.O.U.K.I.R.: Mise en oeuvre de politiques d'ordonnancement temps réel multiprocesseur prouvée. Université de nantes, Thèse de doctorat (2020)

Thanh, Dat N.: Aide à la validation temporelle et au dimensionnement de systèmes temps réel dans une démarche dirigée par les modèles . l'école nationale supérieure de mécanique et d'aérotechnique, Thèse de doctorat (2020)

Maxime, C., Pierre-Emmanuel, H., Anne-Marie, D.: Algorithmes pour l'ordonnancement temps réel multiprocesseur. Journal Européen des Systèmes Automatisés (JESA) (2015)

Frédéric, F., Laurent, G., Damien, M., Serge, M.: Ordonnancement multiprocesseur global basé sur la laxité avec migrations restreintes. Université Paris-Est, Article (2011)

François, D.: Contributions à l'ordonnancement et l'analyse des systèmes temps réel critiques » , Thèse de doctorat , l'école nationale supérieure de mécanique et d'aérotechnique (2010)

Mohammed, N.H. : Résolution exacte du problème de partitionnement de données avec minimisation de variance sous contraintes de cardinalité par programmation par contraintes, Mémoire de maîtrise, Polytechnique Montréal (2020)

Esquirol, P., Lopez, P.: Concepts et méthodes de base en ordonnancement de la production, in Ordonnancement de la production, Information, Commande, Communication », chapitre, p. 25 53 (2001)

Roy, B., Dibon, M.: L'ordonnancement par la méthode des potentiels le programme CONCORD, Automatisme, pp 1–11 (1966)

# A Novel CNN-SVM Hybrid Model for Human Activity Recognition

Imene Charabi, M'hamed Bilal Abidine$^{(\boxtimes)}$, and Belkacem Fergani

Departement of Telecommunications, University of Science and Technology Houari Boumediene, Algiers, Algeria
abidineb@gmail.com

**Abstract.** Human activity recognition (HAR) is a growing field that focuses on using sensor data to automatically detect and analyze human motions. HAR has gained a lot of attention and interest due to its extensive use in many different domains, such as healthcare, sports, security, and many others. In this paper, we propose a hybrid approach for HAR that combines feature extraction using a Convolutional Neural Network (CNN) and classification using a Support Vector Machine (SVM). The preprocessing step is performed using Linear Discriminant Analysis (LDA). LDA transforms the input data from its original high-dimensional space to a new, lower-dimensional space, resulting in enhanced separation between different classes. By passing the transformed features to the CNN, the network can extract more relevant and discriminative features compared to directly using the raw sensor data. This leads to improved classification and prediction performance. The proposed approach is evaluated on two widely used datasets, achieving an accuracy of 96.30% on the UCI HAR and 95.00% on PAMAP2.

**Keywords:** Human Activity Recognition (HAR) · Convolutional neural network (CNN) · Linear Discriminant Analysis (LDA) · Support Vectors Machine (SVM)

## 1 Introduction

Human Activity Recognition (HAR) aims to interpret human motion based on data inputs. It can be categorized into two main types: sensor-based HAR and vision-based HAR [1]. Vision-based HAR specifically refers to the recognition of human activities when the input data consists of captured images or recorded video sequences. On the other hand, sensor-based HAR relies on data collected from smart sensors, such as accelerometers and gyroscopes. Sensor-based HAR has emerged as a preferable approach due to its advantages over vision-based methods. Sensors provide more accurate recognition capabilities, overcoming limitations related to fixed camera positions, privacy concerns, and enabling continuous monitoring and real-time feedback [2]. Moreover, the availability of sensors in everyday devices, such as smartphones and smartwatches, coupled with their low cost and low power consumption, has made sensor-based HAR more accessible and practical. The integration of sensors into everyday devices has paved the way for the creation of smart environments, where sensor data can be utilized to understand human

© The Author(s), under exclusive license to Springer Nature Switzerland AG 2024
M. Hatti (Ed.): IC-AIRES 2023, LNNS 984, pp. 265–273, 2024.
https://doi.org/10.1007/978-3-031-60629-8_27

behavior in real-time. This has led to the adoption of HAR in various domains, including healthcare and sports. In healthcare, HAR is employed for various activities such as patient monitoring [3] and fall detection [4]. Furthermore, in the realm of sports, HAR plays a pivotal role in monitoring and analyzing the activities and movements performed by athletes, providing valuable insights and performance optimization opportunities [5].

Considering the significance of Human Activity Recognition (HAR) in all these domains, we aim to contribute to this field through our research paper. We introduce a novel hybrid approach based on sensor data for Human Activity Recognition. The proposed approach begins with the application of linear discriminant analysis (LDA) to the input data, which enhances the separability and discriminative power of the features among different activity classes. Subsequently, the transformed features obtained through LDA are fed into a convolutional neural network (CNN) for further feature extraction. The CNN exploits its capability to extract meaningful and discriminative features from the transformed input. Finally, the extracted features from the CNN are utilized as input to a support vector machine (SVM) classifier, which classifies the activities into their respective classes. In Section 2 of this paper, a discussion on the previous state-of-the-art in the field of HAR will be presented. Section 3 will provide a detailed explanation of our proposed method. In Sect. 4, we will present the results obtained from our method and compare them with baseline approaches.

## 2   Related Works

In this section, we discuss recent hybrid models relevant to our approach. Luwe et al. (2022) in [6], proposed a hybrid model called 1D-CNN-BILSTM, combining 1D CNN layers for feature extraction and a BILSTM layer for capturing long-range dependencies. Their model achieved recognition rates of 95.48%, 94.17%, and 100% on UCI-HAR, Motion Sense, and Single Accelerometer datasets, respectively. Aboo and Ibrahim (2022) in [7], implemented a PCA-CNN-LSTM hybrid model that incorporated PCA for dimensionality reduction. They compared its performance to variations such as CNN-LSTM without PCA, CNN with PCA, and LSTM with PCA. The CNN-LSTM model achieved the highest performance with up to 96.8% accuracy on the UCI-HAR dataset without requiring PCA. Shuvo et al. (2020) in [8], presented a CNN-SVM hybrid model for HAR. They transformed accelerometer sensor data into spectrogram images and utilized a pre-trained VGG16 model to extract features. An SVM classifier was then used for classification. Athavale et al. (2021), in [9] employed a two-stage learning process in their CNN-SVM based approach. They used a random forest algorithm to classify activities into static and moving categories, followed by SVM for static activities and 1D CNN for moving activities. In the next section, we present our proposed methodology for HAR using a CNN-SVM hybrid approach preceded by LDA.

## 3   Proposed Method

Our proposed approach consists of three main steps as shown in Fig. 1. Firstly, Linear Discriminant Analysis (LDA) is applied as a preprocessing step to the raw sensor data. LDA applies a linear transformation to the data. This transformation maximize the separation between classes (between-class scatter) and minimize the separation within each class (within-class scatter). By doing so, LDA generates new features that capture the essential information that distinguishes one class from another. The new features serve as the input to the CNN. This enhanced discriminability allows the CNN to extract more relevant features.

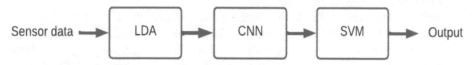

**Fig. 1.** Main steps of the porposed LDA-CNN-SVM approach.

CNNs leverage the process of convolution, which involves using a set of matrices called kernels to scan over the input data. As the kernels traverse the data, they perform a dot product operation between the kernel weights and the corresponding values in the input data, resulting in feature maps. Each feature map captures a specific learned feature. The architecture of the proposed CNN consists of three one-dimensional convolutional layers with kernel sizes of 64, 128, and 512, respectively. All the convolutional layers utilize the Rectified Linear Unit (ReLU) activation function, adopt a kernel size of 5 and 2 strides. Following each convolutional layer, a one-dimensional max pooling layer with a pool size of 2 is applied to reduce the spatial dimensions of the extracted features while retaining their important characteristics. Dropout layers with a rate of 0.2 are incorporated after each pooling layer, aiming to prevent overfitting. Additionally, a batch normalization layer is applied before the dropout in the last layer. A flatten layer is then employed to transform the multidimensional feature maps into a flattened representation. Finally, the softmax layer of the CNN is replaced by an SVM classifier. The SVM classifier performs multiclass classification on the flattened extracted features obtained from the CNN. The Fig. 2 illustrates the layers of the hybrid CNN-SVM model used in this method.

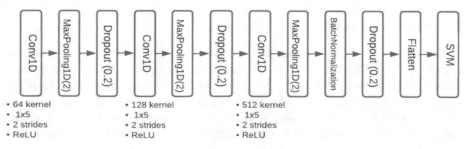

**Fig. 2.** Layers of the proposed CNN-SVM hybrid model.

# 4  Exprimental Results

## 4.1  Datasets

The datasets we utilized in this research are the UCI HAR dataset by Anguita et al. (2013) in [10] and PAMAP2 Reiss and Stricker (2012). Table 1 represents the activities performed in the respective datasets.

The UCI HAR dataset consists of recordings from a Samsung Galaxy S II smartphone worn by 30 individuals. It captures 3-axial angular velocity and 3-axial linear acceleration data at a rate of 50 Hz. Preprocessing includes noise reduction and sampling with sliding windows of 2.56 s (50% overlap), resulting in 128 readings per window. A Butterworth filter separates linear acceleration into body acceleration and gravity components. The dataset contains 561 variables calculated from the time and frequency domains of each window. It was divided randomly into 70% training and 30% test sets.

The PAMAP2 dataset comprises over 10 h recordings of 18 activities performed by 9 subjects (1 female, 8 males). The activities include 12 activities following a specific protocol and 8 optional activities. To collect the data, the subjects wore three Inertial Measurement Units (IMUs) on their dominant arm, dominant ankle, and chest. These IMUs captured tri-axial acceleration, gyroscope, and magnetometer measurements at a sampling rate of 100 Hz.

**Table 1.**  Activities performed in the used datasets.

| Dataset | Activities |
|---|---|
| UCI HAR | WALKING, WALKING_UPSTAIRS, WALKING_DOWNSTAIRS, SITTING, STANDING, LAYING |
| PAMAP2 (protocol) | walking, running, cycling, Nordic walking, ascending stairs, descending stairs, vacuum cleaning, ironing, rope jumping, lying, sitting, standing |

## 4.2  Experimental Details

The implementation of the approach was performed using the Keras library version 2.9.0 in Python 3.11.2. The hybrid CNN-SVM model is end to end trainable. The training process involved 50 epochs, with a batch size of 32 and the utilization of the Adam optimizer for optimization. The remaining hyperparameters were set to their default values. The UCI HAR dataset was divided by the authors by default as mentionned above, and in the case of the PAMAP2, we applied a 30% test and 70% train split.

## 4.3  Results

The training accuracy and loss were tracked throughout the training process for UCI HAR and PAMAP2 dataset, as shown in Fig. 3 and Fig. 4, respectively. The training process shows a steady improvement in both loss and accuracy over the epochs.

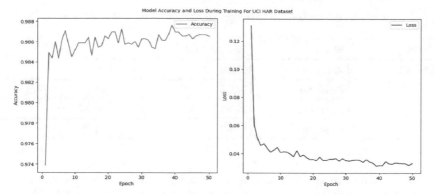

**Fig. 3.** Model accuracy and loss during training for UCI HAR dataset.

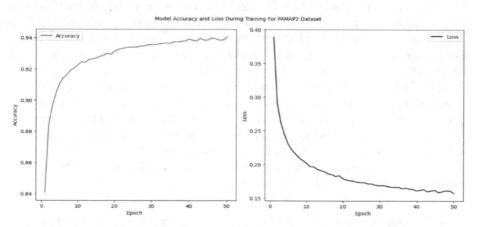

**Fig. 4.** Model accuracy and loss during training for PAMAP2 dataset.

To assess the impact of LDA preprocessing, we conducted tests on previously unseen data. We compared two models: the LDA-CNN-SVM model, which utilized LDA as a preprocessing technique, and the CNN-SVM model without LDA preprocessing. The results of this comparison for each of the datasets are reported in Table 2 and Table 3.

**Table 2.** Performance of the proposed method on the UCI HAR dataset.

| Dataset | Metric | Without LDA (%) | With LDA (%) |
|---------|--------|-----------------|--------------|
| UCI HAR | Accuracy | 95.48 | 96.30 |
| | Precision | 95.68 | 96.51 |
| | Recall | 95.44 | 96.28 |
| | F1 score | 95.52 | 96.35 |

**Table 3.** Performance of the proposed method on the PAMAP2 dataset.

| Dataset | Metric | Without LDA (%) | With LDA (%) |
|---------|--------|-----------------|--------------|
| PAMAP2 | Accuracy | 91.46 | 95.00 |
| | Precision | 91.41 | 94.39 |
| | Recall | 89.31 | 93.57 |
| | F1 score | 89.98 | 93.93 |

The evaluation results demonstrate that the LDA-CNN-SVM approach achieves superior performance compared to the CNN-SVM model without LDA preprocessing for both datasets. Specifically, the LDA-CNN-SVM model achieved an accuracy of 96.30%, outperforming the CNN-SVM model's accuracy of 95.48% on the UCI HAR dataset. Moreover, significant improvements were observed across precision, recall, and F1 score metrics. For the PAMAP2 dataset, the LDA-CNN-SVM model achieved an accuracy of 95.00%, surpassing the CNN-SVM model's accuracy of 91.46%. Similarly, there were notable enhancements in the other metrics.

The obtained results provide evidence of the impact of incorporating LDA as a preprocessing step in enhancing the overall classification performance of our proposed approach for HAR. By applying LDA to the raw sensor data, we are able to enhance the discriminativity of the features, enabling the CNN model to capture more informative and relevant patterns related to the activities being recognized.

Figure 5 illustrates the normalized confusion matrices obtained from the model's predictions on the test sets of the UCI HAR dataset and the PAMAP2 dataset.

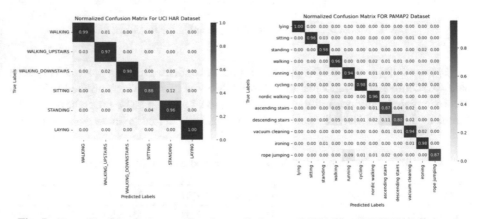

**Fig. 5.** Normalized confusion matrices for predictions on PAMAP2 and UCI HAR Datasets.

For the UCI HAR dataset, dynamic activities (WALKING, WALKING_UPSTAIRS, and WALKING_DOWNSTAIRS) achieved high accuracies of 99%, 97%, and 98% respectively. Similarly, the static activities of STANDING and LAYING demonstrated excellent performance. However, there was a misclassification rate where approximately

12% of SITTING instances were mistakenly classified as STANDING, likely due to their similarity.

Regarding the PAMAP2 activities, most activities such as lying and cycling were accurately classified. However, there were instances of misclassification between ascending stairs and descending stairs. This misclassification may be due to the similarity in movements involved in both activities.

Additionally, We have observed that for both datasets, the activity 'laying' is the easiest to classify, with all instances being correctly identified. This can be attributed to the stationary position and minimal movement involved in this activity, making it easily distinguishable from other activities that involve more dynamic movements.

Furthermore, we conducted a comparative analysis of our proposed method with state-of-the-art methods on both datasets, using accuracy as the primary evaluation metric. The results of this comparison are presented in Table 4 and Table 5.

**Table 4.** Performance comparison of other models for UCI HAR.

| Dataset | Approach | Accuracy (%) |
| --- | --- | --- |
| UCI HAR | 1D CNN [12] | 94.79 |
| | RS-1D-CNN [13] | 95.40 |
| | 1D CNN-BiLSTM [6] | 95.48 |
| | **Proposed** | **96.30** |

We can observe that the proposed method outperforms the other state-of-the-art methods for both datasets. This demonstrates the effectiveness of our approach in accurately predicting the activities in the dataset. The combination of the three components, namely LDA, CNN, and SVM, has contributed to achieving high performance in activity recognition compared to the baseline models. By leveraging the unique capabilities of each component, our model demonstrates competitive accuracy and effectiveness in recognizing and classifying activities.

**Table 5.** Performance comparison of other models for PAMAP2 dataset.

| Dataset | Approach | Accuracy (%) |
| --- | --- | --- |
| PAMAP2 | CNN [14] | 91.00 |
| | CNN-BiLSTM [15] | 94.29 |
| | CNN-LSTM [15] | 92.81 |
| | **Proposed** | **95.00** |

## 5  Conclusion

This paper introduces a novel hybrid approach for Human Activity Recognition (HAR) that combines Linear Discriminant Analysis (LDA) as preprocessing step, Convolutional Neural Networks (CNN) for feature extraction, and Support Vector Machine (SVM) for classification. The proposed methodology achieves high accuracy rates of 96.30% on the UCI HAR dataset and 95.00% on the PAMAP2 dataset, outperforming state-of-the-art approaches. Additionally, the results demonstrated the beneficial impact of LDA as a preprocessing technique in enhancing the overall performance of the approach.

This approach shows promising potential for further enhancement, aiming to expand its applicability beyond the UCI HAR and PAMAP2 datasets and across various domains. To achieve this, our future plans involve fine-tuning hyperparameters, including learning rate, batch size, and SVM kernels, to optimize performance. Additionally, we will explore transfer learning techniques to enhance the generalizability of the approach across diverse datasets and activities.

## References

1. Dang, L.M., Min, K., Wang, H., Piran, M.J., Lee, C.H., Moon, H.: Sensor-based and vision-based human activity recognition: a comprehensive survey. Pattern Recogn. **108**, 107561 (2020)
2. Serpush, F., Menhaj, M.B., Masoumi, B., Karasfi, B.: Wearable sensor-based human activity recognition in the smart healthcare system. Comput. Intell. Neurosci. **2022**, 1391906 (2022). https://doi.org/10.1155/2022/1391906
3. Gul, M.A., Yousaf, M.H., Nawaz, S., Rehman, Z.U., Kim, H.: Patient monitoring by abnormal human activity recognition based on CNN architecture. Electronics **9**(12), 1993 (2020). https://doi.org/10.3390/electronics9121993
4. Kozina, S., Gjoreski, H., Gams, M., Luštrek, M.: Efficient activity recognition and fall detection using accelerometers. In: Botía, J.A., Álvarez-García, J.A., Fujinami, K., Barsocchi, P., Riedel, T. (eds.) Evaluating AAL Systems Through Competitive Benchmarking. CCIS, vol. 386, pp. 13–23. Springer, Heidelberg (2013). https://doi.org/10.1007/978-3-642-41043-7_2
5. Shahar, N., Ghazali, N., Asâari, M., Swee, T.: Wearable inertial sensor for human activity recognition in field hockey: influence of sensor combination and sensor location. J. Phys. Conf. Ser. **1529**(2), 022015 (2020)
6. Luwe, Y.J., Lee, C.P., Lim, K.M.: Wearable sensor-based human activity recognition with hybrid deep learning model. Informatics **9**(3), 56 (2022). https://doi.org/10.3390/informatics9030056
7. Aboo, A.K., Ibrahim, L.M.: Human activity recognition using a hybrid CNN-LSTM deep neural network. Webology **19**(1), 6786–6798 (2022)
8. Shuvo, M.M.H., Ahmed, N., Nouduri, K., Palaniappan, K.: A hybrid approach for human activity recognition with support vector machine and 1D convolutional neural network. In: Proceedings of the 2020 IEEE Applied Imagery Pattern Recognition Workshop (AIPR), pp. 1–5. Washington DC, DC, USA (2020)
9. Athavale, V., Kumar, D., Gupta, S.: Human action recognition using CNN-SVM model. Adv. Sci. Technol. **105**, 282–290 (2021)

10. Anguita, D., Ghio, A., Oneto, L.: A public domain dataset for human activity recognition using smartphones. In: A European Symposium on Artificial Neural Networks, Computational Intelligence and Machine Learning. Proceedings of the 21th International European Symposium on Artificial Neural Networks, Computational Intelligence and Machine Learning, Bruges, pp. 437–442 (2013)
11. Reiss, A., Stricker, D.: Introducing a new benchmarked dataset for activity monitoring. In: Proceedings of the 16th International Symposium on Wearable Computers, Newcastle, UK, pp. 108–109 (2012)
12. Ronao, R., Charissa, A., Cho, S.-B.: Human activity recognition with smartphone sensors using deep learning neural networks. Expert Syst. Appl. **59** (2016)
13. Ragab, M.G., Abdulkadir, S.J., Aziz, N.: Random search one dimensional CNN for human activity recognition. In: Proceedings of the 2020 International Conference on Computational Intelligence (ICCI), Bandar Seri Iskandar, Malaysia, pp. 86–91 (2020). https://doi.org/10.1109/ICCI51257.2020.9247810
14. Wan, S., Qi, L., Xu, X., Tong, C., Gu, Z.: Deep learning models for real-time human activity recognition with smartphones. Mob. Netw. Appl. **25**, 743–755 (2020). https://doi.org/10.1007/s11036-020-01623-4
15. Challa, S.K., Kumar, A., Semwal, V.B.: A multibranch CNN-BiLSTM model for human activity recognition using wearable sensor data. Vis. Comput., 1–15 (2021). https://doi.org/10.1007/s00371-021-02124-9

# A Decision-Making Model for Self-adaptation of Cyber-Physical Systems: Application to Smart Grids

Ouassila Hioual[1,2]([✉]), Arridj Elwouroud Sassi[1], and Walid Djaballah[1]

[1] Mathematics and Computer Science Department, Abbes Laghrour University, Khenchela, Algeria
hioual_ouassila@univ-khenchela.dz
[2] LIRE Laboratory, Abdelhamid Mehri University (Constantine2), Constantine, Algeria

**Abstract.** Cyber-physical systems (CPSs) are systems that combine physical components with computational elements, working together in real-time to achieve specific goals. These systems are widely utilized in various areas, such as industrial automation, autonomous vehicles, and smart infrastructure. In this study, we focus on the field of smart grids, which are advanced electrical power networks featuring intelligent capabilities. We recognize the significance of self-adaptation in CPS, particularly in the context of smart grids. The proposed method involves developing a decision making model for self-adaptation in CPS specifically tailored to the smart grid domain. We employ machine learning techniques, such as neural networks, along with feedback mechanisms and reconfiguration strategies, targeting four key aspects of the smart grid: voltage, current, temperature, and consumption/production. The aim is to enable the system to, automatically, adjust to fluctuations in electricity demand and supply, as well as changes in network conditions. This involves granting the network the ability to independently monitor, analyze, and act upon real-time data, leading to the optimization of electrical energy production, distribution, and consumption. We evaluate the performance of our proposal through various scenarios. Thus, the simulation results demonstrate the effectiveness of our method.

**Keywords:** Cyber-physical systems · Self-adaptif · Smart Grid · Prediction · Decision Making

## 1 Introduction

In the last few decades, a convergence of new systems known as "cyber-physical systems" has taken place. These systems represent a combination of the virtual world of computer systems and the physical world of objects and processes. Indeed, they integrate various elements such as computer components, sensors, actuators, and others to interact with the physical environment. Furthermore, recent advancements in technology have significantly improved the efficiency of cyber-physical systems. Thus, they are utilized in numerous domains of our everyday life, such as healthcare, energy, transportation, robotics, smart electrical grids, and many more.

© The Author(s), under exclusive license to Springer Nature Switzerland AG 2024
M. Hatti (Ed.): IC-AIRES 2023, LNNS 984, pp. 274–289, 2024.
https://doi.org/10.1007/978-3-031-60629-8_28

With the advent of artificial intelligence, electrical grids are moving towards becoming smart grids. These modern systems leverage cutting-edge information and communication technologies to enhance the performance of conventional electrical networks. Their primary objective is to improve electricity distribution efficiency, reliability, sustainability, and overall electrical system security. In such cyber-physical systems (CPSs), self-adaptation plays a crucial role as it enables the system to, automatically, adjust its behavior in response to changes in its environment. Hence, self-adaptation refers to the inherent ability of a system to autonomously adapt and modify itself in reaction to environmental variations or internal disruptions. Furthermore, in the context of smart electrical grids, self-adaptation plays an even more significant role than in other cyber-physical systems applications. It allows for the automatic adjustment of the system's operation based on changes in electrical network conditions (voltage, production, consumption, intensity, etc.). These adjustments are achieved using mechanisms and algorithms such as machine learning, optimization, feedback, and reconfiguration, which regulate and modify the system's behavior or parameters.

The objective of this paper is to propose a decision support method that enables the dynamic adaptation of a smart electrical grid to the instantaneous needs and changes in its environment, without human intervention. Indeed, self-adaptation is crucial in our study to optimize energy efficiency and maintain the stability of the smart electrical grid. We have considered five key aspects, which are production, consumption, intensity, temperature, and voltage. The proposed method aims to ensure that the system maintains its performance, stability, and optimal functioning, even in the presence of unexpected changes.

The rest of this paper is organized as follows: Sect. 2 offers an introduction to the fundamental concepts of self-adaptation in cyber-physical systems (CPSs) in general, focusing specifically on smart grid environments. Section 3 conducts a thorough review and comparative analysis of existing work in this area. In Sect. 4, we introduce our proposed method and provide a comprehensive explanation of its components and functioning. Section 5 presents the results of our experiments. Finally, in Sect. 6, we conclude the paper and outline potential avenues for future research.

## 2 Self-adaptation in Cyber Physical Systems

Managing a CPS is a challenging task due to the complexity of hardware architectures and the constant evolution of the components that comprise it. As a result, it faces several challenges. Among these problems or challenges, we can mention:

- **Self-configuration:** which is the ability to reconfigure, automatically and dynamically, a system in response to changes. This can include installation, integration, removal, and composition/decomposition of system elements (El Ballouli 2019).
- **Self-protection:** which is defined as the ability to detect security breaches and recover the system's state in case of an attack. This includes recovering from existing and anticipated attacks (El Ballouli 2019).
- **Self-adaptation**, which represents the capability of a system to, automatically, adjust to changes in its environment.

The last decades have been characterized by a growing recognition of the significance of self-adaptation in CPSs. The scientific literature on self-adaptation systems has been extensive, and this subject remains at the core of research and development activities. Indeed, the scientific community has dedicated significant efforts to the study of this field. The objective is to discover innovative approaches to make the fundamental principles of self-adaptation theory and practice more comprehensible. Therefore, when the program fails to achieve expected results or when there is an opportunity to enhance its functionalities or performance, self-adaptation systems are capable of assessing and modifying their own behavior (Macías-Escrivá et al. 2013).

The convergence of computer and physical elements in CPSs allows for seamless and implicit integration. Therefore, adjustment, made possible through feedback loops, is crucial for managing uncertain operational conditions within CPSs (Kamel 2016). Figure 1 illustrates the main concerns regarding self-adaptation in CPSs (Muccini et al. 2016). Furthermore, self-adaptation raises significant concerns. It involves the ability to adjust automatically in response to various changes, whether they originate from within or outside the system. These changes can encompass failures, errors, disturbances, environmental variations, or fluctuations in demand. To achieve this self-adaptation capability, algorithms and machine learning mechanisms are employed (Muccini et al. 2016). Among the application domains of self-adaptation in CPSs, smart grids are noteworthy. Smart grids utilize self-adaptation to efficiently manage the production and distribution of electricity. Hence, energy production can be automatically adjusted based on demand. Furthermore, smart grids can optimally integrate renewable energy sources and respond to outages or distribution issues. Since this paper focuses on self-adaptation within distribution networks (smart grids), we will elaborate on their management in a subsequent subsection.

## 2.1 Self-adaptation in Smart-grids: Uses and Techniques

In our study, we are interested in self-adaptation within electrical networks, particularly in smart grids. Self-adaptation in smart grids involves enabling electrical networks to automatically adjust to variations in electricity demand and supply. In order to utilize this technology, it is essential to establish real-time monitoring and control systems, including the installation of smart sensors, the use of dedicated energy management software and the integration of optimization algorithms. These systems allow gathering data about energy production, consumption, and storage, and making real-time decisions to optimize the network's operation.

Several techniques can model electrical network behaviors and make real-time decisions. These later optimize energy production and consumption. Among these techniques, we can mention (Houssin et al. 2020) (Guérard 2014): Neural Networks, Genetic Algorithms, Multi-Agent Systems, Bayesian Networks, Decision Trees and Markov Models.

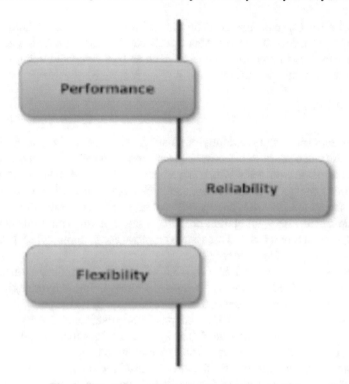

**Fig. 1.** The main concerns of self-adaptation in CPSs

## 2.2 Management of Electricity Distribution Networks

The management of distribution networks is a key aspect of self-adaptation in smart grids. It aims to optimize the real-time utilization of energy within the distribution network based on changing needs. There are several techniques for managing distribution networks, which enable smart grids to automatically adapt to changing energy consumption needs and ensure energy quality for consumers. In the following, we will address a few of them:

- **Voltage and Current Regulation:** Smart grids can be equipped with voltage and current control devices to maintain the quality of electric current and prevent issues such as overvoltages and undervoltages.
- **Load Balancing**: Load balancing involves distributing energy consumption across different sections of the network to prevent outages. There are load management algorithms that can be used to optimize real-time load distribution. It should be noted that the choice of the load distribution model depends on several factors such as network characteristics, cost and quality of service objectives, and available resources for implementation.
- **Fault Detection**: Smart grids can be equipped with fault detection systems. This can include detecting overloads, equipment failures, and more.

- **Power Quality Control**: Power quality is an important aspect for both consumers and energy producers. Therefore, smart grids can be equipped with a set of power quality controllers to monitor parameters such as frequency and voltage in order to maintain them within defined limits.

## 3  Related Works

Ensuring the reliability and continuity of smart grid operations is crucial. To achieve these goals, it is necessary to effectively assess and manage unforeseen and ongoing changes. As a result, numerous efforts have been made in the field of smart grids to propose various methods, architectures, and self-adaptation models to address these continuous changes. In this section, we will present some of existing research.

In (Aladdin et al. 2020), the authors proposed a self-contained multi-agent system for demand response in Smart Grids. This system employs autonomous agents to coordinate agent responses and minimize energy costs.

The authors of (Sang et al. 2022) proposed a self-adaptive optimization strategy for energy management in microgrids. The proposed strategy employs hybrid reinforcement learning to optimize energy production and consumption in real-time.

In (Radhakrishnan et al. 2016), an adaptive and distributed energy management system for Smart Grids was introduced. The system employs fuzzy logic to make real-time decisions and adapt to changes in network conditions.

In the work presented in (Moghaddam et al. 2022), the authors introduced a self-adaptive hierarchical control system for distributed energy resources in smart grids. The system utilizes optimization algorithms to adjust control parameters based on network conditions and optimization objectives.

In (Mu et al. 2019), the authors proposed an adaptive particle swarm optimization algorithm for intelligent management of electrical networks. The algorithm adapts, automatically, to environmental changes and multiple objectives such as cost reduction, energy consumption optimization, and emissions minimization.

The previously mentioned works employ various methods and approaches as follows: the first work focuses on the Internet of Things (IoT) and utilizes multi-agent systems, emphasizing consumption. The second work primarily relies on reinforcement learning, concentrating on production and consumption aspects. The third work predominantly employs fuzzy logic for consumption. The fourth work relies on a self-adaptive hierarchical control system, utilizing optimization algorithms for consumption. Lastly, the fifth work employs an adaptive particle swarm optimization algorithm for consumption.

In summary, it is evident that the mentioned works successfully leverage machine learning techniques, showcasing their significant promise in this field. Additionally, it is clear that a decision-making system plays a crucial role in this domain. The table bellow (Table 1) provides a summary of the five works.

**Table 1.**  A comparative table of related works

| Year | Research work | Adressed Issues | | | | | Used techniques |
|---|---|---|---|---|---|---|---|
| | | Production | Voltage | Current | Consumption | Temperature | |
| 2021 | (Aladdin et al. 2020) | − | − | − | − | + | Multi Agent Systems IoT |
| 2020 | (Sang et al. 2022) | + | − | − | + | − | Reinforcement learning |
| 2021 | (Radhakrishnan et al. 2016) | − | − | − | + | − | Fuzzy logic |
| 2021 | (Moghaddam et al. 2022) | − | − | − | + | − | Optimization algorithms |
| 2011 | (Mu et al. 2019) | − | − | − | + | − | Adaptive particle swarm optimization algorithm |
| 2023 | **Our proposal** | + | + | + | + | + | Feedback mechanisms, neural networks, and reconfiguration mechanisms |

## 4  Proposed Method

Managing a cyber-physical system is a challenging task due to the complexity of hardware architectures and the constant evolution of the components that comprise it. The objective of this work is to propose a decision support method to dynamically adapt CPSs to instantaneous needs and changes (without human intervention). We particularly focused on smart grids, which require efficient management and optimization of energy production, distribution, and consumption.

### 4.1  Our System Architecture

Figure 2 depicts the overall architecture of our system, explaining its functioning. Firstly, data from sensors is received and stored in the storage database (SD). Then, the system uses this data to predict target data. For each aspect (detailed in the subsection below), we have developed a prediction model using neural networks. After each prediction, the SD is updated. Subsequently, the system makes decisions based on the predictions and implements necessary actions accordingly.

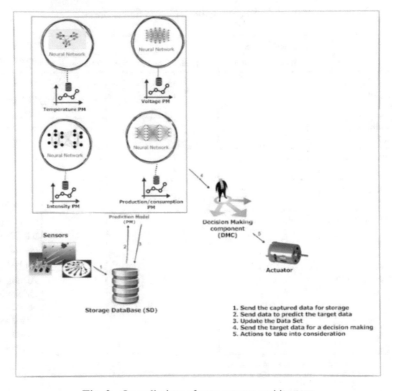

**Fig. 2.** Overall view of our systems architecture

## 4.2 Addressed Aspects

In our research, we examined five essential elements to ensure the proper functioning of a smart grid. These five elements are as follows:

- *Electricity production*: involving the conversion of various primary energy sources such as coal, natural gas, oil, nuclear energy, and forms of renewable energy into electricity (HOCINE 2019).
- *Consumption*: which is the amount of energy used by consumers to operate a variety of systems, devices, and processes. It can be categorized into various segments, such as household consumption, commercial usage, industrial usage, and so on. *Production* and *consumption* of electrical energy are two interdependent aspects of an energy system's operation.
- *Voltage*: it represents the force exerted by a power source on the charged electrons in an electrical circuit, creating a current that flows through a conducting loop. This current enables electrons to perform various tasks and is measured in Volts (V) (Mazur 2006).
- *Current*: corresponds to the flow of charges through a circuit. It's measured in Amperes (A) and represents the quantity of electric charge passing through a specific part of the circuit in a given unit of time.

- **Temperature:** it is a physical quantity that's assessed and explored using a thermometer, known as temperature in the field of thermometry.

### 4.3 Used Mechanisms

In this section, we present the mechanisms, methods, and functional algorithms we have used to adapt our study's model to the target values of smart grid aspects.

- **Feedback mechanisms**: feedback is a control process that comprises three subprocesses: information acquisition, evaluation, and reaction. While these operations occur sequentially, they happen simultaneously and continuously. The goal is to enable the concerned individual to correct or enhance their approach (Paquette 1987).
- **Machine Learning:** is a process of adjusting the parameters of a system so that it generates a desired response based on a given input or stimulus (Torres-Moreno 1997). In our study, we use neural networks, which are mathematical models inspired by the functioning of biological neurons, to implement self-adaptation in cyber-physical systems.
- **Reconfiguration mechanisms**: leverage adaptation possibilities to ensure and maintain an appropriate level of quality of service, even in the presence of disruptions.

### 4.4 Functioning of Proposed Method

We use Unified Modeling Language (UML) diagrams, in this study, to represent and visualize the functioning of our method. Additionally, we utilized UML diagrams to describe in detail the self-adaptation of each aspect mentioned earlier, such as consumption/production, current, voltage, etc.

According to the proposed method, we have four use cases which are as follows:

- **Regulate energy consumption/production**: In this case, the system, after receiving data from the external environment, can regulate production according to the predicted consumption of the system users (Smart Grid).
- **Adjust temperature**: The Smart Grid system can receive signals to adjust the temperature during specific periods.
- **Adapt voltage**: The Smart Grid system is capable of adapting voltage according to external demand and predicted values by the system.
- **Regulate current**: The Smart Grid system can regulate current according to external demand (captured information) and predicted values.

In the following subsections, we will examine each use case individually, explaining its functioning through activity and sequence diagrams.

### 4.5 Functioning of the Use Case: Regulate Energy Consumption/Production

- Activity diagram

According to Fig. 3, our system follows the following steps. Firstly, it initially records the captured consumption values (CAPCV) and the captured production values (CAPPV) from two different types of sensors. The system subsequently generates two predicted values (referred to as "target values") using a prediction model based on a neural network. One of these predicted values is the target consumption value (TARCV), and the other is the target production value (TARPV). In our model, we use the target consumption value (TARCV) for prediction and compare TARPV with TARCV.

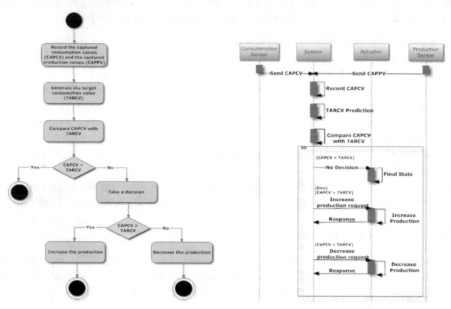

**Fig. 3.** Activity diagram of the "Regulate energy production/consumption" use case

**Fig. 4.** Sequence diagram of the "Regulate energy production/consumption" use case

Next, we compare CAPPV with TARCV. If CAPPV is equal to TARCV, it means that the system is stable. Otherwise, the system needs to make decisions to recover the produced energy or increase production, based on the following two conditions:

If CAPPV is less than TARCV, the system requests an increase in CAPPV until reaching TARCV and sends instructions to the corresponding actuators.

On the other hand, if CAPPV is greater than TARCV, the system requests a decrease in CAPPV towards TARCV, which results in additional recovery of produced energy and subsequent storage.

• Sequence diagram

Among the UML interaction diagrams, we can mention the sequence diagram. It enables the representation of exchanges between different objects and actors of the system based on time. As depicted in Fig. 4, our model comprises three objects: sensors, system, and actuator. They interact with each other through the chronological exchange of messages.

### 4.6 Functioning of the Use Case: Adjust Temperature

- Activity diagram

    As shown in Fig. 5, our system follows a sequence of operations. Firstly, the captured temperature value (CAPV) is stored in the storage base (SB). Next, a prediction of the target temperature value (TARV) is made. This value falls within a defined range known as the "system's operational range". Then, CAPV is compared to TARV. If CAPV is equal to TARC, it signifies system stability and no self-adaptation decision is taken. However, if CAPV differs from TARV, the system sends an adaptation request (gradual reduction of CAPV towards TARV) to the appropriate actuator.

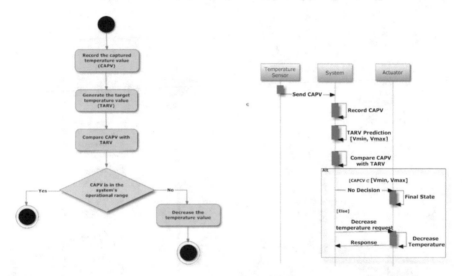

**Fig. 5.** Activity diagram of the "Adjust temperature" use case

**Fig. 6.** Sequence diagram of "Adjust temperature" use case

- Sequence diagram

    As mentioned above, we have three objects: the sensor, the system, and the actuator. Thus, the sequence diagram presented in Fig. 6 provides a detailed explanation of the temperature adjustment process by describing the sequential exchange of messages between these three objects.

    The main steps of this scenario are as follows:

- Step 1: The sensor sends CAPV to the system.
- Step 2: The system records CAPV in the storage base (SB).
- Step 3: The system proceeds to predict TARV based on CAPV. The prediction of TARV is done within the operational range of the system [Vmin, Vmax].
- Step 4: Comparison between CAPV and TARV. If CAPV is within the interval [Vmin, Vmax], then the system is stable (no decision needed). Otherwise, the system must make a decision to decrease CAPV towards TARV and send it to the actuator for execution.

### 4.7 Functioning of the Use Case: Adapt voltage

- Activity diagram

As depicted in Fig. 7, our system follows a sequence of operations. The first two steps of recording and prediction are identical to the previous case. There is only a difference regarding the predicted values and the prediction model used.

It is worth noting that each use case has its own prediction model. Therefore, we have CAPV and TARV. The target or predicted value (TARV) falls within a range defined as the "system's operational range". Then, we compare CAPV with TARV. If CAPV falls within the operational range of the system, then our system is stable. If not, the system requests adaptation by either *increasing* or *decreasing* CAPV towards TARV.

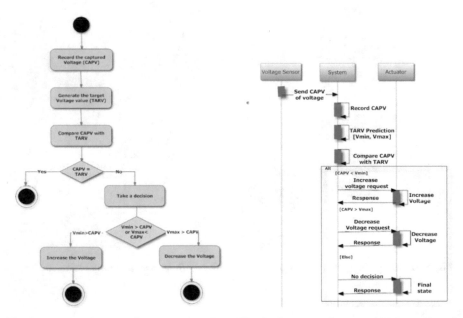

**Fig. 7.** Acivity diagram of "Adapt voltage" use case

**Fig. 8.** Sequence diagram of "Adapt voltage" use case

- Sequence diagram

The operation of this case is detailed in the diagram presented in Fig. 8. This diagram highlights the exchange of messages among the three objects mentioned in the previous subsections. The main steps are as follows:

- Step 1: The appropriate sensor sends CAPV to the system.
- Step 2: The system records CAPV in the storage base (SB).
- Step 3: The system predicts TARV, which should fall within the operational range of the system [Vmin, Vmax].
- Step 4: When comparing CAPV and TARV, if the value of CAPV is lower than Vmin, the system requests an increase in CAPV towards TARV. Similarly, if CAPV is higher

than Vmax, the system requests a decrease in the captured value towards TARV. In both scenarios, the system communicates its decision to the actuator for execution. If neither of these situations occurs, the system remains in its stable state.

### 4.8 Functioning of the Use Case: Regulate Current

The regulation of intensity works in a similar manner to the voltage adaptation, as described in subsections G. However, there is a difference in the operational ranges (intervals) of each aspect.

## 5 Results and Experimentation

The experiments are performed on a machine with an Intel(R) Core i5 7300 H CPU @ 2.60 GHz to 2.70 GHz. It is, also, equipped with 16GB main memory and running 64-bit Windows 10. In these experiments, we utilized the PyCharm development environment.

PyCharm (Islam 2015) is an integrated development environment (IDE) specifically designed for the Python programming language. It provides a wide range of features to facilitate Python development, including code completion, debugging, project management, version control, unit testing, and much more. PyCharm offers a user-friendly and intuitive interface, making it a popular choice among Python developers (Python 2021). We used a specific set of libraries, which are: tkinter, time, csv, random, numpy, tensorflow, pandas and matplotlib.

In our work, we propose a method based on neural networks to create a model for predicting the five aspects: energy consumption, energy production, temperature, intensity, and voltage. To achieve this, we used a dataset generated in Python to collect and organize data in the form of a CSV file (smart_grid_data.csv), based on real values. We assume that each aspect has its own sensor. These sensors are used to measure their values, which should fall within a certain range. For simulating the sensors, we utilize the "uniform" function from the random library. The generated values are added to a list, which is then saved into our dataset "smart_grid_data.csv".

### 5.1 Explanation of the Prediction Model

The model helps predict energy consumption using the LSTM model (Long Short-Term Memory). The latter is trained using the prepared dataset. The dataset is preprocessed and divided into two subsets: the first is reserved for training and the second for testing. The LSTM model is built with LSTM layers and a dense layer, and then it is trained on the training subset data. Predictions are made on the test subset data and are inversely scaled back to the original range. Real and random values are generated for demonstration. The real values are adjusted according to the values obtained from the prediction. If the predicted value is greater than the actual random value, then the predicted value is taken into consideration, otherwise the real value is retained.

## 5.2   Construction and Training of LSTM Model

A sequential model is created, and LSTM layers are added, with 50 units in each LSTM layer. A dense layer with one unit is added to produce the final output. The model is compiled using the *Adam optimizer* and the *mean_squared_error* loss function to minimize the mean squared error during training. The model is trained on the training data (*X_train*, *y_train*) with 10 epochs and a batch size of 32.

## 5.3   Test Data Preparation and Predictions

The test data are prepared similarly to the training data by creating input sequences (*X_test*) and output sequences (*y_test*). The input data *X_test* are reshaped to match the expected structure of the LSTM model. Predictions are made on the test data (*X_test*) using the trained model. The predictions are reversed using the *inverse_transform* method of the scaler to bring them back to the original range.

## 5.4   Adjustment of Real Values and Obtained Results

Random real values are generated using the *generate_random_values* function for demonstration purposes. These values are used for demonstration purposes. The real values are adjusted by comparing each prediction with the corresponding real value.

## 5.5   Some Illustrative Scenarios

The results obtained for each aspect are depicted below in the following figures: Figs. 9, 10, 11, 12, 13, 14, 15 and 16.

- *Scenario 1*

For the graph in Fig. 9, it represents the ranges of energy production variations. The blue curve represents the real production values, varying within the range of 10,000 to 15,000 watts. The orange curve represents the adjusted production values, varying within the range of 15,000 to 22,000 watts. For the graph illustrated in Fig. 10, it corresponds to the energy production over time at t + 5 s.

- *Scenario 2*

The graph in Fig. 11 illustrates the ranges of temperature variations. The blue curve pertains to the actual temperature values, ranging from 30 to 50 degrees. The orange curve represents the adjusted temperature values, ranging from 30 to 35°.
Moving on to the graph in Fig. 12, it corresponds to the results for the temperature aspect at the time instant t + 5 s.

- *Scenario 3*

The graph in Fig. 13 depicts the results for the intensity aspect. The blue curve represents the actual intensity values, ranging from 50 to 100 amperes. Meanwhile, the graph in Fig. 14 illustrates the intensity results at the time instant t + 5 s.

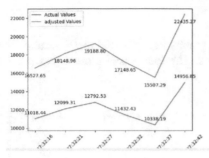

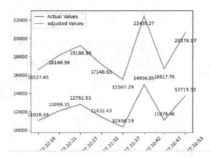

**Fig. 9.** Model result for energy production at time t

**Fig. 10.** Model result for energy production at time T = t + 5 s

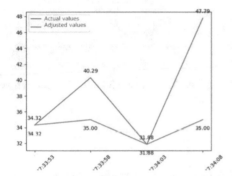

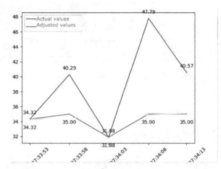

**Fig. 11.** Model result for temperature aspec at time t

**Fig. 12.** Model result for temperature aspec at time t + 5 s

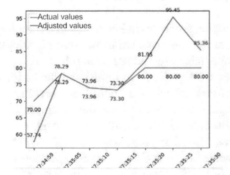

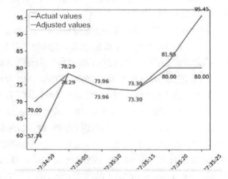

**Fig. 13.** Model result for intensity aspec at time t

**Fig. 14.** Model result for intensity aspec at time t + 5 s

- *Scenario 4*

In the same way as the other graphs, the graphs in Figs. 15 and 16 pertain to the current aspect at different time instants of 5 s. The blue curve represents the actual voltage values, ranging from 200 to 250 V.

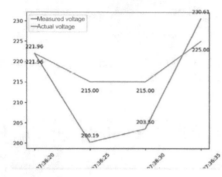 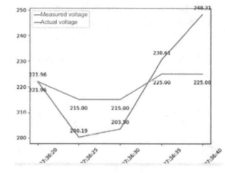

**Fig. 15.** Model result for current aspec at time t        **Fig. 16.** Model result for current aspec at time t + 5 s

# 6   Conclusion

The objective of this work is to propose a decision support method that enables the dynamic adaptation of a cyber-physical system to the instant needs and changes in its environment, without human intervention. This work aims to ensure that the system maintains its performance, stability, and optimal operation, even in the presence of unexpected changes.

In our work, we have taken into consideration five essential elements to ensure the proper functioning of a smart network. These five elements are as follows: electricity production and consumption, electrical voltage, intensity, and temperature.

We used mechanisms, methods, and learning algorithms to implement the various steps of the proposed method, namely: *feedback*, which is a control process. In our context, we used it for information acquisition, evaluation, and reacting to continuous changes that our system may encounter; *neural networks*, which allowed us to adjust the parameters of our system so that it generates a desired response based on a given input; and *reconfiguration mechanisms*, which enabled us to exploit adaptation possibilities to ensure and maintain an appropriate level of service quality, even in the presence of disruptions.

The results obtained confirm both the validity and feasibility of our method. Furthermore, these results demonstrate that the use of our method enables efficient automation of system management and decision-making in response to unforeseen changes. As future work, we plan to continue our exploration of the operation of various components of our architecture, particularly the decision-making component. Furthermore, we intend to test the proposed method using real datasets.

# References

Aladdin, S., El-Tantawy, S., Fouda, M.M., Eldien, A.S.T.: MARLA-SG: Multi-agent reinforcement learning algorithm for efficient demand response in smart grid. IEEE access **8**,
210626–210639 (2020)

El Ballouli, R.: Modeling self-configuration in Architecture-based self-adaptive systems (2019)

Guérard, G.: Optimisation de la diffusion de l'énergie dans les smarts-grids (2014)

Hocine, D.R.T.: Production de l'Energie Electrique. Univ Hassiba Benbouali Chlef Fac Technol
Département d'Electrotechnique 2020 (2019)

Houssin, M., Combettes, S., Gleizes, M-P,, Lartigue, B.: SANDMAN: un système auto-adaptatif
pour la détection d'anomalies dans le flux des données des bâtiments intelligents. In: Rencontres
des Jeunes Chercheur· ses en Intelligence Artificielle (RJCIA 2020@ PFIA) (2020)

Islam, Q.N.: Mastering PyCharm. Packt Publishing Ltd. (2015)

Kamel, L.: Gestion d'énergie dans un réseau intégrant des systèmes à source renouvelable (2016)

Macías-Escrivá, F.D., Haber, R., Del Toro, R., Hernandez, V.: Self-adaptive systems: A survey
of current approaches, research challenges and applications. Expert Syst. Appl. **40**(18), 7267–
7279 (2013)

Mazur, G.A.: Digital multimeter principles. Cengage Learning (2006)

Moghaddam, M.T., Rutten, E., Giraud, G.: Hierarchical control for self-adaptive iot systems
a constraint programming-based adaptation approach. In: HICSS 2022-Hawaii International
Conference on System Sciences, pp 1–10 (2022)

Mu, C., Zhang, Y., Jia, H., He, H.: Energy-storage-based intelligent frequency control of microgrid
with stochastic model uncertainties. IEEE Trans Smart Grid **11**(2), 1748–1758 (2019)

Muccini, H., Sharaf, M., Weyns, D.: Self-adaptation for cyber-physical systems: a systematic
literature review. In: Proceedings of the 11th International Symposium on Software Engineering
for Adaptive and Self-managing Systems, pp 75–81 (2016)

Paquette, G.: Feedback, rétroaction, rétroinformation, réponse... du pareil au même. Commun.
Langages **73**(1), 5–18 (1987)

Python, W.: Python. Python Releases Wind 24 (2021)

Radhakrishnan, B.M., Srinivasan, D., Mehta, R.: Fuzzy-based multi-agent system for distributed
energy management in smart grids. Int J Uncertainty, Fuzz. Knowl.-Based Syst. **24**(05), 781–
803 (2016)

Sang, J., Sun, H., Kou, L.: Deep reinforcement learning microgrid optimization strategy
considering priority flexible demand side. Sensors **22**(6), 2256 (2022)

Torres-Moreno, J-M.: Apprentissage et généralisation par des réseaux de neurones: étude de
nouveaux algorithmes constructifs (1997)

# An Overview of Flying Ad-Hoc Networks

Belghachi Mohammed[✉]

Computer Science Department, Faculty of Science Exact, University Tahri Mohamed of Bechar, Bechar, Algeria
Belghachi.mohamed@univ-bechar.dz

**ABSTRACT.** This survey focuses on exploring the concept of Flying Ad-Hoc Networks (FANETs) and their potential impact on various industries. FANETs involve the deployment of a network of unmanned aerial vehicles (UAVs) or drones that can communicate, coordinate, and exchange data in a dynamic and decentralized manner. This survey aims to gather insights and opinions regarding FANETs, including their network architecture, communication protocols, applications, challenges, and ongoing research efforts. The survey participants' responses will contribute to a better understanding of FANETS and their implications for the future. The survey aims to gather valuable insights and opinions regarding the usage, challenges, and potential benefits of FANETs. By analyzing the survey responses, this research seeks to provide a thorough understanding of FANETs, including their network architecture, communication protocols, applications, and ongoing research efforts. The findings of this survey will contribute to the existing knowledge base and shed light on the current state and future prospects of FANETs.

**Keywords:** Flying Ad-Hoc Networks · unmanned aerial vehicles · network architecture · communication protocols · applications · challenges · future prospects

## 1 Introduction

Flying Ad-Hoc Networks (FANETs) are wireless communication networks formed by a group of unmanned aerial vehicles (UAVs) or drones. FANETs enable communication, coordination, and collaboration among drones, allowing them to exchange information and perform cooperative tasks. Unlike traditional ad-hoc networks, FANETs involve dynamic and mobile nodes, which are the drones. These drones act as both communication devices and network nodes. They establish wireless links with each other to form a self-configuring and self-organizing network without relying on a fixed infrastructure or centralized control. FANETs have the potential to revolutionize various industries by enabling applications such as surveillance and monitoring, disaster management, delivery services, environmental monitoring, and more. They can enhance situational awareness, increase operational efficiency, and provide remote sensing capabilities in challenging or inaccessible environments. To enable communication in FANETs, drones

M. Hatti (Ed.): IC-AIRES 2023, LNNS 984, pp. 290–303, 2024.
https://doi.org/10.1007/978-3-031-60629-8_29

utilize wireless communication technologies such as Wi-Fi, Bluetooth, or specialized protocols. They exchange data, control commands, and other information required for collaborative tasks. FANETs employ specialized routing protocols that are designed to adapt to the dynamic network topology, account for limited bandwidth and resources, and optimize communication efficiency among the drones. The importance and benefits of FANETs will continue to expand as the technology matures, regulations evolve, and new applications are discovered. However, it's essential to address challenges such as safety, privacy, and airspace regulations to ensure responsible and secure integration of FANETs into various industries. FANETs are an emerging technology that involves the deployment of a network of unmanned aerial vehicles (UAVs) or drones for communication, coordination, and data exchange (Fig. 1) [1, 2].

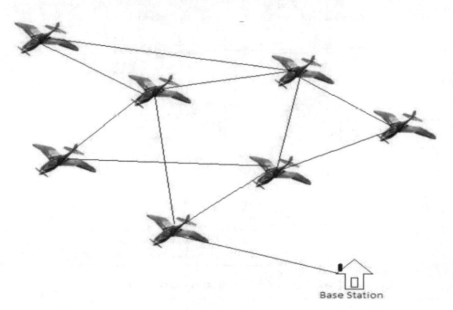

**Fig.1.** Flying Ad-Hoc Networks

These networks hold great potential in various industries, ranging from transportation and surveillance to agriculture and disaster management. The purpose of this survey is to gather insights and opinions regarding FANETs, their applications, challenges, and potential benefits. Your participation will contribute to a better understanding of this technology and its implications for the future. The survey consists of a series of questions that cover different aspects of FANETs, including their network architecture, communication protocols, applications, success stories, challenges, and ongoing research and development efforts. Your responses will be anonymous and used for research purposes only.

## 2  Fanets Architecture

The network architecture of Flying Ad-Hoc Networks (FANETs) involves the interaction and coordination of multiple unmanned aerial vehicles (UAVs) or drones to form a self-configuring and self-organizing wireless network [3, 4]. Here is an explanation of the key components and their roles within the FANET architecture (Fig. 2):

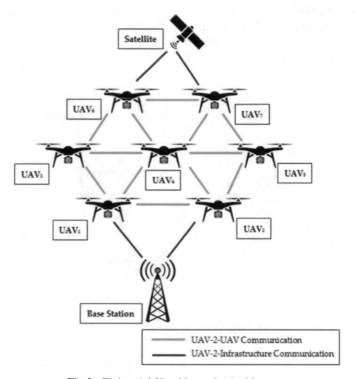

**Fig.2.** Flying Ad-Hoc Networks Architecture

- Drones

   The drones are the main nodes in the FANET network. Each drone is equipped with communication capabilities and functions as both a communication device and a network node. Drones are typically equipped with wireless communication technologies such as Wi-Fi, Bluetooth, or specialized protocols to establish connections with other drones in the network. They can gather and process data, execute tasks, and exchange information with other drones for collaborative operations.

- Ground Stations

   Ground stations serve as base stations or access points for the FANET network. They are fixed installations located on the ground that provide connectivity to the drones. Ground stations serve as gateways for drones to connect to other networks or external systems, facilitating data exchange beyond the FANET network. They can also serve as command and control centers for monitoring and managing the operations of the drones.

• Communication Protocols

FANETs employ communication protocols specifically designed to meet the unique requirements of dynamic aerial networks. These protocols enable the establishment and maintenance of wireless links between drones, facilitating data exchange and coordination among them. The protocols used in FANETs address challenges such as node mobility, intermittent connectivity, bandwidth constraints, and efficient routing in dynamic network topologies.

• Network Formation and Management

FANETs are self-configuring and self-organizing networks. The drones autonomously form and maintain the network without relying on a fixed infrastructure or centralized control. Network formation algorithms enable drones to discover and establish connections with nearby drones, thereby creating a dynamic network topology. The network management mechanisms include monitoring the connectivity status, adapting to changes in the network topology, and optimizing network performance.

• Routing and Data Exchange

Routing protocols in FANETs determine the paths and forwarding decisions for data transmission among drones. These protocols consider factors such as changes in network topology, energy efficiency, quality of service requirements, and avoidance of communication interference. Drones exchange data, control commands, and other information necessary for collaborative tasks, facilitating cooperative operations and distributed decision-making.

The network architecture of FANETs enables drones to communicate, coordinates, and collaborates effectively, facilitating various applications such as surveillance, disaster management, delivery services, and more. The autonomous nature of FANETs allows for flexible and adaptive networking in dynamic and challenging environments.

# 3   Applications of Fanets

FANETs have a wide range of applications across various industries. Here is an exploration of different applications of FANETs (Fig. 3) [5, 6]:

• Surveillance and Monitoring

FANETs can be used for aerial surveillance and monitoring in areas that are difficult to access using ground-based systems. They enable real-time monitoring of large areas, enhancing situational awareness for security, law enforcement, and border control. FANETs provide valuable data for the surveillance of critical infrastructure, crowd monitoring at events, and wildlife tracking.

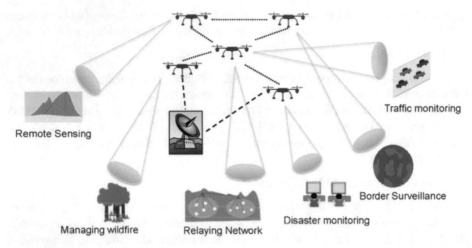

Remote Sensing

Traffic monitoring

Managing wildfire     Relaying Network

Disaster monitoring

Border Surveillance

**Fig.3.** Flying Ad-Hoc Networks Application

- Disaster Management and Search and Rescue

FANETs assist in disaster management by providing aerial assessments of disaster-stricken areas, helping to identify hazards and plan rescue operations. They facilitate search and rescue missions by quickly covering large areas, providing live video feeds, and assisting in locating missing persons or survivors. FANETs enhance coordination and communication among emergency response teams, thereby improving efficiency and response times.

- Delivery Services

FANETs can revolutionize delivery services by enabling autonomous package delivery using drones. They overcome traffic congestion and infrastructure limitations, enabling faster and more efficient delivery in both urban and remote areas. FANETs provide a cost-effective and environmentally friendly alternative for last-mile delivery and logistics.

- Communication and Connectivity

FANETs play a crucial role in establishing communication links in remote or disaster-stricken areas where traditional infrastructure is damaged or nonexistent. They provide temporary connectivity for emergency communication, enabling data transfer, voice communication, and coordination among relief teams. FANETs contribute to extending wireless coverage and bridging the digital divide in underserved or remote region.

- Agriculture and Environmental Monitoring

FANETs offer valuable applications in precision agriculture, as they can monitor crop health, irrigation needs, and pest control. They enable efficient and targeted use of resources, leading to optimized farming practices and increased crop yields. FANETs assist in environmental monitoring by collecting data on pollution, deforestation, wildlife habitats, and climate patterns, thereby aiding in conservation efforts.

● Infrastructure Inspection and Maintenance

FANETs enable the efficient inspection of critical infrastructure, such as power lines, pipelines, and bridges. They reduce the need for manual inspections, improve safety, and detect potential issues promptly. FANETs provide high-resolution imagery, video feeds, and sensor data to assess the structural integrity of infrastructure.

● Entertainment and Events

FANETs add a new dimension to entertainment events by incorporating synchronized drone displays and aerial light shows. They create captivating visual experiences, choreographed movements, and stunning aerial formations. FANETs offer unique perspectives for photography, videography, and live broadcasting, enhancing media coverage of events.

These are just a few examples of the diverse applications of FANETs. As the technology continues to advance, new applications and use cases are continually being explored and developed, expanding the possibilities and impact of FANETs across various industries.

## 4 Challenges and Limitations

FANETs face several challenges that need to be addressed for their successful deployment and operation. Here are some key challenges faced by FANETs [7, 8]:

● Communication and Connectivity

Limited Bandwidth: FANETs operate in a shared wireless medium with limited bandwidth, which can result in congestion and decreased communication performance. Interference and Signal Loss: Drones flying in close proximity can cause interference and signal loss, which can negatively impact communication reliability. Dynamic Network Topology: The mobility of drones leads to frequent changes in the network topology, necessitating efficient routing protocols and adaptive communication mechanisms.

● Mobility and Navigation

Collision Avoidance: Drones must navigate and avoid collisions with each other, as well as with static obstacles such as buildings or trees. Autonomous Navigation: Developing reliable and robust algorithms for autonomous navigation in complex environments is a challenge that ensures safe and efficient drone movements. Battery Life and Range: Drones have limited battery life and operational range, which requires careful energy management and planning for extended missions.

● Safety and Regulations

Airspace Regulations: FANETs need to comply with aviation regulations and airspace restrictions to ensure safe integration with other aircraft and avoid conflicts. Collision Risk with Manned Aircraft: The coexistence with manned aircraft presents challenges in terms of collision avoidance, airspace coordination, and maintaining appropriate separation distances. Safety and Emergency Procedures: FANETs require robust safety measures, including fail-safe mechanisms, emergency landing procedures, and protocols for handling system failures.

- Scalability and Network Management

Scalability: As the number of drones in a FANET increases, the network management and coordination become more complex, necessitating the use of efficient protocols to handle large-scale deployments. Resource Allocation: Optimizing resource allocation, such as bandwidth and energy, becomes challenging in a dynamic and distributed network environment. Network Security: FANETs require robust security mechanisms to safeguard against unauthorized access, cyber-attacks, and the potential hijacking of drones.

- Data Processing and Management

Data Fusion and Processing: Efficiently processing and fusing data collected by multiple drones is a challenging task that requires real-time data analysis, decision-making, and information sharing. Storage and Bandwidth Constraints: Limited storage capacity and bandwidth in drones impose constraints on data transmission and management. This necessitates the use of intelligent data compression and prioritization techniques.

Addressing these challenges requires ongoing research and development efforts, collaboration between industry and regulatory bodies, and the formulation of standards and best practices for the safe and efficient operation of FANETs. Overcoming these challenges will pave the way for the widespread adoption and realization of the full potential of FANETs in various industries.

## 5   Fanet Routing Protocols

Routing protocols specifically designed for FANETs address the unique characteristics and challenges of dynamic aerial networks. These protocols aim to provide efficient and reliable communication among drones in the network. Here's an overview of some routing protocols specifically designed for FANETs [9, 10]:

- Flying Ad-Hoc Network Vector (FANV)

FANV is a routing protocol specifically designed for FANETs. It takes into account the mobility of drones, considering factors such as speed, altitude, and direction. FANV aims to optimize routing paths based on the current and predicted positions of drones, thereby improving efficiency and adaptability in dynamic aerial networks.

- Proactive Cooperative Opportunistic Routing (PCOR)

PCOR is a proactive routing protocol developed for FANETs. It uses proactive routing tables to establish routes in advance. PCOR incorporates cooperative opportunistic routing, where drones work together to transmit packets in order to enhance network performance, throughput, and reliability. It dynamically adjusts routing decisions based on the available opportunities for cooperative packet forwarding.

- Airborne Internet Routing Protocol (AIRP)

AIRP is a proactive routing protocol specifically designed for aerial networks, including FANETs. It establishes and maintains routing tables in advance on each drone. AIRP considers the three-dimensional nature of the airspace, taking into account altitude and flight levels when making routing decisions. It focuses on minimizing the end-to-end delay and maintaining stable routing paths in aerial networks.

- Vector-Based Routing Protocol (VBRP)

VBRP is a reactive routing protocol developed for FANETs. It utilizes the concept of vectors to represent the direction and speed of drone's movements. VBRP establishes routes on demand when communication is required between drones. It adapts the routing decisions based on the changing positions and velocities of the drones.

- Zone-Based Hierarchical Link State (ZHLS)

ZHLS is a hierarchical routing protocol designed for FANETs. It divides the airspace into zones and assigns a zone manager to each zone. ZHLS uses hierarchical link-state information to establish and maintain routes. It enables efficient routing and scalability in large-scale FANETs by reducing control overhead and optimizing route calculation.

These routing protocols for FANETs focus on the unique challenges posed by dynamic aerial networks, such as mobility, three-dimensional routing, cooperation, and efficient resource utilization. They aim to enhance the performance, reliability, and adaptability of communication in FANETs, enabling efficient data exchange and coordination among drones. Ongoing research continues to explore and refine these protocols, considering the evolving needs and advancements in FANET technology.

## 6 Regulatory Considerations

The regulatory framework surrounding FANETs is an important aspect to ensure safe and responsible operation of these aerial networks. As FANETs involve unmanned aircraft operating in shared airspace, they must comply with aviation regulations and guidelines to mitigate risks and ensure the safety of both the drones and other airspace users. Here's an examination of the regulatory framework surrounding FANETs [11, 12]:

- Aviation Authorities and Regulations

Civil Aviation Authorities (CAAs): Each country typically has a CAA responsible for regulating civil aviation activities, including FANET operations. Existing Regulations: FANETs are subject to existing aviation regulations that govern unmanned aircraft operations, such as flight rules, airspace restrictions, and pilot certification requirements. Unmanned Aircraft Systems (UAS) Regulations: Many countries have specific regulations for UAS, which may include requirements for registration, pilot licensing, flight restrictions, and operational limitations.

- Airspace Integration and Separation

Air Traffic Management (ATM): FANETs need to integrate into the existing ATM systems to ensure the safe separation from manned aircraft and coordination of airspace usage. Detect and Avoid (DAA) Systems: FANETs may be required to have DAA systems that enable drones to detect and avoid other aircraft, either through onboard sensors or cooperative communication.

- Flight Authorization and Permissions

Flight Permissions: FANET operators may be required to obtain flight authorizations or permissions from aviation authorities or designated agencies before conducting operations, especially in controlled airspace or sensitive areas. Geofencing and No-Fly Zones: Regulatory frameworks may establish geofencing and no-fly zones to restrict drone operations near sensitive locations, such as airports, military installations, or crowded areas.

- Privacy and Data Protection

Data Privacy: FANET operators may need to comply with data protection and privacy regulations, especially when capturing images or collecting personal data during operations. Surveillance Regulations: In certain jurisdictions, there are specific regulations in place to govern the use of drones for surveillance purposes. These regulations ensure compliance with privacy laws and restrictions on gathering certain types of information.

- Certification and Training

Certification Requirements: Pilots operating FANETs may be required to obtain specific certifications or licenses, depending on the weight and capabilities of the drones used. Training and Education: Aviation authorities may establish training requirements for FANET operators and pilots to ensure they have the necessary knowledge and skills to operate drones safely and responsibly.

- Future Regulatory Developments

Evolving Regulations: Regulatory bodies are continuously evaluating and updating their regulations to keep pace with advancements in drone technology, including Free Flight Autonomous Network (FANET) systems. Remote Identification and Tracking: Many countries are working on regulations for the remote identification and tracking of drones to enhance accountability and security in the airspace.

It's important for operators and users of FANETs to stay informed about the specific regulations and guidelines in their respective jurisdictions. Compliance with the regulatory framework is essential to ensure the safe integration of FANETs into the airspace and to maintain public trust in the technology.

## 7   Research and Development

Research and development (R&D) efforts for Flying Ad-Hoc Networks (FANETs) are crucial for advancing the technology, addressing challenges, and exploring new opportunities. Here are some key areas of research and development in FANETs:

## A. FANETs and Networking

The field of FANETs continues to see active research and development efforts aimed at addressing various challenges and exploring new opportunities. Here's an overview of some of the latest research and development efforts in the field of FANETs [13, 14]:

### 1. *Communication and Networking:*

- 5G and Beyond: Researchers are exploring the integration of FANETs with 5G and future wireless communication networks to enhance connectivity, improve data transmission rates, and support a larger number of connected drones.
- Software-Defined Networking (SDN): SDN principles are being investigated for FANETs, enabling centralized control, dynamic network reconfiguration, and efficient resource allocation.
- Machine Learning for Communication: Machine learning techniques are being applied to optimize communication protocols in FANETs, including adaptive routing, interference management, and spectrum allocation.

### 2. *Navigation and Collision Avoidance:*

- Sensor Fusion and Computer Vision: Researchers are developing advanced sensor fusion techniques that combine data from various sensors (e.g., cameras, LiDAR, radar) to improve navigation, obstacle detection, and collision avoidance capabilities of drones in FANETs.
- Cooperative Collision Avoidance: Ongoing research focuses on cooperative algorithms and communication protocols that enable drones to exchange information and collaborate to avoid collisions, enhancing safety in FANETs.
- Swarm Intelligence: Researchers are investigating swarm intelligence algorithms inspired by natural systems to enable efficient coordination, obstacle avoidance, and distributed decision-making among a large number of drones.

### 3. *Energy Efficiency and Sustainability:*

- Energy Harvesting: Research efforts are exploring energy harvesting techniques, such as solar power and kinetic energy harvesting, to extend the flight time and range of drones in FANETs.
- Energy-Aware Routing and Resource Management: Optimization algorithms and protocols are being developed to minimize energy consumption, optimize routing paths, and manage resources efficiently in FANETs, considering the limited onboard power resources of drones.
- Sustainable Drone Design: Researchers are focusing on developing lightweight and energy-efficient drone designs, utilizing eco-friendly materials and propulsion systems to reduce the environmental impact of FANETs.

### 4. *Autonomous Operations and Swarm Behavior:*

- Autonomous Navigation: Ongoing research aims to improve the autonomy of drones in FANETs, including advanced path planning algorithms, obstacle detection and avoidance, and robust decision-making capabilities.

- Swarm Behavior and Coordination: Researchers are studying and developing algorithms inspired by swarm behavior to enable drones in FANETs to exhibit cooperative behavior, self-organization, and task allocation, enhancing their collective capabilities.

5. *Applications and Use Cases:*

- Emergency Response and Disaster Management: Research is focused on developing FANETs for efficient and rapid response in emergency situations, including disaster assessment, search and rescue operations, and delivery of medical supplies.
- Environmental Monitoring: FANETs are being explored for environmental monitoring applications, such as air quality measurement, wildlife tracking, and ecosystem analysis, leveraging the mobility and aerial capabilities of drones.
- Urban Air Mobility (UAM): Research efforts are underway to integrate FANETs with emerging UAM systems, addressing challenges related to air traffic management, infrastructure integration, and safe operation in urban environments.

These research and development efforts in FANETs are driving innovation, addressing challenges, and expanding the potential applications of this technology. Collaborative efforts between academia, industry, and regulatory bodies are crucial to advancing the field and realizing the full potential of FANETs in various domains.

## B. FANETs and Emerging Technology

Recent advancements, novel solutions, and emerging technologies are shaping the future of Flying Ad-Hoc Networks (FANETs), opening up new possibilities and addressing existing challenges. Here are some key developments that could shape the future of FANETs [15, 16]:

- 6G and Beyond

As 5G networks continue to evolve, research and development efforts are already underway for the next generation of wireless communication, including 6G. The integration of FANETs with future wireless networks will bring enhanced connectivity, higher data rates, ultra-low latency, and massive device connectivity, enabling new applications and services in FANETs.

- Artificial Intelligence (AI) and Machine Learning (ML)

AI and ML techniques are being increasingly applied to FANETs, enabling autonomous decision-making, adaptive routing, intelligent resource management, and collaborative behavior among drones. ML algorithms can learn from real-time data to optimize network performance, predict drone behavior, and enhance overall efficiency and reliability in FANETs.

- Blockchain and Distributed Ledger Technology

Blockchain technology holds potential for enhancing the security, privacy, and trustworthiness of FANETs. It can enable secure and tamper-proof data exchange, decentralized identification and authentication, and transparent transaction tracking. Blockchain-based solutions may facilitate secure communication, resource sharing, and coordination among drones in FANETs.

- Edge Computing and Fog Computing

Edge computing and fog computing technologies are gaining attention in FANETs to address the limitations of onboard processing capabilities and limited bandwidth. These paradigms enable offloading computation and data storage to edge devices, ground stations, or nearby cloud resources, reducing latency and enabling real-time data analysis and decision-making in FANETs.

- Swarming and Collective Intelligence

Swarming techniques inspired by collective behavior in nature are being explored for FANETs. Swarms of drones can exhibit collaborative behavior, self-organization, and distributed decision-making, enabling efficient task allocation, adaptive routing, and cooperative data gathering. Swarming in FANETs holds potential for applications like surveillance, disaster response, and environmental monitoring.

- Advanced Sensing and Perception

Advancements in sensing technologies, such as improved cameras, LiDAR, radar, and other onboard sensors, enhance the perception capabilities of drones in FANETs. Higher-resolution imaging, 3D mapping, and improved object detection and tracking enable better situational awareness, obstacle avoidance, and environmental monitoring in FANETs.

- Hybrid Aerial Platforms

Hybrid aerial platforms, combining the capabilities of fixed-wing and multirotor drones, are being developed for FANETs. These platforms offer longer flight endurance, increased payload capacity, and the ability to transition between vertical and horizontal flight, enabling a wider range of applications and expanding the operational capabilities of FANETs.

- Improved Battery Technologies and Energy Harvesting

Ongoing research focuses on improving battery technologies to enhance the flight time and endurance of drones in FANETs. Additionally, energy harvesting techniques, such as solar power, kinetic energy harvesting, or wireless charging, hold promise for extending the range and autonomy of drones, reducing the reliance on limited onboard power sources.

These recent advancements, novel solutions, and emerging technologies demonstrate the dynamic nature of FANETs and their potential to revolutionize various industries and applications. Continued research, collaboration, and innovation in these areas will shape the future of FANETs, making them more efficient, reliable, and versatile in addressing societal and industrial needs.

The research and development efforts in FANETs involve collaboration between academia, industry, and regulatory bodies to address technical challenges, improve performance, and unlock the potential of this technology in various domains. Continuous innovation and advancements in these areas will shape the future of FANETs, enabling safer, more efficient, and versatile operations in the airspace.

# 8 Conclusion

FANETs have the potential to make a significant impact and offer promising future prospects in various domains. The future prospects of FANETs are promising, but certain challenges need to be addressed, such as regulatory frameworks, airspace integration, safety, and privacy concerns. As technology advances and research and development efforts continue, FANETs have the potential to transform industries, improve efficiency, enhance connectivity, and enable new services and applications. Continued collaboration between academia, industry, and regulatory bodies will be key to realizing the full potential of FANETs and ensuring their safe and responsible integration into our daily lives. Overall, the survey results emphasize the importance of continued exploration, innovation, and collaboration to fully unlock the potential of FANETs. By addressing the challenges, leveraging emerging technologies, and promoting responsible deployment, FANETs can shape the future of various industries, revolutionize existing practices, and create new opportunities. Once again, we appreciate your participation and contribution to this survey. Your insights have been valuable in advancing our understanding of FANETs.

# References

1. Al-Hourani, A., Kandeepan, S., Jamalipour, A.: Modeling air-to-ground path loss for low altitude platforms in urban environments. IEEE Trans. Wireless Commun. **13**(2), 527–537 (2014)
2. Scholten, H., Heijenk, G.: Performance analysis of self-organizing TDMA for aerial communication networks. In: Proceedings of the 2011 IEEE GLOBECOM Workshops (pp. 57–61). IEEE (2011)
3. Boubendir, Y., Laurent-Maknavicius, S., Leconte, M.: A comprehensive survey on unmanned aerial vehicle communication channels. IEEE Commun. Surv. Tutorials **20**(4), 2924–2946 (2018)
4. Mozaffari, M., Saad, W., Bennis, M., Debbah, M.: A tutorial on UAVs for wireless networks: applications, challenges, and open problems. IEEE Commun. Surv. Tutorials **21**(3), 2334–2360 (2019)
5. Singh, D., Panda, S.K.: Flying ad hoc networks: challenges, applications, and future directions. Ad Hoc Netw. **102**, 101999 (2020)
6. Osseiran, A., et al.: Scenarios for 5G mobile and wireless communications: the vision of the METIS project. IEEE Commun. Mag. **52**(5), 26–35 (2014)
7. Vuran, M.C., Akyildiz, I.F.: Flying ad-hoc networks (FANETs): a survey. Ad Hoc Netw. **8**(5), 427–444 (2010)
8. Lohan, E.S., Vinel, A.: UAS as flying base stations: applications, challenges, and approaches. IEEE Commun. Surv. Tutorials **19**(4), 2500–2523 (2017)
9. Iasmin, M., de Albuquerque, C.V.N.: Survey on communication protocols for unmanned aerial vehicles. IEEE Commun. Surv. Tutorials **19**(2), 1123–1152 (2017)
10. De Oliveira, R.C.C., de Carvalho, T.C.M.B., Gomes, R.L., de Albuquerque, C.V.N., Loureiro, A.A.F.: Routing protocols for unmanned aerial vehicles: a survey. J. Netw. Comput. Appl. **168**, 102784 (2020)
11. Khalid, F., Raza, M.A., Tahir, F.A., Khan, I.A.: Swarm intelligence in UAV-assisted wireless networks: challenges, solutions, and future directions. IEEE Wirel. Commun. **28**(1), 136–142 (2021)

12. Saad, W., Bennis, M., Chen, M., Vasilakos, A.V.: A vision of 6G wireless systems: applications, trends, technologies, and open research problems. IEEE Network **34**(3), 134–142 (2019)
13. Li, Z., Zhang, J., Chen, M.: A survey on unmanned aerial vehicle communication networks: communication-theoretic view. IEEE Commun. Surv. Tutorials **22**(1), 700–728 (2020)
14. Saeed, A., Javaid, N., Ahmed, M., Alrajeh, N.: UAV-assisted data collection in wireless sensor networks: a survey. IEEE Commun. Surv. Tutorials **22**(4), 2259–2291 (2020)
15. Haider, S., Pandey, S., Matolak, D.W.: A survey of air-to-ground propagation channel modeling for unmanned aerial vehicles. IEEE Commun. Surv. Tutorials **19**(1), 245–262 (2017)
16. Liu, B., Yang, P., Wang, H., Huang, J., Chen, Y.: UAV communication networks: opportunities and challenges. IEEE Commun. Mag.Commun. Mag. **59**(2), 50–56 (2021)

# Materials in Renewable Energetic Systems

# Interconnections Among Renewable Energy Use and Financial Progress: Empirical Evidence from 12 Arab Countries

Salaheddine Sari-Hassoun[1](✉), Mohammed Seghir Guellil[2], and Samir Ghouali[3]

[1] LEPPESE Laboratory, University Centre of Maghnia, Maghnia, Algeria
salaheddine.sarihassoun@cumaghnia.dz, salah.poldeva08@gmail.com
[2] MCLDL Laboratory, University Mustapha Stambouli of Mascara, Mascara, Algeria
m.guellil@univ-mascara.dz
[3] Department of Electrotechnics, Faculty of Sciences and Technology, Mustapha Stambouli University of Mascara, Mascara, Algeria
s.ghouali@univ-mascara.dz

**Abstract.** In order to achieve the objectives of sustainable development, energy transition, and energy security, renewable energy serves as a crucial pillar. In this study, we examine the relationship between renewable energy consumption and the financial development index for 12 Arab countries from 1990 to 2021. We apply the cross-sectional dependence, panel unit root and panel cointegration tests on the variables of renewable energy consumption per capita, and domestic credit to the private sector per capita. We establish with panel model fixed effect that there is a significant and positive relationship at the level of 1% between renewable energy consumption and the financial development index. According to the Dumitrescu-Hurlin panel causality test, there is a unidirectional causal relationship running from financial development index to renewable energy consumption.

**Keywords:** Renewable Energy · Sustainable Development · Financial Development Index

## 1 Introduction

Any country relies heavily on energy, as evidenced by the fact that more than 80% of it comes from fossil fuels. (Hassoun & Mekidiche, 2019). Since the ground-breaking study by (Kraft & Kraft, 1978), which sparked attention to the link between energy use and GDP growth, energy has certainly become a key component of economic success. Rapid urbanization, technological advancement, climate change, and GDP growth have all led to a greater reliance on fossil fuels (Pata, 2018; Gao & Zhang, 2021). According to Suki et al. (2022) research, the importance of sustaining and effectively governing the energy sector is emphasized as a critical prerequisite for achieving long-term and sustainable GDP growth. With sufficient energy supply, a nation can profit from higher productivity as well as other economic advantages like greater competitiveness, unmet needs, and new value creation (Abbass et al., 2022).

© The Author(s), under exclusive license to Springer Nature Switzerland AG 2024
M. Hatti (Ed.): IC-AIRES 2023, LNNS 984, pp. 307–312, 2024.
https://doi.org/10.1007/978-3-031-60629-8_30

According to Kihombo et al. (2022), FD has a significant impact on energy usage and economic expansion. Shahbaz et al. (2017) said that FD is correlated with cutting-edge technologies that reduce energy consumption, which in turn depends on GDP growth (Danish & Ulucak, 2021; Komal & Abbas, 2015). Boufateh & Saadaoui (2020) observed that reform, financial openness, structural modifications, financial crises, energy prices, and inflation affect both the financial and energy sectors. There have been many studies on the relationship between FD and energy consumption in developing nations (Nkalu et al., 2020; Ma & Fu, 2020; Destek, 2018; Sadi Ali et al., 2015; Chang, 2014). Danish & Ulucak (2021) established that a more efficient financial system could lend money to individuals, businesses, and governments to buy products that consume more energy.

However, very few studies, particularly in developing and Arab nations, have examined the relationship between FD and REC. In this regard, we are going to focus on the connection between REC and FD in 12 Arab countries. (Algeria, Egypt, Libya, Morocco, Tunisia, Iraq, Kuwait, Saudi Arabia, United Arab Emirates, Lebanon, Yemen, and Lebanon). Therefore, this paper will attempt to address the nature of the association between REC and FD in 12 Arab countries during 1990–2021.

The study may contribute to the literature on energy in two ways, taking into account the groups of chosen countries. Therefore, (i) only a few researches have looked into the relationship between FD and REC in previous investigations. On the other hand, (ii) Other than a small number of studies, there is no research in Arab nations looking at how REC affects FD or vice versa.

## 2  Literature Review

Studies on the relationship between REC and FD have been scarce despite the small body of literature that exists because of how quickly technology and the economy are evolving. It follows that research in the areas of energy and finance is crucial to achieving sustainable development. A few publications looked into how FD affected REC. There are some studies that established a positive relationship between REC and FD, Eren et al. (2019) in India, Somoye et al. (2022) in Nigeria. Anton & Afloarei Nucu (2020) examined the association between the demand for RE and the level of FD in 28 European Union member nations over the years 1990–2015 using a panel fixed effects model. According to the research study's findings, REC benefited from the growth of the bond market, banking industry, and capital markets. Zeren & Hizarci (2021) discovered a long-run positive cointegration between FD and biomass energy consumption (REC). Kevser et al. (2022) investigated the relationship between GDP growth, FD, and biomass energy consumption for 15 African countries during the period 1993–2017. They established with DH causality that there is a positive and bidirectional relationship between GDP growth and biomass energy consumption as well as between FD, and biomass energy consumption. The impacts of biomass energy on each country's GDP growth and FD may vary depending on how each country approaches renewable energy sources. Furthermore, because financial markets have different dynamics and conditions in each country, the effects of biomass energy may change by nation.

Peng et al. (2022) used the panel cointegration method to examine the impacts of environmental technologies, environmental regulations, economic growth, and financial development on REC in 29 OECD nations from 1996 to 2018. According to the analysis's results, FD, economic growth, and environmental technology have a positive impact on REC. Fang et al. (2022) employed panel data techniques to analyses the impacts of economic growth, FD, and green innovation on REC in 15 countries covered by the Belt and Road Initiative between 1998 and 2019. According to the study's findings, REC is positively impacted by economic growth, FD, and green innovation.

## 3  Data, Model and Methodology

### 3.1  Data

In this paper, we used annual statistics data for REC and FD collected from 12 Arab nations for the years 1990 to 2021 are used to generate a balanced panel.

Table 1 shows that FD is represented by domestic credit to the private sector per capita (constant 2015 US$), while the data for REC was obtained from the British Petroleum database portal and evaluated. The models that follow were created within the framework of Destek (2018) and Zeren & Hizarci (2021) studies that looked at the connection among REC and FD.

**Table 1.** Information on the dataset

| Variables | Unit | Researchers using the variables | Sources |
|---|---|---|---|
| Renewable Energy Consumption (REC) | Terajoule per capita | **Chica-Olmo et al. (2020); Oluoch et al. (2021); Anton and Nucu** (2020) | British Petroleum, Sustainable Energy for All database, and International Renewable Energy Agency |
| Financial Development Index (FD) | Domestic credit to the private sector per capita (constant 2015 US$) | **Tang and Tan (2014)** | International Monetary Fund, and International Financial Statistics |

### 3.2  Model

Theoretical foundation and the analysis of the empirical literature give us all the variables for the model. Zeren and Hzarc (2021), and Kevser et al. (2022) conducted studies that served as the basis for the following equations. While in this paper, we will use renewable energy consumption instead of biomass energy consumption.

$$FD_{it} = a_{0i} + a_1 REC_{it} + v_{it} \qquad (1)$$

## 4  Empirical Results and Discussion

### 4.1  The Panel Estimation

The table shows the panel estimation with random and fixed effect as follow (Table 2):

**Table 2.**  The panel estimation

| Variables | Coefficient | Prob |
|---|---|---|
| C | 106.297 | 0.8200 |
| REC | 3597095.88*** | 0 |

**Cross-Section Fixed Effects**

| Country | Coefficient | Country | Coefficient | Country | Coefficient |
|---|---|---|---|---|---|
| Algeria | 181.187 | Tunisia | −10158.82 | United Arab Emirate | 3497.339 |
| Egypt | −5779.240 | Iraq | −6357.856 | Jordan | −42.224 |
| Libya | −2487.411 | Kuwait | 17807.81 | Lebanon | −697.557 |
| Morocco | −2013.446 | Saudi Arabia | 6453.548 | Yemen | −403.329 |

Done on EViews 12 by the researchers.
Note: "***", "**", "*'refers to the confidence interval at 99%, 95%, 90% level.

The estimation shows that there is a significant relationship at the level of 1% between REC and FD. The outcomes show that REC has a positive significant influence on FD of these Arab countries. After analysing the country-specific analytical results and cross-section fixed effects, REC has a negative impact on FD for Egypt, Libya, Morocco, Tunisia, Iraq, Jordan, Lebanon and Yemen. Then again, REC has a positive impact on FD for Algeria, Kuwait, Saudi Arabia, and United Arab Emirate. In this situation, a few countries will need to formulate new energy laws and make choices to boost the growth of the RE sector. These policies may support the reduction of environmental deterioration.

### 4.2  Dumitrescu–Hurlin Panel Causality

The causality relationships of the series with cointegration relationships are analyzed with the Dumitrescu and Hurlin (2012) causality test. The following table display the panel causality result (Table 3):

The results indicate a one-way causal relationship between FD and REC, indicating the need for a change in the financial markets that will lead to a more effective composition of the amount, volume, and distribution of loans made to the private sector, which will trigger an economic situation. It should not be forgotten, nonetheless, that Arab countries financial systems are some or less mediocre than those in rich nations, making it impossible for them to quickly achieve financial development and boost renewables.

**Table 3.** Dumitrescu–Hurlin panel causality test

| Null Hypothesis | W-Stat | Zbar-Stat | Prob |
|---|---|---|---|
| FD does not homogeneously cause REC | 5.029*** | 4.17*** | 0 |
| REC does not homogeneously cause FD | 1.893 | −0.409 | 0.6819 |

Note: "***", "**", "*"refers to the confidence interval at 99%, 95%, 90% level.

## 5 Conclusion

Three points serve as a concise summary of the paper's key results. Firstly, there is a positive and significant association between REC and FD for panel data. These findings approve that we can accept the first hypothesis of the study. Secondly, the outcomes from the panel model with fixed effects indicate that the countries that have a negative sign misused RE in their financial and economic systems, while the others have a positive sign use renewable to boost their financial and economic systems with an optimal energy-saving policy. Thirdly, DH panel causality test shows that a there is a one-way causality running from FD to REC.

## References

Abbass, K., Song, H., Khan, F., Begum, H., Asif, M.: Fresh insight through the VAR approach to investigate the effects of fiscal policy on environmental pollution in Pakistan. Environ. Sci. Pollut. Res. **29**(16), 23001–23014 (2022). https://doi.org/10.1007/s11356-021-17438-x

Anton, S.G., Afloarei Nucu, A.E.: The effect of financial development on renewable energy consumption. A panel data approach. Renew. Energy **147**, 330–338 (2020). https://doi.org/10.1016/j.renene.2019.09.005

Boufateh, T., Saadaoui, Z.: Do Asymmetric financial development shocks matter for CO2 emissions in Africa? A nonlinear panel ARDL–PMG approach. Environ. Model. Assess. **25**(6), 809–830 (2020). https://doi.org/10.1007/s10666-020-09722-w

Chang, S.C.: Effects of financial developments and income on energy consumption. Int. Rev. Econ. Financ. **35**, 28–44 (2014). https://doi.org/10.1016/j.iref.2014.08.011

Chica-Olmo, J., Salaheddine, S.H., Moya-Fernández, P.: Spatial relationship between economic growth and renewable energy consumption in 26 European countries. Energy Econ. **92** (2020). https://doi.org/10.1016/j.eneco.2020.104962

Danish, Ulucak, R.: A revisit to the relationship between financial development and energy consumption: is globalization paramount? Energy **227** (2021). https://doi.org/10.1016/j.energy.2021.120337

Destek, M.A.: Financial development and energy consumption nexus in emerging economies. Energy Sources Part B **13**(1), 76–81 (2018). https://doi.org/10.1080/15567249.2017.1405106

Dumitrescu, E.I., Hurlin, C.: Testing for granger non-causality in heterogeneous panels. Econ. Model. **29**(4), 1450–1460 (2012). https://doi.org/10.1016/j.econmod.2012.02.014

Eren, B.M., Taspinar, N., Gokmenoglu, K.K.: The impact of financial development and economic growth on renewable energy consumption: empirical analysis of India. Sci. Total Environ. **663**, 189–197 (2019). https://doi.org/10.1016/j.scitotenv.2019.01.323

Fang, G., Yang, K., Tian, L., Ma, Y.: Can environmental tax promote renewable energy consumption?—An empirical study from the typical countries along the belt and road. Energy **260** (2022). https://doi.org/10.1016/j.energy.2022.125193

Gao, J., Zhang, L.: Does biomass energy consumption mitigate $CO_2$ emissions? The role of economic growth and urbanization: evidence from developing Asia. J. Asia Pac. Econ. **26**(1), 96–115 (2021). https://doi.org/10.1080/13547860.2020.1717902

Hassoun, S.E.S., Mekidiche, M.: Investigating the link amongst the main macroeconomic factors, economic growth and crude oil price in Algeria (2019). 109 مجلة الباحث الإقتصادي. https://doi.org/10.35391/1894-007-011-007

Kevser, M., Tekbaş, M., Doğan, M., Koyluoglu, S.: Nexus among biomass energy consumption, economic growth, and financial development: evidence from selected 15 countries. Energy Rep. **8**, 8372–8380 (2022). https://doi.org/10.1016/j.egyr.2022.06.033

Kihombo, S., Vaseer, A.I., Ahmed, Z., Chen, S., Kirikkaleli, D., Adebayo, T.S.: Is there a trade-off between financial globalization, economic growth, and environmental sustainability? An advanced panel analysis. Environ. Sci. Pollut. Res. **29**(3), 3983–3993 (2022). https://doi.org/10.1007/s11356-021-15878-z

Komal, R., Abbas, F.: Linking financial development, economic growth and energy consumption in Pakistan. Renew. Sustain. Energy Rev. **44**, 211–220 (2015). https://doi.org/10.1016/j.rser.2014.12.015

Kraft, J., Kraft, A.: On the relationship between energy and GNP. J. Energy Dev. **3**(2), 401–403 (1978). http://www.jstor.org/stable/24806805

Ma, X., Fu, Q.: The influence of financial development on energy consumption: worldwide evidence. Int. J. Environ. Res. Pub. Health **17**(4) (2020). https://doi.org/10.3390/ijerph17041428

Nkalu, C.N., Ugwu, S.C., Asogwa, F.O., Kuma, M.P., Onyeke, Q.O.: Financial development and energy consumption in Sub-Saharan Africa: evidence from panel vector error correction model. SAGE Open **10**(3) (2020). https://doi.org/10.1177/2158244020935432

Oluoch, S., Lal, P., Susaeta, A.: Investigating factors affecting renewable energy consumption: a panel data analysis in Sub Saharan Africa. Environ. Challenges **4** (2021). https://doi.org/10.1016/j.envc.2021.100092

Pata, U.K.: Renewable energy consumption, urbanization, financial development, income and CO2 emissions in Turkey: testing EKC hypothesis with structural breaks. J. Clean. Prod. **187**, 770–779 (2018). https://doi.org/10.1016/j.jclepro.2018.03.236

Peng, G., Meng, F., Ahmed, Z., Oláh, J., Harsányi, E.: A path towards green revolution: how do environmental technologies, political risk, and environmental taxes influence green energy consumption? Front. Environ. Sci. **10** (2022). https://doi.org/10.3389/fenvs.2022.927333

Sadi Ali, H., Bin Yusop, Z., Siong Hook, L.: Financial development and energy consumption nexus in Nigeria: an application of autoregressive distributed lag bound testing approach. Int. J. Energy Econ. Policy **3** (2015). http://www.econjournals.com

Shahbaz, M., Van Hoang, T.H., Mahalik, M.K., Roubaud, D.: Energy consumption, financial development and economic growth in India: new evidence from a nonlinear and asymmetric analysis. Energy Econ. **63**, 199–212 (2017). https://doi.org/10.1016/j.eneco.2017.01.023

Somoye, O.A., Ozdeser, H., Seraj, M.: Modeling the determinants of renewable energy consumption in Nigeria: evidence from autoregressive distributed lagged in error correction approach. Renew. EnergyEnergy **190**, 606–616 (2022). https://doi.org/10.1016/j.renene.2022.03.143

Suki, N.M., Suki, N.M., Afshan, S., Sharif, A., Meo, M.S.: The paradigms of technological innovation and renewables as a panacea for sustainable development: a pathway of going green. Renew. Energy **181**, 1431–1439 (2022). https://doi.org/10.1016/j.renene.2021.09.121

Zeren, F., Hizarci, A.E.: Biomass energy consumption and financial development: evidence from some developing countries. Int. J. Sustain. Energ. **40**(9), 858–868 (2021). https://doi.org/10.1080/14786451.2021.1876689

# Integration of Renewable Energy Systems in Building: Case of Solar Thermal Collectors and Photovoltaic Panels for the Tertiary and Residential Sector in Algiers

Tizouiar Ouahiba(✉)

Architecture and Environment Laboratory LAE, Polytechnic School of Architecture and Urbanism EPAU, Algiers, Algeria
otizouiar@yahoo.com, o.tizouiar@epau-alger.edu.dz

**Abstract.** Solar energy can be used in different ways, from thermal to photovoltaic (PV), in different sectors, including the residential one. This energy can be used by solar thermal collectors or PV panels, which convert it into heat energy used either to heat or to generate electricity.

This study defines not only the different types of renewable energy systems and some examples of dimensioning and performance calculations for some real buildings. But above all, it highlights the technical requirements for integrating these systems into buildings from an aesthetic and architectural point of view. This is based on analysis of integration of those systems in a residential building, and in an educational building, then on a state of art about the subject.

Like all components of a building, such as the door and the window, solar collector or photovoltaic panel must be considered as an element of the architectural composition. To achieve this, the designer must integrate it into the design of a façade, roof and envelope, while preserving its specific formal or functional characteristics.

**Keywords:** Solar energy · aesthetics · envelope · thermal collectors · Building-Integrated Photovoltaic (BIPV)

## 1 Introduction

This article focuses, in one side, on the reduction of energy consumption and the development of energy solutions in buildings, while replacing fossil fuels with renewable energies, notably solar energy. This choice is motivated by the immense potential for solar energy in Algeria and by the fact that the building sector poses this consumption problem which is of the order of 47% of the energy produced (Fig. 1), and is responsible for 25% of the greenhouse gas emissions, The Increase in demand from the "Households and others" sector (6.2%) is going from 22.1 M Toe in 2020 to 23.4 M Toe in 2021, driven by the residential sub-sector (4.4%) and the Tertiary and other sub-sector (12.3%) [4]. In order to provide solutions to both the problems of preserving energy resources of fossil

© The Author(s), under exclusive license to Springer Nature Switzerland AG 2024
M. Hatti (Ed.): IC-AIRES 2023, LNNS 984, pp. 313–323, 2024.
https://doi.org/10.1007/978-3-031-60629-8_31

origin and the environmental challenges as well as the climate changes that result from them. Algerian state has developed several regulatory, institutional, financial, incentive and fort systems for energy efficiency in housing. However, the development of measures for such a policy is not enough; it is necessary to ensure the application and impact of such measures.

On the other side, this article particularly focuses on architectural integration of photovoltaic solar systems, which can be installed differently for more architectural aesthetics, namely installation on the roof, the facade or in canopy as an element of protection and decoration. The presence of a solar collector on a building is justified by a functional necessity, but which must meet certain technical constraints and must necessarily be the subject of an aesthetic treatment.

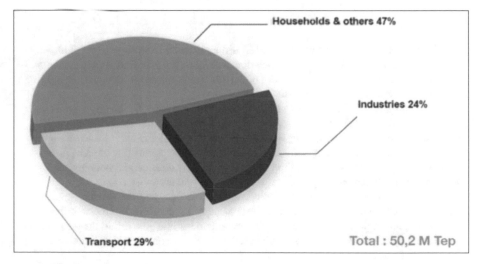

**Fig. 1.** Structure of final energy consumption by sector of activity (source: [4])

## 2 Problematic

The main problems that arise for the integration of renewable energy in residential or tertiary buildings are the following: **"What are the architectural and technical requirements for the integration of solar collectors and photovoltaic panels into buildings in order to meet both energy needs and preserve aesthetics?"**.

The objectives of this study are: Firstly the design of buildings that are energy efficient and thermally comfortable, mainly the replacement of conventional heat production in residential buildings with thermal solar energy, this by evaluating the efficiency of such a system.

Then, contribute for resolving design defects to improve comfort conditions and energy consumption. This by combining the bioclimatic aspect to naturally reduce energy, with the integration of a thermal photovoltaic hybrid system to power tertiary educational equipment. Based on actual parameters of the installation site, a hybrid system dimensioning method while thinking about aesthetics through good architectural integration.

Finally, see what is developed elsewhere on the basis of a state of art on the aesthetic aspect in the BIPV.

## 3 Cases Study

This article is based on two academic case studies that we have supervised [6, 8] then on cases studied in several research articles [1–3, 5, 7, 9, 10].

### 3.1 Case of Solar Thermal Collectors Integrated in a Collective Residential Building

Verification of the feasibility and integrity of an energy saving system with regard to heating was carried out in the case of an existing collective residential building (Fig. 2). The study [8] consists of evaluating the heating needs of an apartment of type (F3), which is part of this building located in the city of Blida, 45 km southwest of Algiers. The thermal simulation is carried out using TRNSYS 16 software taking into account surface of solar collectors, orientation due south, inclination corresponding to the latitude of the city 36% and also dimensioning of the solar trajectory. The chosen system is a solar water heater with circulator and exchanger outside the storage reservoir.

**Fig. 2.** Exterior appearance of the building

## 3.2 Case of PVT Hybrid Panels Integrated into a Tertiary Building

The choice of mathematical high school of Kouba comes from its status as energy-consuming equipment in terms of electricity and heating. The quantity of monthly energy consumed is more than 200,000 kwh and the gas consumption is more than 2500 m³, this results in high costs reaching more than 320,000 Da. The study [6] focused on the block reserved for the secondary cycle represented in (Fig. 3) because it remains the only occupied and accessible block (150 students and 50 staff members). This building is made up of seven volumes of different heights connected and articulated together.

Also, the choice of the hybrid solar system (Fig. 4) comes down to its availability in Algeria, it is essentially made up of a UDES 50 type photovoltaic module in monocrystalline technology and a galvanized steel absorber, with a thickness of 3.25 cm, glued below the photovoltaic module. A heat transfer fluid can circulate inside this absorber to extract the heat stored and not converted by the solar cells.

**Fig. 3.** Mathematical high school of Kouba, secondary cycle block (google earth picture)

**Fig. 4.** The model of Hybrid photovoltaic thermal panel adopted (photo taken at the CDER Renewable Energy Research Center of Bousmail)

## 4  Methodological Approach

For the first case, the research is based on the use of a thermal dynamic simulation (TRNSYS 16) of an existing home with a conventional heating system, then on the evaluation and comparison of energy needs before and after the integration of solar thermal, as well as the evaluation of effect on Carbone dioxide emissions.

Concerning the second case, passive modifications based on the site parameters and the use of the simulation software (ECOTECT) are first proposed. This is to naturally reduce the energy requirement in the building and ensure the comfort of occupants while making the most of solar radiation and natural air circulation.

Subsequently and for better efficiency, active solutions are integrated through the application of a device that meets both demands (electricity and solar heat). It concerns hybrid PVT panels which require laborious prior dimensioning based on in-depth knowledge of the solar resource and the electrical and thermal efficiency of the PVT.

The dimensioning begins by evaluating consumption of the different loads (for this case the average daily consumption is 241,622 Wh/day and air conditioning consumes more than half of the total electrical energy consumed). Then by the definition of the incidence angle which corresponds to the plane formed between photovoltaic panel and light rays, and at the end by the use of equation formulas to calculate necessary surface and number of panels, as well than choosing the operating voltage according to the peak power of the photovoltaic field in watts.

In parallel with the dimensioning, technical requirements and possibilities of architectural integration are proposed so as not to affect the aesthetics of these buildings. In this sense and in addition to the tow cases, basing on several researches, we synthetize other types of BIPV in different contexts.

## 5 Results and Discussion

### 5.1 Energy Consumption Results for the Residential Buildings

The first thermal simulation of the apartment equipped with natural gas heating informs us about the quantity of energy consumed which is of the order of 5298.24 KWH in relation to an outside temperature which varies between 4 and 18 °C in winter. However, the second simulation carried out after the integration of solar collectors with a surface area of $10\,m^2$ and a storage reservoir of $1\,m^3$ allows for lower heating energy requirements than the first, namely 2637.2 KWH of solar energy, compared to only 538.8 KWH of energy from natural gas heating, and this allows us to see an energy saving of 83% in gas energy compared to solar energy and a saving of 60% compared to the totality of energy consumed before and after using the solar heating system.

This economic study [8] of solar heating confirms the efficiency of such an installation, on energy management in the residential sector, and also its positive impact on the environment, namely the reduction in carbon dioxide emissions (Fig. 5).

### 5.2 Energy Consumption Results for the Tertiary Buildings

**The bioclimatic analysis** explains that the flow of energy passing through the wall is pretty significant, especially at the level of the windows due to their exaggerated transmission coefficient and their high glazing rate; which causes enormous losses and therefore high consumption of heating and air conditioning. The use of wall insulation and double glazing to reduce this consumption is necessary. Another parameter is the high altitude of the site witch is open so heat loss is favored by the strong exposure to winds.

The height of the neighboring blocks (on-call housing) is greater than the width of the street and therefore the project presents problems related to lighting and heating. This blocking extends during the cold period, dating from the end of September until the

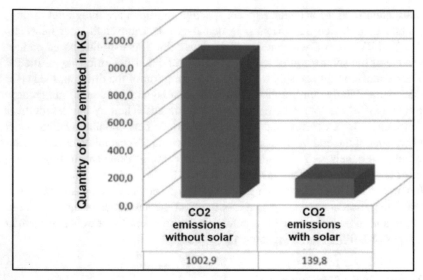

**Fig. 5.** Carbone dioxide emissions before and after the integration of solar collectors (source:[8])

month of March - simulated by ecotect [6]-which requires the use of artificial lighting and heating. Whereas in the heat season from March to June, the height of the sun is high so the accommodation does not create shade on the building, which creates overheating indoor and requires the use of air conditioning to cool.

The presence of corridors, despite their role of protection against bad weather, they create a barrier against light and solar rays and create shaded areas on the facade. Adding to this, spaces requiring significant light and solar input as classrooms, office, media library, they have the most unfavorable orientation (North and northwest).

**The dimensioning results** for photovoltaic installation on the roof for the three blocks (1, 2 and 3) making up the building are summarized in the table (Table 1):

**Table 1.** Dimensioning results of the photovoltaic installation (source: synthesized by the author on the basis of [6])

|  | Element of the photovoltaic system | | | | | |
|---|---|---|---|---|---|---|
|  | Block 1 | | Block 2 | | Block 3 | |
| Peak Power Pc (Wc) | 25552.8Wc | | 3024 Wc | | 18547.2 Wc | |
| Inclination $\propto$ (°) | 36 | | 36 | | 36 | |
| Nomber of modules | Total | 508 | Total | 60 | Total | 368 |
| Electric energy produced by the PVT / year | | | | | 330,673 KWh/day | |
| Thermal energy produced by the PVT / year | | | | | 1267,582 KWh/day | |
| Photovoltaic energy consumed in the building | | | | | 241622 Wh/day | |
| Thermal energy consumed in the building | | | | | 303,4 kWh/day | |

Thermal production is higher due to the thermal efficiency of the PVT panels which is very high (0.46), unlike the electrical production which is low due to the low electrical efficiency of the PVT panels (0.12). This implies that the captured solar irradiation is not converted in its entirety into electrical and thermal production.

Photovoltaic cells produce electricity, but they also produce heat; in a classic system, this heat is not only lost, but it also reduces the efficiency of the photovoltaic cells, which become lazy when the temperature rises. When operating, a standard photovoltaic panel produces 80% heat and only 20% electricity. In a hybrid solar collector, the fluid circulating in the thermal part absorbs the heat created, and therefore cools the photovoltaic cells, thus their efficiency is improved, however heat losses are only reduced and not eliminated categorically.

The cost effectiveness for this hybrid system at the high school level can be reached in around **15 years, i.e. 6 billion DA (total cost of the system) = 15 years \* 4,000,000 DA (annual cost of EC)**. Despite the performance of this system and its ability to reduce energy consumption, it remains necessary to carry out research aimed at improving the lifetime of its components for better cost effectiveness [6]. We consider too that cost effectiveness could be achieved by using the panels as **components of envelope** and not as additions, requiring a double investment, this is why we proceed to developpe in the following architectural integration aspect.

### 5.3  Architectural Integration of Solar Collectors for the Both Cases

The calculated surface area of solar collectors for the first case is 100 m$^2$, installed on the flat terrace which has a surface area of 186 m$^2$, with south-facing orientation and an inclination corresponding to the latitude of Blida city (around 36°). They are placed at a distance of 1.5 m from the acroterium for maintenance and safety reasons; the choice of this type of integration is the most common for an existing collective building.

Solar collectors can be integrated differently for more architectural aesthetics. Apart from the integration on the roof, they can be installed on the facade, taking a part of the stairwell in the middle or as an canopy for each windows of apartments, as element of protection and decoration at the same time. However, to keep the same surface area of the panels, for powering the building with energy, it is necessary to keep a certain number of panels on the terrace, as shown in the figures in (Table 2).

For the high school, the calculated surface area of hybrid panels is about 392 m$^2$, with a south-facing orientation and an inclination corresponding to the latitude of Algiers (around 36°), proposed to be installed on the flat terrace with 450 m$^2$ surface area. The panels are also placed at a distance of 1.5 m from the acroterium for maintenance and safety reasons.

For more aesthetics, the PVT can be integrated on the south facade by taking the part of the stairwell in the middle or as a canopy for each class as an element of protection and decoration; At the same time, we can also exploit the corridors located on the south side.

**Table 2.** Possibilities of solar systems integration on the tow cases (source: synthesized by the author on the basis of [6] and [8])

## 5.4 State of Art About the Impact of BIPV on Building Aesthetics

Several studies have addressed the aspect of BIPV integration from an architectural point of view [1–3, 5, 7, 9, 10], some relate to acceptance and feasibility, others to the approaches and integration tools, and others concern innovation in order to combine energy efficiency and aesthetics.

In our recent research [10] basing on an interview with diffrents actors of building, results indicate that (12.5%) of them integrate renewable energies, except that the costs of the facilities remain high and the depreciation periods relating to this investments are lengthened. (87.5%) made no attempt to integrate renewable energies into their buildings. Because this integration is initially very costly, in addition it requires a very developed techniques and knowledge. In another research [5] The main lessons that can be drawn from a comparative study for the BIPV sector are that this sector needs to be reviewed and developed from an international perspective and that the level of environmental concern in society seems to be higher, thus offering opportunities for BIPV implementation. Also, Companies will need to think about their personal values, strengths and strategies, which could lead to different outcomes.

According to [9] Solar PV manufacturing companies that have emphasized aesthetics in their transactions have gradually risen to the hearts of consumers and architects.

The future of green buildings and renewable energy systems in modern architecture depends in part on the diversification of the aesthetic values of photovoltaic materials. It is argued that the choice of BIPV over its BAPV counterpart is primarily based on aesthetics. Although cost and effectiveness are important, aesthetics provides a magnet that attracts potential consumers. That aesthetics plays a fundamental role in the consumer's purchasing process. It is recommended that when it comes to BIPV, aesthetics be treated as an important aspect by PV manufacturers and installers. Fundamental design principles and elements must be adopted to ensure overall beauty.

Integration should be implicated in the first phase of design. The analysis carried out by certain authors [7] indicates a good level of possibilities for integrating photovoltaic modules into architecture and these vary depending on the geometric characteristics of the roof. Applications on solid and glazed roofs are also specific. These modules are used in glazed roofs and generally serve as highly integrated elements of the building architecture, as they are a component of the building rather than a separate installation. The impact of photovoltaic modules on the aesthetic appearance is more linked to the body of the building in the case of applications on solid roofs and to the interior space in the case of applications on glass roofs. Compared to solid roofs, the use of photovoltaic modules in glass roofs offers greater opportunities to take advantage of these elements while shaping the thermal and visual environment.

About tools of integration, Researchers of the study [9] discuss the primacy of aesthetics using design elements and principles as useful tools in the adoption of BIPV. Essential design elements such as color, shape and texture were covered. as well as design principles such as variety,balance, rhythm, contrast and proportion. However, it is important to balance aesthetics, efficiency and functionality so as not to sacrifice the energy production capacity of photovoltaics for the sake of beauty.

The paper [3] aims to demonstrate how decision-making in the design phase of the PV design can take advantage of two tools that allow the designer to evaluate the use of solar irradiation and the impact of shading in different conditions of orientation and inclination: the solar abacus and the shading masks.. The integration of solar energy as a power source should be viewed as an added requirement and should be included in project commitments. To optimize energy generation and minimize shading losses, informed decisions must be made during the design phase. Due to the simplicity of interpretation of the adopted tools, both architects not specialized in solar energy and customers can understand the decision-making process and the resulting losses from each project choice. Thus helping to decide the most appropriate positioning and inclination of the PV modules.

The irradiation and shading analysis are crucial in the design process of PV systems integrated into buildings. By using useful design tools, architects and engineers can design buildings that are not only aesthetically pleasing but also energy-efficient and sustainable.

Regarding innovation in BIPV, Net-Zero Energy Buildings can be acheived by integrating photovoltaic (PV) building materials, throughout the building skin, which act simultaneously as construction materials and energy generators. Currently, architects and builders are inclined to design a building using BIPV modules due to the limited colors available, namely, black or blue, which gives a monotonous look to building.

Therefore, there is an growing demand/need to develop modern, aesthetically pleasing BIPV green energy products for use of architects and the construction industry [2]. To address current issues related to achieving aesthetically pleasing BIPV modules that pave the way for achieving net-zero energy buildings (NZEB), the authors of [2] have discussed an advanced technical approach based on the development of high-definition colored PV modules with adequate image contrast and a power conversion efficiency (PCE) exceeding 85% of that of a bare PV panel will boost the use of BIPV products and increase their market values in near future.

According to [1], technological advancement in Building Integrated Photovoltaics (BIPV) has converted the building façade into a renewable energy-based generator, the challenge however, is that architectural design objectives sometimes conflict with energy performance, such as the provision of view and daylight versus maximum power output. Theirs findings indicate that BIPV façade customization can be carried out with significant advantages which include: flexibility and applicability at an elemental and compositional level. Versatility in development of both custom BIPV products and custom BIPV integration schemes. Multiple type strategies in single or combined scenarios can be used to achieve objectives. Increase in power output and performance is possible.Based on some research, those authors find that areas such as daylighting, self-cleaning photovoltaic glazing, aesthetics using color, shape or forms, concentrating BIPV, perovskite-based solar cells and solar trees are one of the emerging areas of innovation.

Combining other research, they analyze the design impact of different types of BIPV facades, as curtain wall/cladding systems that can be integrated as single facades. Solar glazing and windows applied as windows and glazing panels for view or daylight. Exterior devices (parasols, sunshades, spandrels, balconies, parapets) which allow PV modules of different shapes to be installed. Innovative envelope systems (double skin facades, active skins, movable facade parts) with possibilities for integrating advanced aesthetic polymer technologies.

# 6  General Conclusion

This article aims to make managers, decision-makers and building designers aware of the importance of taking into consideration the energy side but also the aesthetic side of our cities and our buildings, this through the right choice of solar systems and their proper integration.

Photovoltaic technology has ecological qualities because the finished product is non-polluting, silent and does not cause any disturbance to the environment, such as solar thermal collectors and hybrid PVT panels. Despite their performance and ability to reduce energy consumption, they remain inconvenient for large-scale projects, given the enormous investment cost and the inability to attract gains due to the limited lifetime of their components, which requires more applied research aimed at improving the latter for better cost effectivness.

Solar systems reduce energy consumption in the long term, provided that they renews their components, carries out the correct dimensioning and inclination before installation, and thinks about improved cost effectivness by designing the panels as parts of the components of envelope and not as an addition or a seperate elements.

On sloping or flat roofs, energy gains must be optimised throughout the year according to use, but also aligned or symmetrical with the building's components. Integration can also take several ways on an architectural and decorative level, namely installation on a sloping roof or terrace roof, on a facade or in an calopy or even at the level of the corridors which has a great impact on the aesthetic value of the building.

As part of future research, complementary research could be considered in order to calculate the contribution of different manners of integrating PVT into building, this is in relation to energy consumption rates, and the comparison of these possibilities for optimizing the choice from an aesthetic, functional and energetic point of view.

# References

1. Attoye, D.E., Tabet Aoul, K.A., Hassan, A.: A review on building integrated photovoltaic façade customization potentials. Sustainability **9**(12), 2287 (2017). https://doi.org/10.3390/su9122287
2. Basher, M.K., Nur-E-Alam, M., Rahman, M.M., Alameh, K., Hinckley, S.: Aesthetically appealing building integrated photovoltaic systems for net-zero energy buildings. Current status, challenges, and future developments—a review. Buildings **13**(4), 863 (2023). https://doi.org/10.3390/buildings13040863
3. Zomer, C., Fossati, M., Machado, A.: Designing with the sun: finding balance between aesthetics and energy performance in building-integrated photovoltaic buildings. Solar Compass **6** (2023). https://www.sciencedirect.com/science/article/pii/S2772940023000140
4. Energy Report 2021, edition 2022. Algerian Democratic and Popular Republic. Ministry of Energy and Mines, P(48). https://www.energy.gov.dz/Media/galerie/bilan_energetique_2021_63df78f2b775e.pdf
5. Osseweijer, F.J.W. van den Hurk, L.B.P., Teunissen, E.J.H.M., van Sark, W.G.J.H.M.: CA comparative review of building integrated photovoltaics ecosystems in selected European countries. Renew. Sustain. Energy Rev. **90**, 1027–1040 (2018). https://www.sciencedirect.com/science/article/pii/S1364032118300716
6. Habbani, K.: Impact de l'énergie solaire sur la consommation énergétique dans le bâtiment Intégration des capteurs hybrides PVT au Lycée des mathématiques à Kouba, Master memory in architecture at Polytechnic School of Architecture and Urbanism of Algiers, Supervised by Tizouiar, O., 80 p. (2019)
7. Marchwiński, J.: Architectural analysis of photovoltaic (pv) module applications on non-flat roofs. Original Paper Acta Sci. Pol. Architectura, 1–10 (2023)
8. Moulay, R.: Intégration des systèmes d'économie d'énergie dans l'habitat en Algérie: cas des capteurs solaires thermiques. Master Memory in Architecture at Polytechnic School of Architecture and Urbanism of Algiers, Supervised by Tizouiar, O., 95p. (2015)
9. Awuku, S.A., Bennadji, A., Muhammad-Sukki, F., Sellami, N.: Myth or gold? The power of aesthetics in the adoption of building integrated photovoltaics (BIPVs). Energy Nexus **4** (2021). https://www.sciencedirect.com/science/article/pii/S2772427121000218
10. Tizouiar, O., Boussoualim, A.: The approach used for comfortable and high efficient buildings in sustainable cities by different actors. In: Hatti, M. (ed.) Renewable Energy for Smart and Sustainable Cities. ICAIRES 2018. LNNS, vol. 62, pp. 3–13. Springer, Cham (2019). https://doi.org/10.1007/978-3-030-04789-4_1

# Study and Simulation of MSM Photodetector Based on the $Al_xGa_{1-x}N$/GaN Structure

N. Hafi[1], A. Aissat[1,2(✉)], Mohamed Kemouche[1], Nesrine Bakalem[1], and D. Décoster[2]

[1] Faculty of Technology University of Blida 1, Laboratory LATSI Faculty of Technology, University of Blida 1, 09000 Blida, Algeria
sakre23@yahoo.fr

[2] Institute of Electronics, Microelectronics and Nanotechnology (IEMN), UMR CNRS 8520, University of Science and Technology of Lille 1, Avenue Poincare, BP 60069, 59652 Villeneuve d'Ascq, France

**Abstract.** In this paper, we present the structure of an MSM $Al_{0.25}Ga_{0.75}N$/GaN photodetector operating in ultraviolet. The structure studied is made from the alloy of two binary semiconductor materials GaN and AlN. Modeling and simulation were carried out by Atlas-Silvaco software. The component characteristics dark current, photocurrent and reactivity were simulated and optimized. The optimal structure corresponds to the aluminum concentration of 25%. A dark current around 3.47 nA under 1 V bias was obtained. Then we changed the polarization to 10 V the dark current achieved is equal to 0.24 µA. Also, the transmission and reflection coefficients were studied. The optimal responsivity reaches 0.28 A/W at a wavelength $\lambda = 250$ nm and a polarisation of 10 V.

**Keywords:** Materials · Semiconductors III-V-N · UV Photo-Detector · Responsivity · Detection

## 1 Introduction

III-V nitride compound materials (InN, GaN, AlN) and their alloys (AlGaN, InGaN) exhibit interesting physical properties such as high electron mobility, high carrier saturation velocity, high thermal stability and conductivity, direct and tunable band gap, of high optical absorption coefficient [1]. Recently, GaN and its alloys have been materials of choice for the fabrication of UV detectors in the wavelength range 200–370 nm, because UV photodetectors fabricated to date have a wide range of applications, including engine control, lithography aligners, astronomy, solar UV monitoring, secure space-to-space communications, or missile detection, as indicated in a recent review [2].High-quality materials such as AlN, GaN, InN, and their alloys are required in new technologies. Understanding growth mechanisms is critical for improving the quality of materials that control growth conditions and discovering new methods to apply the capabilities

© The Author(s), under exclusive license to Springer Nature Switzerland AG 2024
M. Hatti (Ed.): IC-AIRES 2023, LNNS 984, pp. 324–331, 2024.
https://doi.org/10.1007/978-3-031-60629-8_32

of modern growth [5]. Photodetectors (PD) are widely used in optoelectronic applications, particularly in communication [3]. MSM (metal, semiconductor, and metal) photodetectors are commonly used in UV detection. The Schottky contact, which is characterized by a significant barrier between the metal and the semiconductor, offers excellent performance [4]. Since the contacts in the MSM arrangement are rectified and the material has a high resistivity, the dark current is very low. Moreover, the maximum internal gain, the low noise, the low dark current, and the reduced parasitic capacitance have been optimized [6, 7]. In this work, a photodetector MSM with the structure $Al_{0.25}Ga_{0.75}N/GaN$ have been designed and simulated. The device simulator Atlas-Silvaco, the current-voltage characteristics in dark and illumination conditions, spectral responsivity, absorption, and transmission of the MSM photodetector were investigated [8–10].

## 2  Device Structure and Simulation Model

The device structure represented in Fig. 1 was simulated using Atlas Silvaco simulator [11]. The AlGaN/GaN MSM layer structure structure consists of a 40 nm thick GaN nucleation layer, 50 nm thick insulator GaN, 4 nm AlGaN layer is n-type with carrier concentration $10^{17}$ $cm^3$, and 6 nm unintentionally-doped AlGaN spacer layer, two layer were simulated with Al content 25%. The photodetector is grown on sapphire substrate of 300 nm thickness by metalorganic chemical vapor deposition (MOCVD). The structure of the MSM consists of a pair of Pt Schottky electrodes contacts with finger is 20 nm wide, a 160 nm spacing band gap and a 2 nm thickness was deposited [11]. Simulations were performed using the two-and three dimensional computer simulators Atlas-Silvaco that solves a set of basic semiconductor equations consisting of the Poisson equation and the continuity and transport equations for electrons and holes [8]. The $Al_xGa_{1-x}N$ band gap energy alters from 3.4 to 6.2 eV as the Al mole fraction x changes. The relationships between the $Al_xGa_{1-x}N$ band gap energy $E_g$ and the Al components $x$ are as follows [12]:

$$E_{g\,AlGaN} = xE_{g\,AlN} + (1 - x)E_{g\,GaN} - bx(1 - x) \tag{1}$$

where

$E_{g\,AlN} = 6.2$ eV, $E_{g\,GaN} = 3.39$ eV, and $b = 1$.
The absorption coefficient of AlGaN can be expressed as [12]:

$$\alpha^2 = \alpha_0^2[E - E_{g_{AlGaN}}] \tag{2}$$

where E is photon energy, $E_{g\,AlGaN}$ is the band gap energy of AlGaN, $\alpha_0 = 10^5$ $cm^{-1}eV^{-1}$.

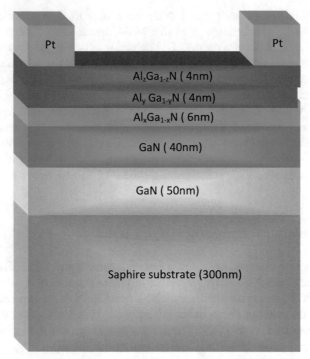

**Fig. 1.**   Structure of $Al_{0.25}Ga_{0.75}N$/GaN MSM Photodetector with Pt electrodes.

## 3   Results and Discussion

The simulated results were created by writing a program for $Al_{0.25}Ga_{0.75}N$/GaN MSM photo-detector in the DECKBUILD window interfaced with ATLAS at room temperature.

Figure 2 shows the dark current of the $Al_{0.25}Ga_{0.75}N$/GaN MSM ultraviolet photo-detector at room temperature 300 K. The dark current of the photo-detector is 3.47 nA under 1 V bias, 62.27 nA under 5 V bias and is 0.24 μA at a bias of 10 V. It can be seen that the dark current increases with the applied reverse bias and doesn't show any effect of saturation. One of the Schottky contacts is forward biased and the other is reverse biased under a bias voltage. As the bias voltage increases, the depletion width of the reverse biased. One of the Schottky contacts is forward biased and the other is reverse biased under a bias voltage. As the bias voltage increases, the depletion width of the reverse biased contact increases while the depletion width of the forward biased contact reduces, and the sum of the two depletion widths increases as well. The I-V curve under the illumination at room temperature is illustrated in Fig. 3. Under the illumination, the photocurrent of the AlGaN/GaN MSM ultraviolet photodetector is 0.20 nA under 1 V bias, is 39.09 μA under 5 V bias and is 1.57 mA under 10 V polarization.

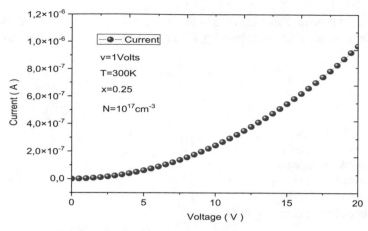

**Fig. 2.** The I–V characteristics of the $Al_{0.25}Ga_{0.75}N$/GaN MSM photodetector in dark condition.

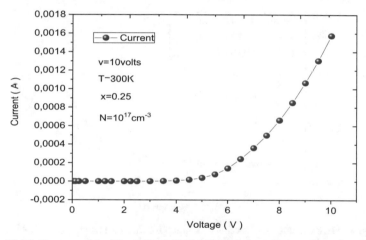

**Fig. 3.** The I–V curve of the $Al_{0.25}Ga_{0.75}N$/GaN MSM photodetector under illumination

The current-voltage characteristics are simulated under dark for different doping of AlGaN layer; results are present in Fig. 4. For a concentrations of $1 \times 10^{16}$, $1 \times 10^{17} cm^{-3}$ and $1 \times 10^{18} cm^{-3}$, we obtain approximately the same dark current about 60nA under 5 V polarization. Figure 5 presents the I-V curve of the photo-detector for uniform doping levels in the AlGaN layer; with various concentrations of $1 \times 10^{17}$, $1 \times 10^{18} cm^{-3}$ and $1 \times 10^{19} cm^{-3}$. We obtain photocurrents of about 39.09 μA, 41.4 μA and 0.14 mA, respectively under 5 V bias voltages. The Schottky contact can be described by taking into account the height of the barrier due to electric field lowering tunnel effects, the presence of an interfacial layer, and carriers combination in the metal semiconductor contact's space charge region [13, 14]. The spectral absorption and transmission of the diode are of interest. Figure 6 shows the transmission and absorption spectrum from 100 to 500 nm. The results indicate that ultraviolet absorption is much greater than transmission. The rejection rate is affected by the Pt/AlGaN contact. The results show that ultraviolet spectral absorption is significantly greater than transmission. When the incident light satisfies the condition: $h \geq E_g$, the spectral responsivity is equal to:

$$R = \frac{q\lambda}{hc}\eta g \tag{3}$$

Where

$q$ is the electron charge.
$\lambda$ is the wavelength of the incident light.
$h$ is the Planck' constant.
$c$ is the vacuum speed of light,
$\eta$ is the quantum efficiency,
g is the photoelectric current gain [15].

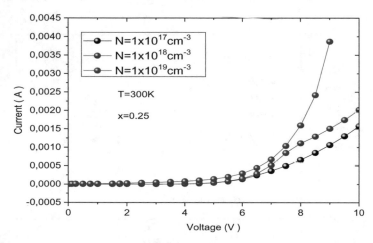

**Fig. 4.** Dark current as a function of voltage for different doping of AlGaN layer

Figure 7 presents the simulated spectral responsivity with a 10V reverse polarization. The maximum responsivity is found at 250 nm and is approximately 0.28 A/W. The detector has a high responsivity in the ultraviolet spectral region, with only a slight increase from 100 to 250 nm. We observe that when the wavelength increases from 100 to 250nm the responsivity varies from 0.19 to 0.28 A/W, that is to say we have a gain of 0.09. Above $\lambda = 250$ nm the responsivity begins to decrease rapidly. When $\lambda$ reaches 400 nm the responsivity becomes constant around the value 0.22 A/W.

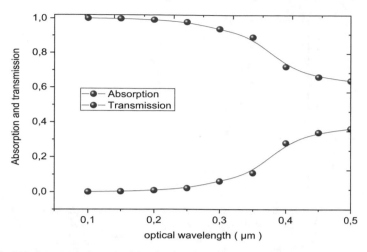

**Fig. 5.** I-V characteristic under illumination for different doping of AlGaN active layer.

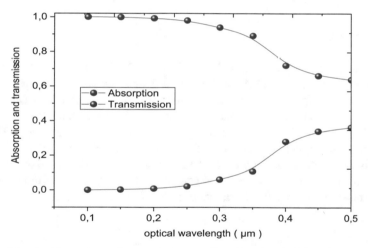

**Fig. 6.** Absorption and transmission as a function of optical wavelength.

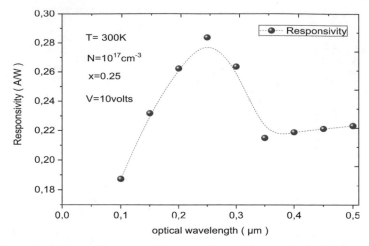

**Fig. 7.** The responsivity spectral of the $Al_{0.25}Ga_{0.75}N/GaN$ MSM photo-detector as a function of wavelength at 10 V polarization.

## 4 Conclusion

We presented numerical simulations of an MSM ultraviolet photo-detector $Al_{0.25}Ga_{0.75}N/GaN$ using the Atlas-Silvaco- software. We have investigated its dark current, photo-current and spectral responsivity was 3.47 nA at 1 V polarization and 0.24 $\mu A$ at 10 V polarization. The photocurrent was 0.20 nA with a 1 V bias, 39.09 $\mu A$ with a 5 V bias, and 1.57 mA with a 10V polarization with $N = 10^{17}cm^{-3}$. The simulated responsivity reaches a maximum of 0.28 A/W near 250 nm at 10V polarization. This study has enabled us to optimize the photodiode structure in order to produce a high performance, fast and low cost component. In the prospect, we will be able to use nanostructures to improve the performance of the photodiode studied.

## References

1. Elbar, M., Alshehri, B.,Tobbeche, S., Dogheche, E.: Design and Simulation of InGaN/GaN p–i–n Photodiodes. Phys. Status Solidi A (2018)
2. Monroy, E., Omnes, F., Calle, F.: Wide-bandgap semiconductor ultraviolet photodetectors. Semicond. Sci. technol **18**, R33–R51 (2003)
3. Chen, M., Hu, L., Xu, J., Liao, M., Wu, L., Fang, X.: ZnO hollow-sphere nanofilm-based high-performance and low-cost photodetector. Small **7**, 2449–2453 (2011)
4. Al Salman, H.S., Abdullah, M.J.: Fabrication and characterization of undoped and cobalt-doped ZnO based UV photodetector prepared by RF-sputtering. J. Mater. Sci. Technol. **29**, 1139–1145 (2013)
5. Masri, P.: Silicon carbide and silicon carbide based structures, the physics of epitaxy. Surf. Sci. Rep. **48**, 1–51 (2002)
6. Monroy, E., Calle, F., Munoz, F., Omnes, F.: AlGaN metal-semiconductor-metal photodiodes. Appl. Phys. Lett. **74**, 3401 (1999)
7. Palacios, T., Calle, F., Monroy, E., Munoz, F.: Submicron technology for III-nitride semiconductors. J. Vac. Sci. Technol. B **20**, 2071 (2002)

8. ATLAS User's Manual, Device Simulation Software, Version 5.20.2. R, SILVACO International, Santa Clara, CA (2016)

9. Aissat, A., El Besseghi, M., Decoster, D.: Optimization of photoswitch constituted of a coplanar line introducing an interdigitated MSM photodetector. Superlattices Microstruct. Elsevier **72**, 245–252 (2014)

10. Zebentout, A.D., Aissat, A., Bensaad, Z., Zegaoui, M., Pagies, A., Decoster, D.: Opt. Laser Technol. **47**(April), 1–3 (2013)

11. Allam, Z., Hamdoune, A., Boudaoud, C.: High-performance solar-blind photodetector based on AlGaN/GaN heterostructure. Proc. IEEE (2014)

12. Pulfrey, D.L., Nener, B.D.: Suggestions for the development of GaN-based photodiodes. Solid-State Electron. **42**(9), 1731–1736 (1998)

13. Hao, G., Chen, X., Chang, B., Fu, X., Mingzhu, Y.Z., Zhang, J.Y.: Optik **125**, 1377–1379 (2014)

14. Poochinda, K., Chen, T.C., Thomas, G., Stoebe, N., Ricker, L.: J. Cryst. Growth **261**, 336–340 (2004)

15. Razeghi, M., Rogalski, A.: Semiconductor ultraviolet detectors. J. Appl. Phys. **79**(10), 7433–7473 (1996)

# Design of a Hybrid Photovoltaic-Thermal Solar Collector: Utilization of High-Efficiency Perovskite-Based Solar Cells

H. Soufi[1(✉)], K. Rahmoun[1], and M. E. A. Slimani[2]

[1] Unité de Recherche Matériaux et Energies Renouvelables URMER, University of Tlemcen, Abou Bekr Belkaid, BP 119, 13000 Tlemcen, Algeria
hadjersoufi1@gmail.com

[2] Theoretical and Applied Fluid Mechanics Laboratory, Department of Energetic and Fluid Mechanics, University of Science and Technology Houari Boumediene (USTHB), 16111 Alger, Algeria

**Abstract.** Recently, perovskite lead-free solar cells (PSCs) are a novel advancement in photovoltaic technology that have attracted attention due to their affordability, potential as a future solution for exceeding cell efficiency limits, PSCs have a poor thermal stability, to overcome this problem using $CsSnI_3$ perovskite material, $CsSnI_3$ perovskite solar cell is environmentally [1], friendly and inexpensive elements with high emission efficiency [2]. In this work, a device simulation of $CsSnI_3$ based solar cells is performed. Glass/FTO/ W $S_2$/$CsSnI_3$/HTL and metal back contact (Au) make up the architecture. SCAPS-1D software is employed in the simulation to examine the effectiveness and performances of this solar cell, we used $MoO_3$ as the hole transport layer (HTL) to identify the $WS_2$/$CsSnI_3$/HTL combinations showed an overall efficiency of 35.08%, FF of 79.98%, $J_{sc}$ of 35.80 mA/cm$^2$ and $V_{oc}$ of 1.28v. The effect of thickness of absorber layer, electron transport layer, hole transport layer and temperature was studied. For the second part, we theoretically implemented the optimized solar cell in an air-based hybrid photovoltaic/thermal (PV/T) solar collector using MATLAB. This approach enables a more versatile utilization of solar energy, allowing for simultaneous generation of electricity and heat in a more efficient manner.

**Keywords:** Perovskite Solar Cell · hybrid photovoltaic/thermal solar collector · Power Conversion Efficiency

## 1 Introduction

Solar energy is one of the most promising forms of renewable energy, inexhaustible, and readily available. It harnesses the power of the sun to convert light energy into heat using a thermal collector and into electricity using a photovoltaic (PV) cell.

Perovskite solar cells are a recent innovation that has generated significant interest due to their high performance and potential for reducing production costs [3]. They have a unique crystalline structure that allows efficient absorption of sunlight across a wide range of wavelengths, making them a promising option for more efficient and affordable solar energy production. However, lead-containing perovskite cells ($APbX_3$) present challenges in terms of long-term durability and stability due to the toxicity of lead, raising concerns for health and the environment [4]. Researchers are actively working on developing alternative perovskites to improve stability and reduce toxicity [5] as $CsSnI_3$ [6].

In this work, we are fully committed to pursuing this path and contributing to the improvement of lead-free perovskite cell performance. A theoretical study is conducted using a one-dimensional solar cell capacitance simulator (SCAPS-1D). The studied solar cell architecture is as follows: glass/FTO/ETL/$CsSnI_3$/HTL/Au. The effect of the thickness of the active, ETL, and HTL layers, as well as the influence of temperature on device performance, are analyzed and studied. Then, we intend to use the previously studied solar cell in a hybrid photovoltaic/thermal (PV/T) solar collector for simultaneous electricity and heat production, along with cooling of the PV cell.

## 2 Simulation Methodology

### 2.1 Device Simulation Details

The investigated cell in this study is a tin-based perovskite p-i-n device, as depicted in Fig. 1. The cell follows the general configuration of glass/FTO/ETL/$CsSnI_3$/HTL/Au. The ETL, or the n part of the cell, is composed of tungsten disulfide ($WS_2$). This material was chosen due to its favorable optoelectronic properties, such as a bandgap ranging from 1.33 to 2.2 eV, high electron transport mobility (approximately $260 \, cm^2 \, V^{-1} \, s^{-1}$) [7], and high transparency. The absorber, or the i part of the cell, is $CsSnI_3$, also known as cesium tin iodide. This material is selected for its efficient light absorption (bandgap of 1.3 eV) and relatively high thermal stability[8]. $MoO_3$ was chosen as representative inorganic HTL material due to its very high hole transport mobility (approximately $100 \, cm^2 \, V^{-1} \, s^{-1}$). It is cost-effective and has a desirable band alignment with the perovskite absorber. The input parameters characterizing the device were compiled from a combination of published experimental and theoretical data, which are provided in a summarized form in Table 1.

In this study, the numerical simulations were performed using the SCAPS-1D (Solar Cell Capacitance Simulator-one dimensional) tool version 3.3.10, developed by Mark Burgelman at the Department of Electronics and Information Systems (ELIS) of the University of Gent, Belgium. SCAPS-1D is widely used among PV researchers due to its ability to closely match simulation results with experimental data.

### 2.2 Description of Studied Collector PV/T Device

We exposed the optimized perovskite solar cell in a photovoltaic/thermal (PV/T) hybrid air collector. The schematic diagram of the studied PV/T collector is illustrated in Fig. 2. The PV/T collector investigated in this study is designed with a monocrystalline photovoltaic module as its core component. It comprises several elements, including a glazing layer, a protective glass layer, photovoltaic cells embedded in an EVA (Ethylene-Vinyl-Acetate) polymer layer, a Tedlar protective layer beneath the PV cells, an air duct for

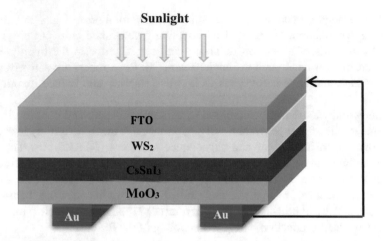

**Fig. 1.** Device structure of the p-i-n primary perovskite solar cell

**Table 1.** Properties of the FTO, ETL, absorber, and HTL material.

| Parameter | FTO [9] | W $S_2$ [10] | CsSnI$_3$ [11] | MoO$_3$[12] |
|---|---|---|---|---|
| Epaisseur (nm) | 100 | 200 | 1000 | 100 |
| Band gap, Eg (eV) | 3.5 | 1.87 | 1.3 | 3 |
| Affinity $\chi$, (eV) | 4 | 4.3 | 3.6 | 2.5 |
| Diélectric permittivity (relative), $\varepsilon$ | 9 | 11.9 | 9.93 | 12.5 |
| Effective density of state at CB, Nc (cm$^{-3}$) | $2.2 \times 10^{18}$ | $1.0 \times 10^{18}$ | $1.0 \times 10^{18}$ | $2.2 \times 10^{18}$ |
| Effective density of state at VB, Nv (cm$^{-3}$) | $1.8 \times 10^{19}$ | $2.4 \times 10^{19}$ | $1.0 \times 10^{19}$ | $1.8 \times 10^{19}$ |
| Mobility of electrons, $\mu_n$ (cm$^2$ V$^{-1}$ s$^{-1}$) | 20 | 260 | 1500 | 25 |
| Mobility of holes, $\mu_h$ (cm$^2$ V$^{-1}$ s$^{-1}$) | 10 | 51 | 585 | 100 |
| Density of n-type doping, ND (cm$^{-3}$) | $2.0 \times 10^{19}$ | $1.1 \times 10^{19}$ | 0 | 0 |
| Density of p-type doping, NA (cm$^{-3}$) | 0 | 0 | $1 \times 10^{16}$-$1 \times 10^{21}$ | $1.0 \times 10^{18}$ |
| Density of defects, Nt (cm$^{-3}$) | $1.0 \times 10^{15}$ | $2.0 \times 10^{11}$ | $1 \times 10^{15}$- $1 \times 10^{19}$ | - |

thermal energy transfer, a metal plate positioned above the insulator, and thermal insulation to minimize lateral heat loss. The rear side of the hybrid solar collector is also accounted for.

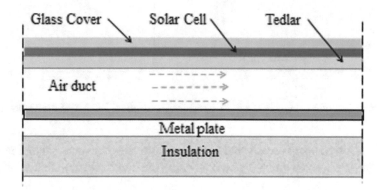

**Fig. 2.** Schematic cross-sectional view of the studied solar PV/T air collector.

## 3    Results and Discussion

### 3.1    Effect of Absorber Layer Thickness

In our simulation, to investigate the effect of the absorber layer thickness on the performance parameters of the photovoltaic cell, we varied the thickness of the active layer between 200 nm and 1200 nm, as illustrated in Fig. 3. The simulation results show that $J_{SC}$ increases with increasing absorber layer thickness and reaches a maximum value of around 35 mA/cm$^2$ for a thickness of 1100 nm. This indicates that a thicker active layer can absorb more photons and increase the concentration of excess photo-generated carriers, the $V_{OC}$ value decreases from 7.789 V to 2.539 V. The FF value decreases from 17% to 11% due to the presence of parasitic resistance as the thickness increases from 200 nm to 300 nm. However, it then increases to a maximum value of 31% for a thickness of 1200 nm. The solar cell efficiency increases with increasing absorber layer thickness, reaching a maximum value of approximately 28% for thicknesses of 1100 nm and 1200 nm.

### 3.2    Effect of External Operating Temperature

During regular operation, solar cells are typically installed on rooftops and are exposed to varying weather conditions, which can impact their stability. It has been observed that the temperature of the devices can be approximately twice as high as the ambient temperature, leading to stability issues. Figure 4 illustrates the change in photovoltaic performance with temperature ranging from 300 to 400 K. The decrease in PCE at higher temperatures can be attributed to a decline in the $J_{SC}$ and $V_{OC}$ parameters, as evidenced

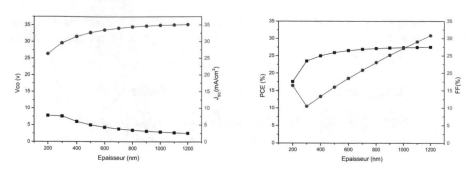

**Fig. 3.** Variation of $V_{OC}$, $J_{SC}$, PCE and FF as a function of the thickness of the absorber layer.

by Eq. (1). This equation reveals that as temperature rises, the dark saturation current ($J_0$) also increases, leading to a decrease in $V_{OC}$, consistent with our simulation results.

$$VOC = \frac{AKT}{e}Ln[\frac{J_{ph}}{J_0} + 1] \tag{1}$$

With increasing temperatures, the internal resistance may decrease as a result of the reduced resistivity of the materials employed. This facilitates improved current flow and minimized energy losses, ultimately leading to an enhancement in the FF.

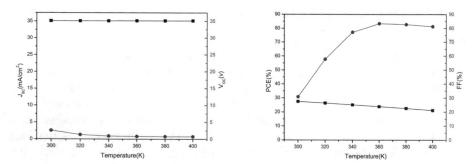

**Fig. 4.** Variation of photovoltaic performance with operating temperature.

### 3.3 Hybrid Collector Performance

The evolution of the electrical and thermal efficiencies of the air-based hybrid solar collector over time is observed. The graph in Fig. 5 depicts the variation of the electrical efficiency of the studied hybrid collector over time. It shows an increase in electrical efficiency in the morning, reaching a peak of approximately 33% at noon, followed by a period of stability in the afternoon until 5 pm. This indicates the influence of solar intensity and the characteristics of the perovskite photovoltaic module, such as its ideal bandgap and broad absorption of the solar spectrum, which facilitate the transport and collection of generated electrons and holes to generate electrical current. Towards the end of the day, there is a progressive decrease in efficiency. This may be attributed to the decrease in solar intensity as the sun sets, as well as operational factors such as shading or conversion losses.

The thermal efficiency gradually increases from 24% to 34% from 7 am to noon, which corresponds to the period of highest solar intensity and ambient temperature. This indicates efficient conversion of solar energy into thermal energy and maximum heat transfer from the PV/T collector to the cooling fluid. After this peak, the efficiency slightly decreases but remains relatively stable around 34% until 6 pm, with slight variations. This demonstrates the consistent performance of the PV/T collector in converting solar energy into thermal energy despite environmental fluctuations. Towards the end of the day, there is a progressive decrease in efficiency.

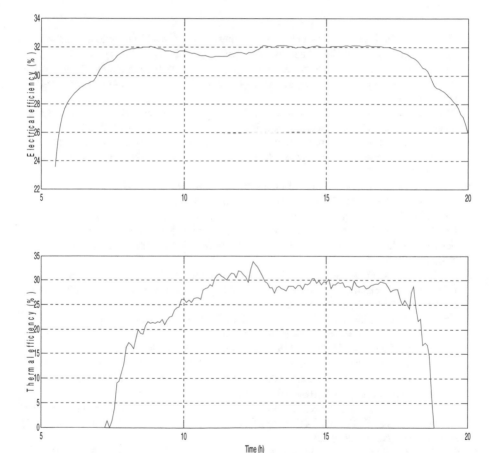

**Fig. 5.** Variation of electrical thermal efficiency and of the PV/T hybrid collector throughout the day.

## 4 Conclusion

In this study, we investigated the impact of absorber thickness and operating temperature using one-dimensional device simulation (SCAPS-1D). The performance of a perovskite solar cell with the architecture Glass/FTO/WS$_2$ /CsSnI$_3$/MoO$_3$/Au was designed and analyzed. Then, We investigated theoretically the optimized solar cell in an air-based photovoltaic/thermal (PV/T) hybrid solar collector using MATLAB. This was done to enhance the cooling efficiency of the photovoltaic module and improve its durability and reliability. The numerical results show that the electrical and thermal reach respectively 33% and 34%.

## References

1. Huaxia Ban, M.W., et al.: Efficient and stable mesoporous CsSnI3 Perovskite solar cells via imidazolium-based ionic liquid additive. Sol. RRL. **6**, 3642–3649 (2022)
2. Kumar, M.H., et al.: Lead-free halide perovskite solar cells with high photocurrents realized through vacancy modulation. Adv. Mater. **26**(41), 7122–7127 (2014). https://doi.org/10.1002/adma.201401991
3. Mohammadi, F.M.: Perovskite solar cell simulation based on perovskite CH 3 NH 3 PbI 3 through SCAPS-1D simulation software. Int. J. Adv. Acad. Stud. **3**(3), 1–2 (2021)
4. Filippetti, A., et al.: Fundamentals of tin iodide perovskites: a promising route to highly efficient, lead-free solar cells. J. Mater. Chem. A **9**(19), 11812–11826 (2021). https://doi.org/10.1039/D1TA01573G
5. Gholipour, S., et al.: Globularity-selected large molecules for a new generation of multication perovskites. Adv. Mater. **29**(38), 1–9 (2017). https://doi.org/10.1002/adma.201702005
6. Heo, J.H., Kim, J., Kim, H., Moon, S.H., Im, S.H., Hong, K.H.: Roles of SnX2 (X = F, Cl, Br) additives in tin-based halide perovskites toward highly efficient and stable lead-free perovskite solar cells. J. Phys. Chem. Lett. **9**(20), 6024–6031 (2018). https://doi.org/10.1021/acs.jpclett.8b02555
7. Hankare, P.P., Manikshete, A.H., Sathe, D.J., Chate, P.A., Patil, A.A., Garadkar, K.M.: WS 2 thin films: opto-electronic characterization. J. Alloys Compd. **479**, 657–660 (2009). https://doi.org/10.1016/j.jallcom.2009.01.024
8. Zhou, Y., Garces, H.F., Senturk, B.S., Ortiz, A.L., Padture, N.P.: Room temperature ' one-pot ' solution synthesis of nanoscale CsSnI 3 orthorhombic perovskite thin films and particles. Mater. Lett. **110**, 127–129 (2013). https://doi.org/10.1016/j.matlet.2013.08.011
9. Ouslimane, T., Et-taya, L., Elmaimouni, L., Benami, A.: Impact of absorber layer thickness, defect density, and operating temperature on the performance of MAPbI3 solar cells based on ZnO electron transporting material. Heliyon **7**(3), e06379 (2021)
10. Lohia, P., Dwivedi, D.K., Ameen, S.: A comparative study of quantum dot solar cell with two different ETLs of WS2 and IGZO using SCAPS-1D simulator. Solar **2**(3), 341–353 (2022). https://doi.org/10.3390/solar2030020
11. Lin, S., et al.: Inorganic lead-free B-γ-CsSnI 3 Perovskite solar cells using diverse electron-transporting materials: a simulation study. ACS Omega **6**(40), 26689–26698 (2021). https://doi.org/10.1021/acsomega.1c04096
12. Li, W., Li, W., Feng, Y., Yang, C.: Numerical analysis of the back interface for high efficiency wide band gap chalcopyrite solar cells. Sol. Energy **180**(January), 207–215 (2019). https://doi.org/10.1016/j.solener.2019.01.018

# Hydrodynamic Lubrication Analysis in Self-lubricating Journal Bearings with Power Law Fluid Model

M. Malki[1]($\boxtimes$), S. Boubendir[2], S. Larbi[3], and L. Boukhalkhal[4]

[1] LTI, Department GMP National High School of Technology Bordj El Kiffan, Algiers, Algeria
maamar.malki@enst.dz

[2] USTHB, BP 32, El Alia, 16111 Bab Ezzouar Algiers, Algeria

[3] LGMD, Polytechnic National School of Algiers, 10, Avenue Hassen Badi, Algiers, Algeria

[4] Dpt GM, Faculty of Science and Technology, University of Djelfa, Djelfa, Algeria

**Abstract.** The effect of non-Newtonian lubricants behavior using power law fluid model of self-lubricating porous journal bearings with sealed ends is analyzed. The nonlinear Reynolds equation is derived by considering both the fluid flow in the porous matrix and the lubricant rheological behavior where Darcy's law and power- law model were used. Governing differential equations were solved numerically using the finite difference method. Static characteristics are obtained by considering three types of lubricants: dilatant $n > 1$, pseudo-plastic $n < 1$ and Newtonian fluids $n = 1$. Obtained results showed that the power law index has important effects on the performance of porous and non-porous bearings. An improvement in the fluid bearing characteristics (load capacity, pressure) is observed for dilatant fluids. The permeability of the porous structure has significant effects on the performance of porous journal bearings of finite length, particularly at higher eccentricity ratios.

**Keywords:** Hydrodynamic · Self-lubricating · Power law model

## 1 Introduction

The porous journal bearing is a special case of fluid bearings used in situations of congestion (lack of space) or inaccessibility for lubrication. They operate quietly and maintenance-free, costing less than similar externally lubricated bearings [1].

The porous bearings are made from sintered metal based on bronze, with the incorporation of lubricant (oil, graphite, etc.) into the pores. In rotation of the bearing, the oil is sucked to lubricate the contact shaft-bush. It is reabsorbed by the bush when stopped. Unlike a conventional fluid bearing, the shaft-bush contact is never dry even at starting. In rotation, the hydrodynamic pressure created will ensure the circulation and supply of oil.

M. Hatti (Ed.): IC-AIRES 2023, LNNS 984, pp. 339–346, 2024.
https://doi.org/10.1007/978-3-031-60629-8_34

The recent studies or most of them conducted on porous bearings [2, 3] were dedicated to the study of hydrodynamic performances. They showed that the increase in permeability leads to a reduction in the performances of the bearings. Notice that the lubricating fluid considered in these studies is Newtonian.

Nevertheless, non-Newtonian effects have been observed in several problems, when high shear rate and high pressure gradient.

Several theoretical developments were established during recent years to describe the behavior of fluids. Most of them have successfully applied the power law [4, 5] in lubrication problems.

In several industrial activities [6], the lubricants used are often described by Ostwald equation. The use of this behavior law is due to the wide coverage in the analysis of lubricants as well as its mathematical simplicity. The exponent "n" of this law is called: structure index. The fluid behaves like a dilating fluid, whereas it corresponds to a pseudo-plastic fluid. It behaves like a Newtonian fluid. Other researchers have [7, 8] extended this study by taking into account thermal effects. They noted, in addition to the previous results, that the power law index and temperature are proportional. This increase is more significant in the case of dilatant fluids.

## 2   Physical and Mathematical Model

The porous journal bearing consists of a shaft and bearing separated by a lubricating film generating a hydrodynamic pressure field, under the effect of the load the two centers never coincide constituting an equilibrium angle (attitude angle). Figure 1 shows the simplified diagram of a sealed self-lubricating porous bearing.

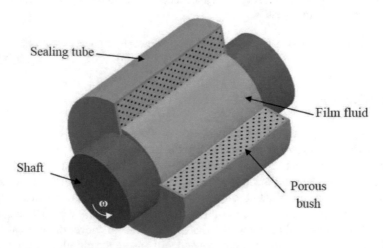

**Fig. 1.** Diagram of a sealed self-lubricating porous bearing.

The thickness of the film is expressed as:

$$h = C + e\,Cos\theta \tag{1}$$

$$\begin{cases} \varepsilon = \frac{e}{C} & \text{(eccenticity ratio)} \\ C = R_c - R_a & \text{(Radial clearance)} \\ \theta = \dfrac{x}{R_a} \end{cases} \tag{2}$$

The boundary conditions on the velocities in the porous bush are:

- On the bush:

$$U_1 = -\frac{K}{\mu}\frac{\partial p^*}{R\,\partial\theta}\bigg|_{R=R_c} \tag{3}$$

$$V_1 = -\frac{K}{\mu}\frac{\partial p^*}{\partial R}\bigg|_{R=R_c} \tag{4}$$

The exponent (*) designates the porous part.

- The velocity components at the interface Shaft-fluid are:

$$U_2 = U|_{h=H} = \omega R_a \tag{5}$$

$$V_2 = V|_{h=H} = \omega\frac{dh}{d\theta} \tag{6}$$

## A. *Power Law Fluid Behavior*

One of the empirical laws frequently used in rheology to characterize incompressible viscous fluids is that of Oswald. Its expression is:

$$\delta = m.\dot{\delta} \tag{7}$$

The coefficient of apparent viscosity, $\mu_a$, relating to this law, has the expression:

$$\mu_a = \frac{m}{\dot{\varepsilon}^{(1-n)}} \tag{8}$$

where: n designates the structure index and "m" the consistency (n and m > 0). The type of fluid depends on the power law index (n).

$$\mu = [2\Pi_\varepsilon]^{\frac{(n-1)}{2}} \tag{9}$$

where: $\Pi_\varepsilon$, is the strain rate tensor. $\delta$ and $\dot{\delta}$, represent the strain and the strain rate respectively.

## B. *Generalized Reynolds Equation*

The nonlinear Reynolds equation for a self-lubricating porous journal bearing using a fluid type power-law has been deduced from the fundamental thin film equations, and it is written in the following form [9]:

$$\frac{\partial}{\partial\theta}\left[\left(\bar{h}^3\bar{G} + \frac{\overline{KChF}}{\mu}\right)\frac{\partial\bar{p}}{\partial\theta}\right] = \frac{\partial}{\partial\theta}\left[\bar{h}(1-\overline{F})\right] - \frac{\overline{K}}{\mu}\frac{\partial\bar{p}^*}{\partial\overline{R}}\Big|_{\overline{R}=1} \tag{10}$$

where:

$$\overline{G} = \int_0^{\bar{h}}\frac{\bar{y}}{\mu}d\bar{y} - \overline{I}_2\overline{F}\;;\; \overline{F} = \frac{1}{\overline{J}_2}\int_0^{\bar{h}}\frac{\bar{y}}{\mu}d\bar{y}\;;\; \overline{I}_2 = \int_0^{\bar{h}}\frac{\bar{y}}{\mu}d\bar{y}\;;\; \overline{J}_2 = \int_0^{\bar{h}}\frac{d\bar{y}}{\mu} \tag{11}$$

The relative component at tangential speed is written in the form:

$$\bar{u} = \frac{\partial\bar{p}}{\partial\theta}\left[\bar{h}^2\left(\overline{I} - \overline{J}\frac{\overline{I}_2}{\overline{J}_2}\right) + \frac{KC}{\mu_b}\left(\frac{\overline{J}}{\overline{J}_2} - 1\right)\right] + \frac{\overline{J}}{\overline{J}_2} \tag{12}$$

With: $\overline{J} = \int_0^{\bar{y}}\frac{\xi}{\mu}d\xi$, $\overline{J} = \int_0^{\bar{y}}\frac{d\xi}{\mu}$

Equations governing the fluid flow are put into dimensionless form on the basis of the following reference quantities:

$$p = \frac{\bar{p}\mu_0\omega R_a R_c}{C^2}\;;\; \overline{R} = \frac{R}{R_c}\;;\; \bar{h} = \frac{h}{C}\;;\; \overline{K} = \frac{KR_c}{C^3}\;;\; \overline{C} = \frac{C}{R_c}\;;\; x = R_c\theta; y = \bar{y}h\;;\; (u,v) = \omega R_a(\bar{u},\bar{v})\;;\; \overline{\mu} = \frac{\mu}{\mu_0} \tag{13}$$

If we neglect the non-Newtonian effect in the porous matrix (constant dynamic viscosity), the Darcy's [10] equation governed the flow in the porous matrix becomes:

$$\frac{1}{\overline{R}}\frac{\partial}{\partial\overline{R}}\left(\frac{\overline{R}}{\mu}\frac{\partial\bar{p}^*}{\partial\overline{R}}\right) + \frac{1}{\overline{R}^2}\frac{\partial}{\partial\theta}\left(\frac{1}{\mu}\frac{\partial\bar{p}^*}{\partial\theta}\right) = 0 \tag{14}$$

The set of boundary conditions used for the resolution of the complex mathematical system are:

- The condition of Swift [11] and Stieber [12] are used for the cavitation zone.
- The boundary conditions of a sealed porous bush.
- The condition of equal pressures at the porous matrix-fluid film at the interface.

## C. *Static Characteristics*

The integration of the pressure at equilibrium gives the attitude angle and the load:

$$\overline{Q} = \frac{Q}{\mu_0\,\omega\,R_b^3 L/C^2} = \sqrt{\overline{Q}_L^2 + \overline{Q}_K^2} \tag{15}$$

$$\tan\phi = -\overline{Q}_L/\overline{Q}_K \tag{16}$$

where:

$$\overline{Q}_L = \int_0^{2\pi} \overline{P}\,Sin\theta\,d\theta \qquad (17)$$

$$\overline{Q}_K = \int_0^{2\pi} \overline{P}\,Cos\theta\,d\theta \qquad (18)$$

## 3   Results Presentation and Discussions

The governing equations given above have been solved numerically using an iterative Gauss-Seidel method ($\Omega = 1.8$) [13]. A calculation code was thus developed.

In the following, we will consider a self-lubrication porous bearing, whose characteristics are as follows:

$\overline{C} = 2.10^{-3}$; $\varepsilon = 0.7$; $R_{ec} = 1.5$. $\mu$ the dynamic viscosity is defined by the power law.

The distribution of the fluid dimensionless pressure, function of the circumferential direction for a relative eccentricity value of 0.7, is represented in Fig. 2, for different values of the structure index for the porous and non-porous bearing cases. The figure shows that the dimensionless pressure increases with the increase of the structure index and decreases with the increase in permeability.

Figure 3 illustrates the variation of the load versus of the eccentricity ratio, for three values of the structure index for the porous and non-porous bearing cases. It is noted the considerable increase of the load with the dilating fluid. This increase becomes significant with increasing relative eccentricity and in the case where $k = 0$ (non-porous bearing).

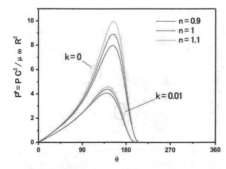

**Fig. 2.** Evolution of dimensionless circumferential pressure function of structure index and permeability.

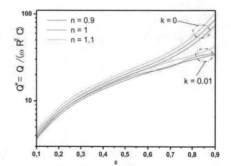

**Fig. 3.** Load variation with relative eccentricity.

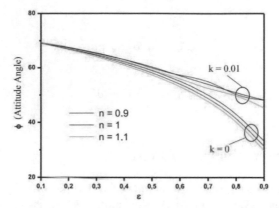

**Fig. 4.** Variation of attitude angle versus the eccentricity ratio.

Figure 4 shows the variation of the attitude angle as function of the eccentricity ratio, for three values of the structure index in the case of porous and non-porous bearing. Notice that the equilibrium angle is inversely proportional to the relative eccentricity. It is also noted that the structure index does not have a big influence on the attitude angle for small relative eccentricities, on the other hand the pseudo-plastic fluids have a bigger attitude angle for large relative eccentricities. We can also be seen that the attitude angle increases with the increase of the permeability.

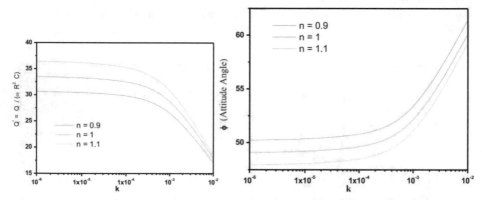

**Fig. 5.** Variation of load versus the permeability for different values of structure index.

**Fig. 6.** Variation of the attitude angle versus the permeability for different structure index values.

The variation of the carrying load versus dimensionless permeability for three values of structure index is shown in Fig. 5 where we notice the proportionality between the load capacity and the power law index and the decrease of the load with the increase in permeability from a critical value of permeability $k = 10^{-4}$.

Figure 6 illustrates the variation of the attitude angle versus the permeability for different values of power law index. We can remark that the attitude angle decreases with the increase of the structure index and increases with the increase of the dimensionless permeability from a critical value $k = 10^{-4}$.

Table 1 summarizes a comparative study of the characteristics of a porous bearing for non-Newtonian or Newtonian fluids cases in the porous matrix. We notice that the effect of the viscosity variation in the porous matrix is negligible.

**Table 1.** Comparison between non newtonian and newtonian effects in the porous medium.

| | n | Non Newtonian Fluids | | Newtonian Fluids | | Difference in % | |
|---|---|---|---|---|---|---|---|
| | | $P_{max*}$ | Q | $P_{max*}$ | Q | $P_{max*}$ | Q |
| K = 0.01 | 0.9 | 4,07 | 17,04 | 4,10 | 17,16 | 1% | 1% |
| | 1 | 4,38 | 17,94 | 4,38 | 17,94 | 0% | 0% |
| | 1.1 | 4,58 | 18,40 | 4,61 | 18,47 | 1% | 1% |

## 4  Conclusion

In this work, the analysis of the lubrication of porous fluid bearings under hydrodynamic regime using a non-Newtonian fluid lubricant following the behavior of the power law has been studied. A porous fluid bearing performance calculation code was developed to determine the static characteristics.

Obtained results showed:

- There is a link between the reduction in the load and pressure, as well as the attitude angle and the permeability of the media constituting these bearings.
- The increasing of structure index increases the pressure and hydrodynamic load and a decrease in the stability angle.
- The increase in the eccentricity ratio causes a decrease in the stability angle and an increase of the load capacity.
- The effects of the non-Newtonian behavior of the fluids, on the characteristics of a hydrodynamic bearing, are negligible for the case of porous bearing.

## References

1. Braun, A.L.: Porous bearings. Tribol. Int. **15**(5), 235–242 (1982)
2. Guha, S.K.: Linear stability performance analysis of finite hydrostatic porous journal bearings under the coupled stress lubrication with the additives effects into pores. Tribol. Int. **43**(8), 1294–1306 (2010)
3. Balasoiu, A.M., Minel, J.B., StefanI, M.: A parametric study of a porous self-circulating hydrodynamic bearing. Tribol. Int. **61**, 176–193 (2013)
4. Chu, H.M., Li, W.L., Chen, M.D.: Elastohydrodynamic lubrication of circular contacts at pure squeeze motion with non-Newtonian lubricants. Tribol. Int. **39**(9), 897–905 (2006)

5. Wu, Y.S., Pruess, K.: Numerical method for simulating non-Newtonian fluid flow and displacement in porous media. Adv. Water Resour. **21**(5), 351–362 (1998)
6. Sivakumar, P., Prakash, R.B., Chhabra, R.P.: Steady flow of power-law fluids across an unconfined elliptical cylinder. Chem. Eng. Sci. **62**(6), 1682–1702 (2007)
7. Nessil, A., Larbi, S., Belhaneche, H., Malki, M.: Journal bearings lubrication aspect analysis using Non-Newtonian fluids. Adv. Tribol. **213**, 1–9 (2013)
8. Boubendir, S., Larbi, S., Bennacer, R.: Numerical study of the thermo-hydrodynamic lubrication phenomena in porous journal bearings. Tribol. Int. **44**(1), 1–8 (2011)
9. Malki, M., Larbi, S., Boubendir, S., Hammoudi, D., Bennacer, R.: Elastic deformation effects on self-lubricating journal bearings using pseudo-plastic lubricants. Tribol. Online **16**(4), 299–308 (2021)
10. Darcy, H.: Les fontaines publiques de la ville de Dijon. Edition Dalmont, Paris, France (1856)
11. Swift, H.W.: The stability of lubricating films in journal bearings. Proc. Inst. Civil Eng. UK **233**, 267–288 (1932)
12. Stieber, W.: Das Schwimmlager. VDI, Berlin, Germany (1933)
13. Christopherson, D.G.: A new mathematical method for the solution of film lubrication problems. Proc. Inst. Mech. Eng. **146**(1), 126–135 (1941)

# Physical Properties of TiO2 Nanotubes Grow in Different Electrolytes by Varying Anodizing Parameters

Henia Fraoucene[1]([✉]), El Hadi Khoumeri[2], and Djedjiga Hatem[1]

[1] Laboratory of Advanced Technologies of Genie Electrics (LATAGE), Faculty of Electrical and Computer Engineering, Mouloud Mammeri University (UMMTO), BP 17 RP 15000, Tizi-Ouzou, Algeria
henia.fraoucene@ummto.dz

[2] Laboratory of Innovative Technologies (LTI), National School of Technology, Algiers, Algeria

**Abstract.** Highly ordered $TiO_2$ nanotubes ($TiO_2$ NTs) were successfully prepared by electrochemical anodization process. Titanium (Ti) foil was anodized at 20V in three different electrolyte; one organic consist on glycerol containing $NH_4F$ and water, and two aqueous solution [$(NH_4)_2SO_4$ and $Na_2SO_4$] containing different amount of $NH_4F$ at varying anodization time. The $TiO_2$ NTs were characterized using Scanning Electron Microscopy (SEM), X ray Diffraction (XRD) and UV-Visible spectrophotometer. SEM images shows homogeneous, regular and aligned $TiO_2$ NTs morphology obtained using organic solution, whereas not well nanotubular structure with lower ordering degree was produced in aqueous solution due to the high chemical dissolution rate of titanium dioxide ($TiO_2$) ensured by the high fluoride ions concentration; and by the extended anodization time. From the XRD analysis, the annealed samples at 450 °C for 3h indicate apparition of a single anatase phase that exhibits a high optical absorption band edge for the $TiO_2$ NTs developed in glycerol electrolyte where the morphological parameters are well improved.

**Keywords:** $TiO_2$ nanotubes · anodization process · electrolyte · properties · nanomaterials

## 1 Introduction

The Titanium dioxide (TiO2) is a semiconductor material that has been actively researched due to its chemical stability, non toxicity and biocompatibility [1, 2]. These unique properties enable it to be utilized as an active component in several devices; specifically, photocatalysis, photovoltaics [3] and biomedicine [4]. In recent decades, research has shown that for these applications; nanotubular structure of titanium dioxide (TiO2 NTs) has a major interest due to their one dimensional (1D) nature, ease of handling and simple preparation.

M. Hatti (Ed.): IC-AIRES 2023, LNNS 984, pp. 347–354, 2024.
https://doi.org/10.1007/978-3-031-60629-8_35

Several methods have been used on the synthesis of TiO2 NTs including the hydrothermal synthesis [5], template assisted [6] and anodic oxidation [7]. Electrochemical anodization was generally privileged due to its simplicity, low cost and efficient process to fabricate TiO2 NTs structures. The anodization process is performed in an electrochemical cell consisting of two electrodes, the counter electrode (the cathode) generally is a platinum bar and the metal (Ti) is employed as an anode where the oxidation process take place to obtain different TiO2 structures such as a compact oxide, a disordered porous layer and highly self-organized nanotubular layer. Assefpour-Dezfuly.[8] in 1984 and Zwilling et al.[9] in 1999, were the first reported the formation of TiO2 porous films by anodization of Ti foil in chromic acid (CA) (0.5 mol 1-1 Cr2O3) solution with and without hydrofluoric acid (HF) (9,5*10–2 mol 1-l). Over the past years, the control over the morphology (diameter, length, thickness,…) was strongly improved by continuously optimizing the anodizing conditions. Qin et al. [10] study the effect of anodization voltage and time on the morphology of TiO2 NTs synthesized from anodization of Titanium foil (Ti) in ethylene glycol (EG) solution containing 0.25 wt% NH4F with 2vol% water. They showed that pore diameter and length of nanotubes increase with increasing the anodization voltage and time respectively. Pasikhani et al. [11] fabricated Highly ordered and aligned TiO2 NTs using two-step anodization process at different voltages. The results showed that due to an increase in the anodization voltage from 40 V to 60 V, the nanotubes length and diameter linearly increase from 6 μm to 10 μm and from 54 nm to 101 nm, respectivelyIn addition to anodization voltage, time and NH4F concentration, the nature of electrolyte strongly affects the morphology and the structure of TiO2 NTs. The first proposition to use HF containing water media were only able to produce 0.5μm long of TiO2 NTs due to the high chemical dissolution of TiO2 in the acidic electrolyte [12]. Recently, Taib et al. [13] synthesized TiO2 NTs by anodization in KOH/fluoride/EG electrolyte at 60V. The result exhibit that the addition of KOH reduces the oxide dissolution process ~ 700 nm long nanotubes in 1 min. In this study, we investigated the effects of anodization time and fluoride ions (F −) concentration on morphological, structural and optical properties of TiO2 nanotubes arrays produced in organic (Glycérol) and aqueous electrolytes (Ammonium Sulfate and Sodium Sulfate) using the anodic oxidation method.

## 2 Experimental

### 2.1 Substrate Preparation

The titanium (Ti) foils (0.25 mm, 99.5% purity, Sigma-Aldrich) were cut into square shape (1.5 cm x 1.5 cm) with a selected work area of 0.6 cm$^2$. The samples were subjected to a final polishing using a rotating felt pad (01 μm) impregnated with alumina until a metallic mirror surface was obtained. The sprayed Ti foils were degreased by sonication in acetone, methanol and 2-Propanol for 10 min respectively, rinsed with ultrapure water and dried in a stream of compressed air before anodization.

### 2.2 Synthesis of TiO$_2$ Nanotubes

To synthezise TiO$_2$ NTs, electrochemical anodization process was carried out using a direct current power supply in a two-electrode electrolytic cell with the pretreated Ti foil as the working electrode and Pt wire as the counter electrode under constant potential (20V) at room temperature. Three types of electrolytes were adopted in this work to investigate the physical properties of $TiO_2$ NTs based on organic and aqueous solutions. The prepared organic glycerol solution containing 1.3 wt% $NH_4F$ and 2 wt% $H_2O$. Two different nature of aqueous solution were used; the first solution based on Ammonium Sulfate (1M $(NH_4)_2SO_4$) and the second was the Sodium Sulfate (1M $Na_2SO_4$) using two different fluoride concentration of 0.5 wt% and 1 wt%. For all experiences, the anodization time varying from 1 to 2 h. After anodization, the samples were soaked in ultrapure water for 10 min and then dried in an oven at 50 °C for 10 min. In order to transform the amorphous crystallographic structure obtained just after electrochemical anodization into crystalline structure, the samples were annealed at 450 °C for 3 h with a heating and cooling rate of 5 °C/min.

## 2.3 Characterization of the Samples

The Morphological characterization of theTiO₂ NTs was investigated using a Scanning Electron Micrographs (SEM). The crystalline phases were characterized by X- Ray Diffraction (XRD) analysis was performed with an X"pert Philips MPD with a Panalytical X'celerator detector using graphite monochromized CuKα radiation ($\lambda = 1.5418°A$) was used. The measurements were performed within the range of 2θ from 20° to 70°. The optical properties were investigated using UV spectrophotometer from 250 to 700 nm.

# 3 Experimental Results

Figure 1 shows Scanning Electron Microscope (SEM) images of $TiO_2$ NTs growing in glycerol organic electrolyte containing 1.3 wt% NH4F and 2wt% H2O at 20V by varying the anodization time. From the SEM images, it is evident that the anodization in glycerol electrolyte solution produced a beautiful structure of self-organized TiO2 NTs with homogeneous and regular distribution along the surface, more ordered, good uniformity, well separated and aligned morphology. In general, the electrochemical anodization process is explained in terms of field-assisted oxidation; since the TiO2 grows on the Ti foil as a result of the movement of ions under the applied electric field-assisted according to the reaction 1 [14]:

$$Ti + 2H_2O \rightarrow TiO_2 + 4H^+ \tag{1}$$

In addition, the fluoride ions present in the electrolyte can chemically dissolve (chemical dissolution) the oxide layer formed on the surface of Ti foil by the formation of soluble fluoride complexes $[TiF6]^{2-}$ assisted by electric field according to the relations 2 and 3. The localized dissolution of oxide resulted in the formation of small holes on the oxide layer and gradually grew into bigger pores with increasing anodization time to form a tube.

$$Ti^{4+} + 6F^- \rightarrow [TiF6]^{2-} \tag{2}$$

$$TiO_2 + 6F^- + 4H^+ \rightarrow [Ti F_6]^{2-} + 2H_2O \tag{3}$$

when the rate of electrochemical etching at the metal/oxide interface equals to the rate of chemical dissolution at the pore/electrolyte interface, the thickness of the nanotubular layer remains unchanged.

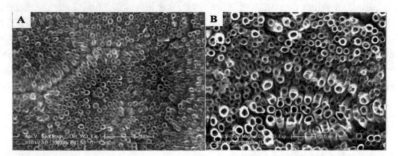

**Fig. 1.** SEM images of TiO$_2$ NTs synthesized at 20V in Glycérol + 1.3wt% NH$_4$F + 2wt% H$_2$O during: 1h (A) and 2h (B).

The morphological parameters of the TiO2 NTs developed in glycerol electrolyte during 1h presented in Fig. 1a, shows a regular nanotubular structure consists of tube arrays with an inner diameter varying between 13–27 nm and an average spacing of 6 nm. By increasing the anodization time to 2h (Fig. 1b), the morphological parameters were significantly improved characterized by an inner diameter varying between 26–53 nm and the walls thickness approximately 9 nm. The morphological difference between the two samples caused by an extended anodization time and the viscosity of the glycerol solution. The viscosity of electrolyte favors the diffusion of the spaces present in the solution and permits the dissolution rate of F- ions during the anodizing process, result on the formation of TiO2 NTs with dissimilar morphological parameters.

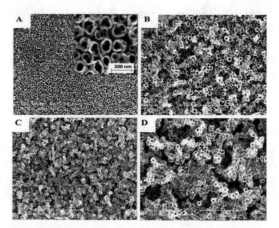

**Fig. 2.** SEM images of TiO$_2$ NTs synthesized at 20V in 1M (NH$_4$)$_2$SO$_4$ containing: 0.5wt%NH$_4$F (A and B) and 1wt%NH$_4$F (C and D) during 1h (A and C) and 2h (B and D).

To investigate the effect of electrolyte on the resulting morphology, a series of anodization experiences were performed using two different aqueous solution including Ammonium Sulfate (1M NH4)2SO4) and Sodium Sulfate (1M Na2SO4) at 20V with varying the anodization time and F- ions concentration. Figures 2 and 3 reveal the SEM images of TiO2 NTs synthesized in different aqueous solutions. A nanotubular structure with an inner diameter approximately 64nm was obtained a constant voltage of 20 V in ammonium sulfate solution containing 0.5wt% NH4F during 1 h of anodization (Fig. 2a). By increasing the anodization time to 2h (Fig. 2B), non uniform and not well nanotubular structure can be clearly observed with a pore diameter of ~ 58 nm. This behavior is due to the partial collapse of the tubes due to the excessive dissolution at the top of tubes by fluoride ions. However, an additional amount of 1wt% NH4F for 1h anodization reveals a remarkable lower ordering degree of TiO2 NTs organization (see in Fig. 2C). For an extended anodization time to 2h (Fif.2D), the nanotubular structure was completely destroyed resultant to the high dissolution rate of TiO2 by the presence of great amount of F- ions in the solution, thus preventing the formation of nanotubes.

For these results, it evident that both F- ions concentration and anodization time, are key parameters to achieve high-aspect-ratio of TiO2 NTs in appropriate electrolyte. Gong et al. [12] synthesized TiO2 TNTs with an average inner diameter of 25–65 nm and tube length shorter than 500 nm by anodization in aqueous solution containing 0.5–3.5% HF which plays the role of source of fluoride ions. Then, we can deduce that fluoride concentration in electrolyte is assumed to play an important role during anodization to carry out the dissolution reaction.

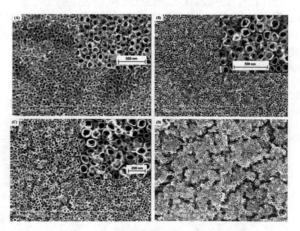

**Fig. 3.** SEM images of TiO$_2$ NTs synthesized at 20V in 1M Na$_2$SO$_4$ containing: 0.5wt% NH$_4$F (A and B) and 1wt% NH$_4$F (C and D) during 1h (A and C) and 2h (B and D).

Figure 3 shows SEM images of TiO2 NTs synthesized at 20V in 1M Na2SO4 with different anodization time and fluoride concentration. Different anodique TiO2 NTs architectures were observed. The films formed at 20 V during 1h of anodization using 0.5wt% NH4F (Fig. 3A) exhibit a nanotubular structure with an inner diameter ranged from 48 to 83 nm and walls thickness about 25nm. For two hour of anodization (see in. Figure 3B), nanotube structure was obtained with pore diameters varying between

43 to 86nm; in (NH4)2SO4 solution the pore diameter measured is about ~ 58 nm with non uniform and not well nanotubular structure was observed (Fig. 2B). However, regular nanotubular structure with high aspect ratio was obtained using glycerol organic solution (Fig. 1B). The difference in apparent morphology under identical conditions (20V, 2h) has led to an important distinction between aqueous and organic electrolytes: fluoride ions in aqueous solutions are much more aggressive than in organic media, which explains the growth of TiO2 NTs in aqueous solutions is limited to lengths of a few microns. By increasing both anodization time and fluoride concentration (2h and 1wt% NH4F respectively), the nanotubular structure obtained is completely destroyed with very low ordering degree of TiO2 NTs (see. Figure 3D) which is explained by the high dissolution rate of oxide formed on the Ti surface according to the reaction 2and 3. From the results obtained, the key to achieve high-aspect-ratio growth of TiO2 NTs is to adjust the dissolution rate of TiO2 by selecting an appropriate electrolyte containing an optimum concentration of fluoride ions to ensure TiO2 NTs growth but minimize dissolution.

Figure 4a and 4b show the XRD patterns of the samples anodized at 20V for 2h in glycerol containing 1.3 wt% NH4F and 2wt% H2O before and after annealing at 450°C for 3 h respectively. The XRD measurement of as-anodized sample (Fig. 4a) revealed that the self-organized TiO2 NTs had an amorphous structure, as only Ti-peaks were shown and it was originated from the Ti foil. The XRD pattern of the sample after annealing at 450 °C showed anatase and Ti peaks. Ti-peaks were present because the substrate information was revealed. The anatase phase peaks appear at $2\theta = 25.60°$, $37.95°$, $48.35°$ and $55.10°$, which can be indexed respectively to the (101), (004), (200) and (211) crystal faces of anatase TiO2 (JCPDS Pattern no 00–021-1272). Figure 4c shows SEM image of annealed TiO2 NTs at 450°C for 3 h; synthesized in fluoride glycerol electrolyte at 20V for 2h. The result indicated that the morphology of TiO2 NTs during the annealing remains unchanged.

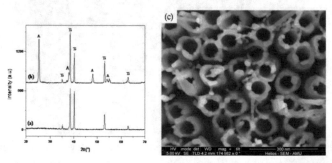

**Fig. 4.** XRD patterns of TiO$_2$ NTs: as-anodized (a), film annealed at 450 °C (b). "A" is Anatase, "Ti" is Titanium.

The optical properties of annealed TiO$_2$ NTs at 450 °C for 3 h synthesized at 20 V in organic/aqueous solutions by varying the anodization parameters, as displayed in Fig. 5. For all TiO$_2$ NTs samples, the strong adsorptions are in the range of 250 to 350 nm which is in good agreement with the comportment of TiO$_2$ film in ultraviolet range. Using glycerol solution, it can be seen from Fig. 5 that the band gap absorption edge increases until 375 nm with increasing the anodization time (2h). This behavior can be attributed to the improvement on surface morphology with formation of homogeneous, regular and more ordered $TiO_2$ NTs (Fig. 1b); which increases the reactionnel surface thus allowing the photons incorporation in tube tops. In addition, apparition of anatase phase at 450 °C in XRD patterns favors better photons absorption due to the width of the band gap of anatase. Furthermore, using aqueous solution to fabricate $TiO_2$ NTs with different anodization parameters; the same behavior was observed for all samples with the absorption intensity is a little lower compared to that of samples developed in the organic solution. This behavior is explained by the fact that the structure is non-uniform and not very nanotubular. From these results, it can be deduced that the morphology of $TiO_2$ NTs plays an essential role in the photons absorption by the improvement of the specific surface, while developing these nanotubes in suitable electrolytes.

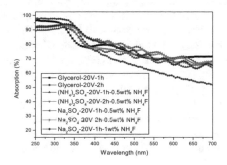

**Fig. 5.** UV-Vis absorption spectra of $TiO_2$ NTs synthesized in different electrolyte.

## 4   Conclusion

The formation of Titanium dioxide nanotubes (TiO2 NTs) in fluoride organic (glycerol) electrolyte and two aqueous solution [(NH4)2SO4 and Na2SO4] containing different amount of NH4F at varying anodization time has been investigated. Different anodiques TiO2 NTs architectures were observed, a regular nanotubular structure was obtained in fluoride organic solution; with an inner diameter varying between 13–27 nm and 26–53 nm for films synthesized for 1h and 2h respectively; assured by a competition between electrochemical etching and chemical dissolution of oxide. Whereas using aqueous solution, the morphology was greatly affected by the fluoride ions concentration and anodization time. For anodization at 20V during 1h in (NH4)2SO4 solution containing 0.5wt% NH4F, a nanotubular structure was produced with tube diameter about 64nm. By increasing anodization time to 2h, non uniform and not well nanotubular structure has developed explained by the excessive dissolution of the tubes at the top by fluoride ions. Moreover, by increasing anodization time and fluoride ions concentration in aqueous solutions, the morphological structure was greatly affected and completely destroyed with very low ordering degree of TiO2 NTs. So, a compromise between the fluoride ions concentration and the anodizing time is required to growth high aspect-ratio of TiO2

NTs in an appropriate electrolyte. From XRD analysis, amorphous structure was confirmed for as-anodized layers; however, anatase phase was obtained for annealed TiO2 NTs films at 450 °C for 3h. The optical absorption band edge of the growth TiO2 NTs in glycerol solution at 20V for 2h was enhanced until 375 nm due to their improvement in morphological structure and formation of anatase phase.

# References

1. Morozova, M., Kluson, P., Krysa, J., Vesely, M., Dzik, P., Solcova, O.: Electrochemical properties of TiO2 electrode prepared by various methods. Procedia Eng. **42**, 573–580 (2012)
2. Reszczyńska, J., et al.: Lanthanide co-doped TiO2: the effect of metal type and amount on surface properties and photocatalytic activity. Appl. Surf. Sci. **307**, 333–345 (2014)
3. Mor, G.K., Varghese, O.K., Paulose, M., Shankar, K., Grimes, C.A.: A review on highly ordered, vertically oriented TiO2 nanotube arrays: fabrication, material properties, and solar energy applications. Sol. Energy Mater. Sol. Cells **90**, 2011–2075 (2006)
4. Van Noort, R.: Titanium: the implant material of today. J. Mater. Sci. **22**, 3801–3811 (1987)
5. Liu, N., Chen, X., Zhang, J., Schwank, J.W.: A review on TiO2-based nanotubes synthesized via hydrothermal method: formation mechanism, structure modification, and photocatalytic applications. Catal. Today **225**, 34–51 (2014)
6. Bavykin, D.V., Friedrich, J.M., Walsh, F.C.: Protonated titanates and TiO2 nanostructured materials: synthesis, properties, and applications. Adv. Mater. **18**, 2807–2824 (2006)
7. Khudhair, D., et al.: Anodization parameters influencing the morphology and electrical properties of TiO2 nanotubes for living cell interfacing and investigations. Mater. Sci. Eng., C **59**, 1125–1142 (2016)
8. Assefpour-Dezfuly, M., Vlachos, C., Andrews, E.H.: Oxide morphology and adhesive bonding on titanium surfaces. J. Mater. Sci. **19**, 3626–3639 (1984)
9. Zwilling, V., Aucouturier, M., Darque-Ceretti, E.: Anodic oxidation of titanium and TA6V alloy in chromic media. An electrochemical approach. Electrochimica Acta **45**, 921–929 (1999)
10. Qin, L., et al.: Effect of anodization parameters on morphology and photocatalysis properties of TiO2nanotube arrays. J. Mater. Sci. Technol. **31**, 1059–1064 (2015)
11. Pasikhani, J.V., Gilani, N., Pirbazari, A.E.: The effect of the anodization voltage on the geometrical characteristics and photocatalytic activity of TiO2 nanotube arrays. Nano-Struct. Nano-Objects **8**, 7–14 (2016)
12. Gong, D., et al.: Titanium oxide nanotube arrays prepared by anodic oxidation. J. Mater. Res. **16**, 3331–3334 (2001)
13. Taib, M.A.A., Razak, K.A., Jaafar, M., Lockman, Z.: Initial growth study of TiO$_2$ nanotube arrays anodised in KOH/fluoride/ethylene glycol electrolyte. Mater. Des. **128**, 195–205 (2017)
14. Xue, Y., Sun, Y., Wang, G., Yan, K., Zhao, J.: Effect of NH4F concentration and controlled-charge consumption on the photocatalytic hydrogen generation of TiO$_2$ nanotube arrays. Electrochim. Acta **155**, 312–320 (2015)

# Analyzing the Generated Power for Different CubeSats Solar Panel Configurations

Hadj-Dida Abdelkader[1,3]([✉]) and Maamar Djamel Eddine[2]

[1] Department of Research and Space Instrumentation DRIS, Satellites Development Centre
CDS, Algerian Space Agency ASAL, Ibn Rochd USTO-MB, Po Box 4065, 31130 Bir El Djir,
Oran, Algeria
ahadjdida@cds.asal.dz
[2] Department of Space Engineering DING, Satellites Development Centre CDS, Algerian Space
Agency ASAL, Ibn Rochd USTO-MB, Po Box 4065, 31130 Bir El Djir, Oran, Algeria
dmaamar@cds.asal.dz
[3] Department of Electrical Engineering, University of Sciences and Technology of Oran
Mohamed Boudiaf USTO-MB BP 1505, El M'naouer USTO, 31130 Bir El Djir, Oran, Algeria
abdelkader.hadjdida@univ-usto.dz

**Abstract.** This paper presents an analysis study of the generated power for different CubeSat solar panel configurations using the Systems Tool Kit (STK) software. CubeSats are small satellites used for various applications, including scientific research, Earth observation, and communication; they are designed to be cost-effective and quick to manufacture, making them ideal for space missions with limited budgets and short development timelines. Solar panels are an essential component of CubeSats as they provide the necessary power to operate the satellite's systems, that their configuration affects the amount of power generated which is increasing since the missions are getting more and more sophisticated and complicated. In this study, the performance comparison of different CubeSats with fixed and/or deployable solar panel configurations was evaluated using STK software. The simulation results show that the solar panel configuration significantly affects the generated power, and the optimal configuration depends on the satellite's orbit and mission requirements. This analysis study can provide valuable insights into the design of CubeSats solar panels systems, contributing to the development of more efficient and reliable CubeSats and optimize their solar power generation to greatly enhance the success of their missions.

**Keywords:** CubeSats · Generated power · Solar panels · Panels configurations · Power analysis · STK software

## 1 Introduction

CubeSats are small satellites with a standard size of $10 \times 10 \times 10$ cm and a maximum weight of 1.33 kg that are designed for a variety of scientific and commercial missions. They typically have limited space and power budgets, which pose significant challenges for their design and operation. One critical component of CubeSats is their solar panels,

© The Author(s), under exclusive license to Springer Nature Switzerland AG 2024
M. Hatti (Ed.): IC-AIRES 2023, LNNS 984, pp. 355–370, 2024.
https://doi.org/10.1007/978-3-031-60629-8_36

which are responsible for generating power to run the satellite's electronics during different mission phases. In this paper, we present an analysis study of the generated power for different CubeSat solar panel configurations using the Systems Tool Kit (STK) software to simulate the power generation of the CubeSats during different mission scenarios. The solar generator model is designed based on the specifications of a commercial CubeSats solar generator, which consists of a set of photovoltaic panels and batteries. The simulation involves modelling the CubeSats orbit and attitude, as well as the solar illumination conditions during the mission. The STK software is used to calculate the solar flux on the photovoltaic panels and the power generated by the solar generator. The generated power is then used to simulate the consumption of the on-board systems during different mission phases [1]. As State-of-the-Art of CubeSats Solar Power Generation on Small-Sats, there were several academic papers and research studies published on the topic of analyzing CubeSats solar power generation. Power generation on CubeSats is a necessity typically governed by a common solar power architecture. As the Small-Sat missions drives the need for lower cost and increased production rates of space solar arrays, the standardization of solar array and panel designs, deployment mechanisms, and power integration will be critical to meet the desire of large proliferated constellations. Here are some research works and papers: [2] This paper proposes a method for power estimation of CubeSats with high-aspect-ratio solar panels, which is based on the analytical modelling of the solar panel system and considers the shadowing effect of the spacecraft body. [3] This paper presents a comprehensive approach to solar power system analysis and modelling for CubeSats, which includes the development of a mathematical model, the integration of simulation tools, and the validation of the results through experimental tests. [4] This paper proposes a framework for predicting power generation from solar panels on CubeSats, which is based on the use of machine learning algorithms and considers factors such as solar panel orientation, temperature, and degradation over time [5]. This paper presents an experimental validation of a solar panel model for CubeSats in low Earth orbit, which is based on the use of a sun sensor, a temperature sensor, and a voltage-current curve tracer. Overall, these papers demonstrate the importance of developing an approach for analyzing CubeSats solar power generation, and highlight the various methods and tools that can be used to achieve accurate results [6].

## 2 Design of the Project Scenario for Analyzing Cubesats Solar Power Generation Under STK Software

### 2.1 Material and Methods

This part aims to create a flowchart for calculating the energy balance of a model with fixed and deployable solar panels for a CubeSat in low orbit. To do this, it is essential to import the microsatellite model from Solid-works to STK and then use the "Solar Panel" tool to draw up an energy balance of a microsatellite model with fixed solar panels and another deployable one. This approach aims to demonstrate the impact of the orientation of solar panels on the maximum power supplied in low orbit: case of microsatellites before going into the detail of the calculations and the prototyping of a model with deployable solar panels [7–9] (Fig. 1).

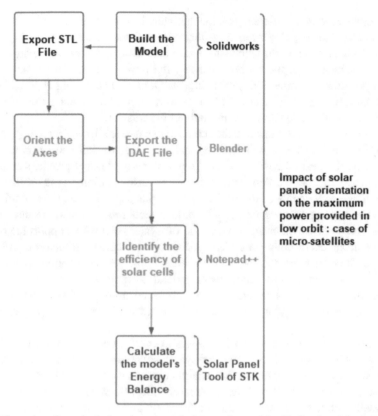

**Fig. 1.** Flow chart for calculating the energy balance of a model with fixed and deployable solar panels.

## 2.2 Theory and Calculation

CubeSats are small, standardized satellites that have revolutionized the way we approach space exploration. These miniature satellites are often used for scientific research, technology demonstrations, and educational purposes. As CubeSats continue to evolve, the need for efficient and reliable power generation becomes increasingly important. Advances in technology have greatly improved the efficiency, size, and flexibility of solar panels, making them an increasingly popular choice for CubeSat power generation. The size and efficiency of solar panels can vary depending on the CubeSat's specific mission requirements. Many CubeSats use off-the-shelf solar panels, while others use custom-made panels that are designed to fit the CubeSat's unique size and shape. The efficiency of solar panels has greatly improved, making them an increasingly popular choice for CubeSat power generation. This increased efficiency means that CubeSats can generate more power from the same size solar panel, allowing for longer mission durations and more advanced scientific instruments. One of the most significant technological advances in CubeSat solar panel power generation is miniaturization. As CubeSats continue to decrease in size, so do their solar panels. CubeSat solar panels have become smaller, lighter, and more efficient, making them ideal for use in even the smallest Cube-Sats. Miniaturization has also led to more flexibility in CubeSat design, allowing for solar panels to be integrated into the CubeSat's structure in unique and creative ways [14, 15]. First part "calculation of the energy balance", the power subsystem consists of three main components: Equipment for generating, storing and distributing electrical energy; A power budget to manage and distribute power; A production/capacity schedule to support satellite power consumption throughout the mission.

These points are interconnected; the choice of use in one allows progress in the other. Selecting efficient hardware will lead to creating a system that can generate enough power to meet the desired power budget. An initial energy budget provides an estimate of how much energy each component will draw at various points in the orbit. This first budget guides a first choice of use of the type of solar panels. If the type of solar panel selected does not meet the desired profile, adjustments must be made to the choice of using one type over the other up to where the subsystem can fully support the mission. Energy requirements of other subsystems may therefore constrain the on-board energy subsystem. The sum of the minimum power consumption of many components through-out the mission defines a minimum power generation required. In order to ensure the success of the mission, it is extremely important to accurately estimate the production and consumption of energy supplied whether by fixed solar panels or deployable panels [10–12].

In this work, we have developed, created and carried out scenarios under the STK software of satellites having fixed and deployable solar panels in orbit around the Earth which makes it possible to calculate the instantaneous power received by the solar panels of these satellites; as well as to evaluate the moment for the flow of information emitted by the satellites and received by the ground station and then the interpretation of the results. Two satellites with fixed solar panels and one with deployable panels in LEO circular orbit, Nadir pointing type of attitude at an altitude of 650 km were realized. A ground station has also been installed comprising an antenna and a receiver allowing the transmission and reception of data by the satellites. An antenna has also been associated with the satellite which will be used to transmit data to the ground station. Once this scenario was realized, we calculated the instantaneous power received by the solar panels, as well as the surface covered by the satellite when it is detected by the ground station. The Solar Panel tool allows us to model the exposure of the solar panels mounted on the satellite over a given time interval, such as an orbit period. The result of the analysis can be used to determine the variable availability of electric power for the operations to be performed by the bus and its payload. To calculate the electrical power captured by solar panels at any given time, the Solar Panel tool applies the following basic power equation [13]:

$$\text{Power} = \text{Efficiency} * \text{Soalr Intensity} * \text{Effective Area} * 1358 \frac{w}{m^2}$$

Efficiency is specified for the solar panel in the satellite model file and varies between 0 and 1, solar intensity varies from shade (0) to penumbra ($0 < i < 1$) to full sunlight (1). The Solar Panel tool calculates solar irradiance over time by animating the timeline and periodically counting the pixels corresponding to the illuminated parts of the considered solar panels. With energy production graphs, the analysis becomes a simple matter of comparing the graphs with the energy balance. This build data can be exported from STK into an Excel.

## 2.3   Construction of CubeSats with Fixed and Deployable Solar Panels

In the design of solar panels of CubeSats, we had used a high-efficiency GaAs-Based solar cells with an efficiency equal to 28% and the number of solar cells used per panel is 14 cells. Figure 2 shows the characteristics and the size of solar cells used in the design of CubeSats under Solid-works software. At first, my objective was to build two types of Cube-Satellites one having fixed solar panels and the other one with deployable solar panels comprising the characteristics illustrated in Table 1 using a construction software named Sketchup or Blender. And then, we will export the CubeSats built on the STK software in Collada format "*.dae", in which we will calculate the power received by the solar panels of different CubeSats. Building accurate models required CAO files of the solar panels, which to date have not been received. The solar panels used in this project were attached to a simple rectangle representing the dimensions of the satellite. Each file was loaded into Blender as an.stl file and assigned to their respective locations on the main satellite bus. It is important to ensure that each panel has the correct coordinate system. As described in the previous RTA, the first step consists in making two 3D models of satellites with a deployable solar panel, and another with fixed solar panels with the Solid-works software as shown in the following figures (Figs. 3 and 4).

**Table 1.**  Characteristics of the two types of Built CubeSats with fixed and deployable solar panels

|  | Parameters |
| --- | --- |
| Size (cm$^3$) | 10 * 10 * 10 |
| Mass (Kg) | 1.3 |
| Altitude (Km) | 650 |
| Orbit (°) | 97 - Circular |
| Antenna type | Dipole |
| Pointing | Earth |

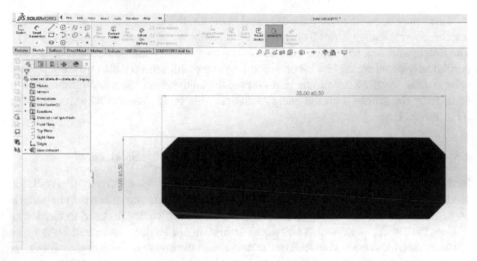

**Fig. 2.** Characteristics and size of solar cells used in the design of CubeSats

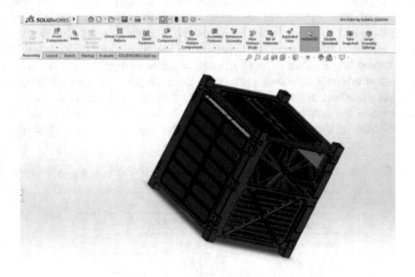

**Fig. 3.** 3D model of the satellite with fixed solar panels

**Fig. 4.** 3D model of satellite with a deployable solar panel.

The second step consists in exporting the CAO file, in the BLENDER software in order to fix the axes of the types of CubeSats as indicated in the following figures. When designing with Solidworks, the solar cells must be separated from the panel. Once the axes are well positioned, we must save the new axes so that our new model can have the same style of coordinate system. The last step is to export our model as a file with a "*.dae" extension (Fig. 5).

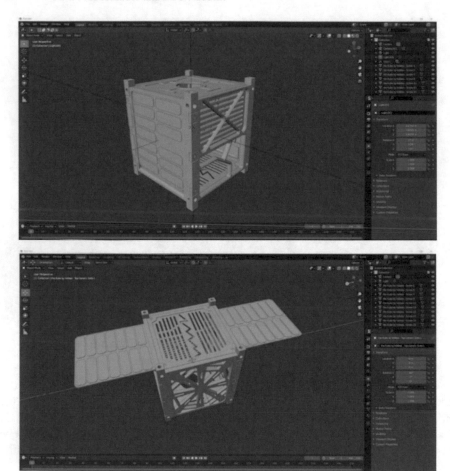

**Fig. 5.** Opening the Solidworks 3D models of CubeSats in BLENDER.

With the model built and saved as a digital asset modification file ("*.dae" needed for STK integration), assign solar panel groups and their respective efficiencies, the "*.dae" file should be edited in NotePad++ according to the following Fig. 6.

The previous figure shows a direct example of changes made to the "*.dae" file. After having developed the Visual scenes part, we inject the lines highlighted in gray. These lines regroup the solar cells, and give an efficiency of 28% (Fig. 7).

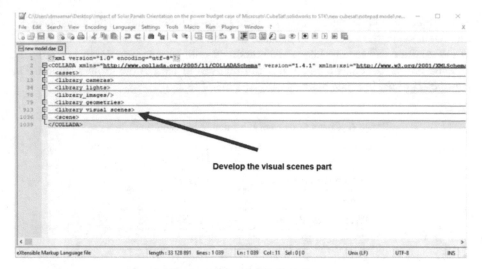

**Fig. 6.** Editing the "*.dae" file.

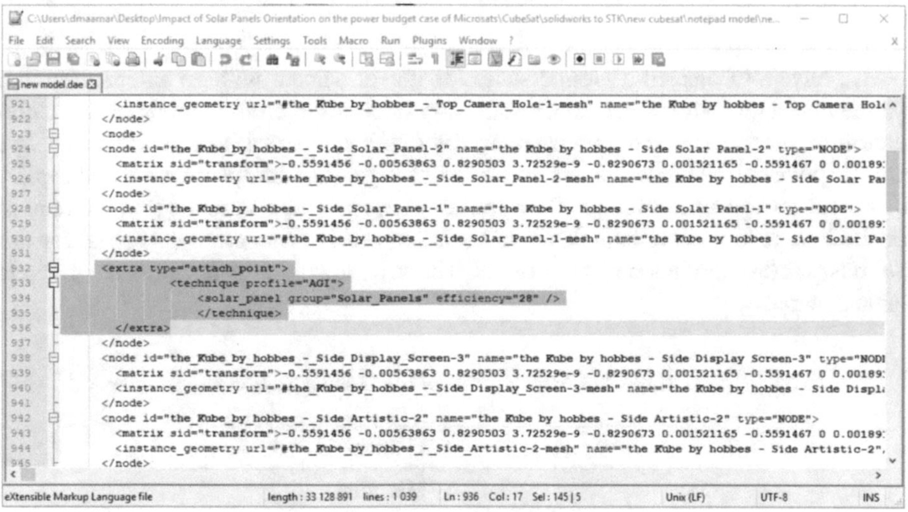

**Fig. 7.** Grouping solar cells and injecting their efficiencies.

## 2.4 Design of the Project Scenario Under the STK Software

The next step is to create a scenario with the STK software. A scenario in STK is an instance of an analytical or operational task that you model with STK. When we select Save from the File menu in STK, we save a scenario. However, unlike other applications which save a single file, STK saves a scenario as a group of files comprising the collection of objects relevant to the scenario. When we save a scenario, the scenario itself is saved as an object, and each object in the scenario is saved individually. The scenario object file extension is "*.sc". Once the scenario has been created, we must change the model of the satellite with the one that we had modified with Notepad++. We had installed an antenna on the CubeSats built for the transmission of data to the ground station. The parameters of the installed antenna are given in Table 2.

**Table 2.** Characteristics of the antenna installed on CubeSats

|                          | Parameters |
| ------------------------ | ---------- |
| Design Frequency (GHz)   | 0,1458     |
| Length (m)               | 0.514      |
| Wate Length Ratio        | 0.2499     |
| Efficiency (%)           | 100        |
| Antenna type             | Dipole     |
| Azimuth (°)              | 0          |
| Elevation (°)            | 90         |

We have also created a ground station that's associated with an antenna that has parameters in line with the antenna installed on satellites in order to minimize the information losses and transmit commands to the satellites, as well as a receiver having a circular polarization which can receive the wave regardless of the direction and capture all the information transmitted by the satellites. After inserting a ground station, we must set the longitude and the latitude so that it is located in Oran city of Algeria, for this we must set the station's own parameters. The ground station parameters are given in Table 3, 4 and 5.

**Table 3.** Characteristics of the Ground Station

|                    | Parameters |
| ------------------ | ---------- |
| Type               | Geodetic   |
| Laltitude (°)      | 35.6       |
| Longitude (°)      | −0.6       |
| Atitude (Km)       | 0.1        |
| Atitude Reference  | WGS84      |

After having determined all the parameters of the project, we began to design these models of CubeSats under the STK software. In Fig. 8, we can observe the detection of the CubeSats by the ground station and we can also begin to manipulate it in order to be able to calculate and determine the instantaneous power received by the sun at the level of the solar panels, or even the instant or the satellite is detected by ground station.

**Table 4.** Characteristics of the antenna installed on Ground Station

|                        | Parameters |
|------------------------|------------|
| Design Frequency (GHz) | 0.1458     |
| Length (m)             | 1          |
| Wate Length Ratio      | 0.4863     |
| Efficiency (%)         | 100        |
| Antenna type           | Dipole     |
| Azimuth (°)            | 0          |
| Elevation (°)          | 90         |

**Table 5.** Characteristics of the receiver installed on Ground Station

|                                          | Parameters           |
|------------------------------------------|----------------------|
| Design Frequency (GHz)                   | 0.4375               |
| Length (m)                               | 1                    |
| Bandwidth (MHz)                          | 0.01                 |
| G/T correspond to receiver noise (dB/k)  | 17.26                |
| Polarization                             | Right-hand Circular  |
| Demodulator                              | BPSK                 |

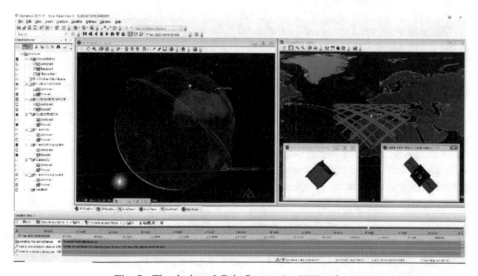

**Fig. 8.** The designed CubeSats under STK software.

# 3   Simulations Results and Discussions

After realizing and simulating the scenario, we must calculate a few quantities to verify that the satellites return data consistent with reality. In our case, we considered that the solar panel which points in the direction of the sun located in the upper face which is operational because the STK software does not manage to detect all the solar panels. Figure 9 shows the trend and graph of the instantaneous power in watts obtained by the solar panels coming from the sun as a function of time under the STK software and the Fig. 10 shows the Area of solar panel illuminated by the sun for the two types of CubeSats. It can be seen that the power received is maximum for a value of 0.7 W for CubeSats with fixed solar panels and 3.6 W for CubeSats with deployed solar panels respectively at certain times when the solar panel is completely illuminated. We also observe that there are slight fluctuations in this power which remains generally constant during this period and this is certainly due to the albedo of the Earth. To check that this graph is coherent, we calculated the full orbital period using Kepler's 3rd law in a theoretical way and we will compare the value obtained with the value present in the graph. According to Kepler's law:

$$T^2 = \frac{4P_i^2}{G(M+m)} * a^3$$

where:

T: Period of the satellite.

G: Universal gravitational constant 6.67 * 10–11 ui.

M: Mass of the Earth (6.0 * 1024 kg).

m: mass of the satellite (1.3 kg).

a: Corresponds to the orbit radius (Earth radius + Satellite altitude).

We have neglected the mass of the satellite because it is much lower than the mass of the Earth, we will have:

$$T^2 = \frac{4\pi^2 * (7028\,e3)^3}{(6.67\,e-11) * (6\,e24)} = 5851\,s = 1\,h\,30\,min$$

After calculation, we have obtained a period T = 1 h 30 min, which is in agreement with the graph, therefore the calculation of the power is correct for Shadow/Day periods. From the graph of the power curve, we can calculate the average power received by the solar panel, this power will correspond to the power stored by the satellite at the level of these batteries to ensure its operation in orbit. It is calculated as follows:

$$P_{recuemoy} = \frac{(P_{recuemax} * T_{on})}{(T_{on} * T_{off})}$$

$P_{received\ max}$: Maximum power received by the solar panel.

$T_{on}$: Time corresponding to the maximum received power.

$T_{off}$: Time corresponding to zero power.

By calculation, we obtain an average received power about 1,9 W by deployed solar panels CubeSat. After putting into practice this scenario, we were able also to calculate certain specific parameters such as the instantaneous power received by the solar panel of

CubeSats, the average power received and the coverage of the moment when CubeSats is detected by the ground station as shown in Fig. 11. This study can be allowed us to see if the satellite powers its entire navigation system thanks in part to the power stored by the solar panels.

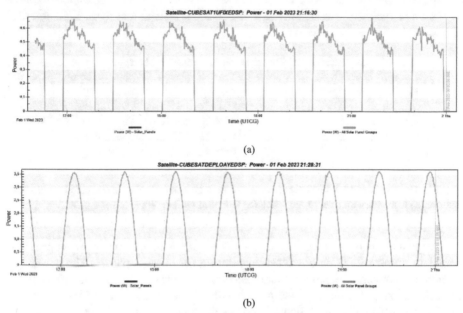

**Fig. 9.** Graph of the instantaneous power obtained by the solar panels of the two types of CubeSats. (a) CubeSat with fixed solar panel. (b) CubeSat with deployable solar panel.

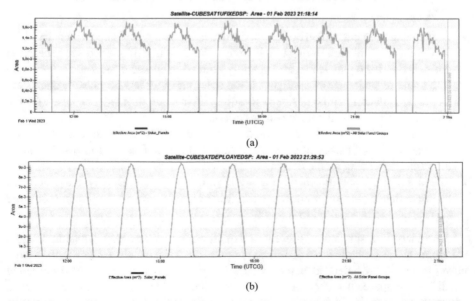

**Fig. 10.** Graph illustrating the Area of solar panel illuminated by the sun for the two types of CubeSats. (a) CubeSat with fixed solar panel. (b) CubeSat with deployable solar panel.

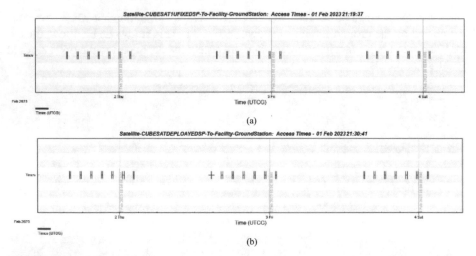

**Fig. 11.** Graph illustrating the coverage of the moment when CubeSats is detected by the ground station. (a) CubeSat with fixed solar panel. (b) CubeSat with deployable solar panel.

The results analysis shows that the CubeSats solar generator is capable of providing sufficient power for the on-board systems during the mission phases. The power generation comparison of deployable and fixed solar panels on CubeSats is influenced by the orientation of the photovoltaic panels with respect to the sun and the distance of the CubeSats from the sun. It depends on several factors such as the orientation of the panels, the angle of incidence of sunlight, the efficiency of the solar cells, and the size of the panels. Here are comparisons and simulations results discussions based on these factors:

- Deployable solar panels can be designed to track the sun, which allows them to maintain optimal orientation throughout the day. This results in higher power generation compared to fixed panels that are typically pointed in a fixed direction.
- The angle of incidence of sunlight on the solar panel affects its power generation. Deployable panels can be designed to adjust their angle of incidence based on the position of the sun, while fixed panels cannot.
- The efficiency of the solar cells used in the panels affects the amount of power generated. The efficiency of solar cells used in deployable and fixed panels can be the same, depending on the design.
- The size of the panels affects the amount of power generated. Deployable panels can be larger than fixed panels since they can fold or unfold as needed, allowing for more surface area to be exposed to sunlight.

The analysis of simulation results based on these factors have shown that deployable solar panels can generate up to 50% more power than fixed panels. However, the actual power generation will depend on the specific design of the panels and the environmental conditions in space. It's important to note that deployable solar panels can be more complex and expensive to design and deploy compared to fixed panels. Simulation results demonstrate also the effectiveness of the STK software-based approach for analyzing the performance of a CubeSat solar generator for on-board power generation and highlight the importance of battery capacity for CubeSats power management, as well as the need for more efficient photovoltaic panels for space applications. The limitations and recommendations for future work are discussed in this paper. This developing approach provides a useful tool for optimizing the design of the solar generator and the on-board power management system for CubeSats missions.

## 4   Conclusions

In conclusion, the development of CubeSats has opened up new possibilities for space exploration and scientific research. However, the power generation capabilities of these miniature satellites are limited due to their small size and weight constraints. Therefore, a detailed analysis of the solar power generation potential of CubeSats is crucial for their successful deployment and operation in space. This paper has proposed a useful method for analyzing CubeSats solar power generation with different solar panel configurations which can be applied to a wide range of CubeSats missions that's considers various factors such as: the CubeSat's orbit, solar panel size and efficiency, and the orientation of the CubeSats relative to the Sun. This analyze approach involves modelling and simulating the power generation performance of the CubeSat, and the results can be used to optimize the CubeSat's design and operation for maximum power generation. The use of the STK software-based approach for power analysis of CubeSats solar generators is a promising area of research that has the potential to greatly enhance the success of CubeSats missions. However, there are still some challenges to be addressed, such as accurately modelling the space environment and improving the accuracy of battery performance predictions. Overall, this analyzing study can help researchers and engineers to improve the power generation capabilities of CubeSats, leading to more efficient and cost-effective space missions.

## References

1. Etchells, T., Berthoud, L.: Developing a power modelling tool for CubeSats. In: Etchells 1, Conference on Small Satellites (2019)
2. Wang, Y., Gong, X., Zhang, W., Liu, Z.: Power estimation for cubesats with high-aspect-ratio solar panels. IEEE Trans. Aerosp. Electron. Syst. **55**(6), 3486–3495 (2019)
3. Franco, C.C., Kumar, A., Han, K., Garcia, A., Dordizadeh, F.: A comprehensive approach to solar power system analysis and modeling for CubeSats. IEEE Trans. Aerosp. Electron. Syst. **56**(5), 3355–3365 (2020)
4. Collins, J.M., Davis, R.M., Spina, T.V.: A framework for predicting power generation from solar panels on CubeSats. J. Aerosp. Inf. Syst. **17**(12), 746–756 (2020)
5. Manrique, J.P., Sarmiento, E.R., Jaramillo, R.E.: Experimental validation of a solar panel model for cubesats in low earth orbit. IEEE Aerosp. Electron. Syst. Mag. **33**(7), 54–61 (2018)
6. Cervone, I., Gatto, A., D'Acquisto, L.: Design and verification of a power management system for cubesat with custom solar panels. In: 14th European Space Power Conference, ESPC (2020)
7. Park, H., Kim, S.M., Lee, S.H.: Analysis of solar panel orientation for CubeSat using genetic algorithm. IEEE Access **8**, 22182–22192 (2020)

8. Porter, J.R., Keating, A.J., Taylor, J.P., Mankins, J.C.: Optimizing CubeSat power system design: defining the optimal power generation-to-storage ratio. In: IEEE Aerospace Conference, Big Sky, MT, USA, p. 111 (2014)
9. Kim, G.-N., et al.: Development of CubeSat systems in formation flying for the solar science demonstration: the CANYVAL-C mission. Adv. Space Res. **68**(11), 4434–4455 (2021)
10. Hernandez, D.T.: CubeSat power system design for high precision, solar observation. A Project Present to the Faculty of the Department of Aerospace Engineering San Jose State University (2015)
11. Kharsansky, A.: Power modelling and budgeting design and validation with in-orbit data of two commercial LEO satellites. In: Proceedings of the 31st Annual AIAA/USU Conference on Small Satellites, SSC17-X-08 (2017)
12. Gonzalez-Llorente, J., Ortiz-Rivera, E.I.: Comparison of maximum power point tracking techniques in electrical power systems of CubeSats. In: Proceedings of the 27th Annual AIAA/USU Conference on Small Satellites, SSC13-WK-4 (2013)
13. Socolovsky, H., Muoz, S., Raggio, D., Bolzi, C.: Development and testing of solar panels for small satellite applications at CNEA. In: Proceedings of the 31st Annual AIAA/USU Conference on Small Satellites, SSC17-P1–18 (2017)
14. Walker, J., Lorentz, T.: Efficient spacecraft power systems for CubeSats. IEEE Aerosp. Electron. Syst. Mag. **32**(3), 30–35 (2017)
15. Kendall, D.: The evolution of CubeSat solar panels. Aerosp. Am.. Am. **55**(5), 40–45 (2019)

# Smart Systems and Communication

# TinyML Model for Fault Classification of Photovoltaic Modules Based on Visible Images

Z. Ksira[1], N. Blasuttigh[2], A. Mellit[1(✉)], and A. Massi Pavan[2]

[1] Renewable Energy Laboratory, University of Jijel, Jijel, Algeria
adel_mellit@univ-jijel.dz
[2] Department of Engineering and Architecture, and Center for Energy, Environment and Transport Giacomo Ciamician, University of Trieste, Trieste, Italy
{nicola.blasuttigh,apavan}@units.it

**Abstract.** In this paper, a Tiny Machine Learning (TinyML) model is developed for fault classification of photovoltaic (PV) modules. A dataset based on visible images of healthy and faulty PV modules has been collected at different locations. The examined defects are: discolored cells, cracked PV modules, bubble formation, bird droppings, dirt accumulation, sand deposit, corrosion, shading effect, and snail trails. The Edge Impulse platform has been used to develop and optimize our TinyML model, which is then integrated onto a low-power microcontroller for a real time application. The simulation results show a good overall classification accuracy of 92% whereas experimental results demonstrate the ability of the developed TinyML model to be deployed for real-world and low-cost applications.

**Keywords:** Photovoltaic · Fault classification · Visible images · TinyML · Microcontroller · Deep learning

## 1 Introduction

Photovoltaic (PV) cells/modules are subject to internal (physical degradation) and external (environmental effects) factors that significantly contribute to a decrease in their performance, stability, and operating lifetime [1]. Recently, researchers have been more concerned in the application of deep learning (DL) in fault detection and diagnosis of PV modules/arrays. For example, in [1] the authors used CNN to classify various types of PV module defects, by obtaining a classification rate of 97%. The well-known You Only Look Once (YOLOv5) and Deep algorithm Residual Network (ResNet) algorithms have been used for fault localization and detection in PV plants. It has been shown that, the YOLOv5 intensely boosts the efficiency of this method, and the detection speed can reach up to 36 images per second [2]. Similar work was performed in [3], where the authors applied three machine learning algorithms and a pre-trained AlexNet CNNs was used to extract features. Similarly, the authors in [4] used a pre-trained VGG16 network for feature extraction and classification. The results showed that the classification results are very promising, with an average accuracy of 95.4%. Other works related to the application of machine learning and deep learning methods for fault detection, localization

© The Author(s), under exclusive license to Springer Nature Switzerland AG 2024
M. Hatti (Ed.): IC-AIRES 2023, LNNS 984, pp. 373–380, 2024.
https://doi.org/10.1007/978-3-031-60629-8_37

and diagnosis of PV modules can be found in [5–7]. Although these methods produce good results in terms of classification rate, the cost-effectiveness of the proposed devices and instrumentation is questioned. In this regard, recent research advancements have led to the deployment of AI algorithms into low-cost and low-power devices. In our earlier work, a fault diagnosis method based on infrared images, deep learning methods and the Edge Impulse platform was developed [8]. To this end, the main goal of this work is to propose an embedded solution for fault classification of PV modules based on visual images. The idea aims at the development of a TinyML classifier and its integration into a low-cost and low-power microcontroller for real-time applications. This solution could be integrated on-board in drones for aerial inspection of large-scale PV plants. To the best of our knowledge, no work related to this subject has been published before.

The rest of the paper is organized as follows: Section II presents the dataset used in this study. The developed method, including the CNN classifier and experimental step is explained in Section III. Section IV provides the simulation and the experimental results of the proposed solution in details. Conclusion and future perspective are given in Section V.

## 2  Database

The dataset was collected from different regions in Algeria, and after several inspections the most common visual defects observed on PV modules are snail trails, corrosion due to overheating cells, sand accumulation, bird droppings, glass breakage, browning and discoloration, dirty, and shading. Table 1 lists the label of these classes. Figure 1 shows an example of augmented visible images. The size of the dataset used is 5000 images, and each class contained 500 images.

**Table 1.** Classes description

| Class | Description |
| --- | --- |
| D1 | Discolored PV modules |
| D2 | Healthy PV modules |
| D3 | Cracked PV modules |
| D4 | Bubble formation on PV modules |
| D5 | Bird droppings on PV modules |
| D6 | Dirt on PV modules |
| D7 | Corrosion |
| D8 | Sand deposit |
| D9 | Shading |
| D10 | Snail trails |

**Fig. 1.** Example of defective PV module images

## 3 Methods

The diagram block of the embedded procedure is given in Fig. 2.

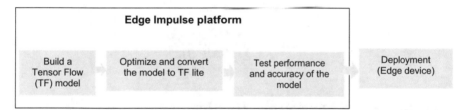

**Fig. 2.** The embedded procedure using Edge Impulse platform

It consists of four main steps: 1) build a TensorFlow (TF) model, 2) optimize and convert the model into a TF lite model, 3) test the performance of the optimized model and 4) integrate the model inside an Edge device and deploy it.

The model was developed under Edge Impulse platform [9]. The selected microcontroller is the Arduino Nano BLE sense [10]. The procedure of the model development using an Edge Impulse can be summarized in the flowchart shown in Fig. 3.

The following metrics are used to evaluate the model's performance:

$$Acc = \frac{\sum_i CM(i, i)}{\sum_i \sum_j CM(i, j)} \tag{1}$$

$$F1 - score = 2 * \frac{(Pre * Rec)}{(Rec + Pre)} \tag{2}$$

where *Pre* is the precision, *Rec* is the recall and *CM* is the confusion matrix.

$$Pre = \frac{CM(i, i)}{\sum_j CM(j, i)} \tag{3}$$

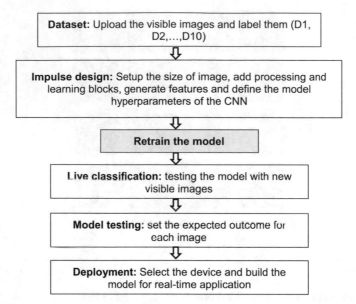

**Fig. 3.** Flowchart of the model development under Edge Impulse platform

$$Rec = \frac{CM(i,i)}{\sum_j CM(i,j)} \tag{4}$$

## 4  Simulation and Exprimental Results

Table 2 shows the obtained hyperparameters of the proposed model. These hyperparameters have been selected after various experiments.

**Table 2.** Hypermeters of the developed model

| Hyperparameters of the CNN model | Name/Value |
| --- | --- |
| Activation function | Adam |
| Number of training cycle | 65 |
| Learning rate | 0.0045 |
| Validation set size | 20% |
| Dropout | 0.3 |
| Number of nodes at the end layer | 10 |

The confusion matrix of the training performance is shown in Fig. 4a. The accuracy is 92% and the loss is 0.24. The F1-score ranges between 84% to 97%. Overall, in terms of accuracy, these results are very promising. Figure 4b shows the confusion matrix of

**Fig. 4.** Confusion matrices during the validation stage: (a) unoptimized (float32) and (b) optimized (quantized Int8)

the validation performance where a slight decrease in the accuracy is observed (87.1%), the ranges of the f1-score is the same, while the uncertain varies between 0% and 11.7%.

In order to validate the proposed model in real-time, Fig. 5a presents an example of a live classification test. An unknown visual image, a PV module affected by snails trails (class D10), was selected to verify the capability of the model to classify the defect. The obtained results are reported in Fig. 5b, where it can be seen that the model can predict the class of defect correctly with a classification rate of 86%.

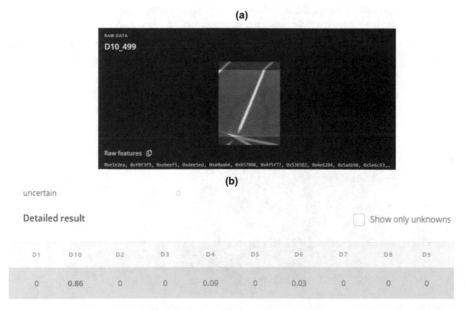

**Fig. 5.** a) Selected image for real-time testing of the model and b) the corresponding classification results.

Despite the increasing development and exponential increase in the performance of microprocessors available on the market, it is always useful if not necessary to reduce the burden and computational load of the employed applications as much as possible. For this reason, an optimization process of the proposed TinyML model was carried out in order to decrease hardware resources and latency. Table 3 compares the non-optimized and optimized models in this regard, where the RAM utilization (kB), flash memory usage (kB), and latency (ms) for each model have been taken into consideration.

**Table 3.** Comparison between optimized and unoptimized models

| Model | RAM usage (kB) | Flash usage (kB) | Latency (ms) |
|---|---|---|---|
| Non-optimized model (float32) | 305.2 | 864.3 | 8,971 |
| Optimized model (quantized Int8) | 124.8 | 300.1 | 1,181 |
| Obtained reduction | **180.4** | **564.2** | **7,790** |

As listed in Table 3, the optimized model RAM, and flash usage have been decreased and a significant drop was observed in latency (7,790 ms). This indicates that the optimized code can run approximately 7 times faster than the nonoptimized code. Once the model was optimized, the last step consisted of building the optimized model and download it into the Arduino Nano BLE sense.

The generated firmware of the optimized model is shown in Fig. 6a. This file is uploaded into the Nano 33 BLE sense. Figure 6b shows the designed prototype, which consists of a low-cost, low-power microcontroller and a low-cost visual camera (OV7670). The total cost of the developed prototype is around 50 €. Figure 6c depicts the experimental result (live classification), where it can be seen that class D10 (snail trails defect on PV module) represents the largest value (61.72%), confirming that the examined PV module is faulty.

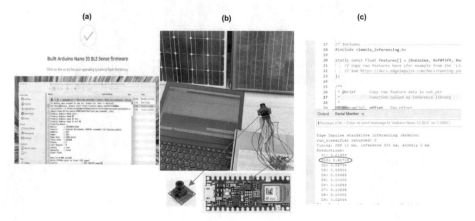

**Fig. 6.** a) The generated firmware file after building the model, b) the designed prototype, and c) experimental results of the executed code in real-time (live classification, case of snail trails).

# 5    Conclusion and Future Perspectives

In this work, a TinyML model for fault classification of PV modules based on visual images was developed. Simulation results demonstrated the ability of the model to identify the type of defect with good accuracy (92%). The designed prototype was evaluated under various tests, and the experimental results report the aptitude of the proposed system to classify the defect with an acceptable accuracy of 77.24%. With this quantization technique (Int8) the size of the model was reduced by up to 30%, the latency was also decreased significantly (7,790 ms) and, however, the accuracy of the implemented model was reduced by only 10%. The developed solution, which is cost-effective and therefore accessible to everyone, can be easily used even by non-experts in the field due to the proposed TinyML-based machine learning process. Moreover, this embedded system can be installed into drones for the aerial inspection of large-scale PV systems, reducing costs, time, and labor resources. With respect to future work, we will improve the experimental accuracy of the embedded TinyML model by using other quantization techniques, increase the size of the dataset, and try to integrate this system onto a drone for automatic inspection of large-scale PV plants.

**Acknowledgements.** This work was supported in part by the "Ministero dell'Istruzione, dell'Universita' e della Ricerca" (Italy) under the Grant PRIN2020–HOTSPHOT 2020LB9TBC. The second author (Dr. A. Mellit) would like to acknowledge support from the ICTP through the Associates Programme (2023–2028).

# References

1. Mellit, A., Kalogirou, S.: Assessment of machine learning and ensemble methods for fault diagnosis of photovoltaic systems. Renew. Energy **184**, 1074–1090 (2022)
2. Li, X., Yang, Q., Lou, Z., Yan, W.: Deep learning based module defect analysis for large-scale photovoltaic farms. IEEE Trans. Energy Convers. **34**(1), 520–529 (2018)
3. Hong, F., Song, J., Meng, H., Wang, R., Fang, F., Zhang, G.: A novel framework on intelligent detection for module defects of PV plant combining the visible and infrared images. Sol. Energy **236**, 406–416 (2022)
4. Sridharan, N.V., Sugumaran, V.: Visual fault detection in photovoltaic modules using decision tree algorithms with deep learning features. Energy Sources, Part A: Recovery, Utilization, and Environmental Effects, pp.1–17 (2021)
5. Gawre, S.K.: Advanced fault diagnosis and condition monitoring schemes for solar PV systems. In: Bohre, A.K., Chaturvedi, P., Kolhe, M.L., Singh, S.N. (eds.) Planning of Hybrid Renewable Energy Systems, Electric Vehicles and Microgrid. Energy Systems in Electrical Engineering, pp. 27–59. Springer Nature Singapore Springer, Singapore (2022). https://doi.org/10.1007/978-981-19-0979-5_3
6. Vlaminck, M., Heidbuchel, R., Philips, W., Luong, H.: Region-based CNN for anomaly detection in PV power plants using aerial imagery. Sensors **22**(3), 1244 (2022)
7. Rahman, M.R., Tabassum, S., Haque, E., Nishat, M.M., Faisal, F., Hossain, E.: CNN-based deep learning approach for micro-crack detection of solar panels. In: 2021 3rd International Conference on Sustainable Technologies for Industry 4.0 (STI), pp. 1–6. IEEE (2021)
8. Mellit, A., Blasuttigh, N., Pavan, A.M.: TinyML for fault diagnosis of photovoltaic modules using edge impulse platform. In: 2023 11th International Conference on Smart Grid (icSmartGrid), pp. 01–05. IEEE (2023)

9.  'Edge Impulse'. https://www.edgeimpulse.com/. Accessed 07 Apr 2023
10. 'Arduino Nano 33 BLE Sense'. Arduino Official Store. https://store.arduino.cc/products/arduino-nano-33-ble-sense. Accessed 07 Apr 2023

# Development of a Hybrid DLDH Fault Detection and Localization Algorithm for Two Types of PV Technologies with Experimental Validation

S. Arezki[1](✉), A. Aissaoui[2], and M. Boudour[1]

[1] Laboratory of Electrical and Industrial System, University of Sciences and Technology (USTHB), Algiers, Algeria
arezki.saliha@gmail.om

[2] Department of Electrical Engineering, University of Tahri Mohamed Bechar, Bechar, Algeria

**Abstract.** The work lies in its ability to detect and locate photovoltaic (PV) (DLD) faults, which is essential for diagnosing and monitoring the efficiency of a PV field. The paper focuses on presenting the DLD algorithm, providing details on the approach and implementation of the diagnostic algorithm based on the fault detection and localization flowchart. To enhance diagnostic accuracy and fault detection, a hybrid diagnostic method is employed, combining fuzzy logic with the thresholding method. Additionally, fault location diagrams for monocrystalline and polycrystalline PV modules are developed. The algorithm developed allows for the differentiation of defects with the same signature (behavior). This is achieved by introducing a range of symptoms and applying an intelligent method alongside the "threshold method" to distinguish between faults with identical symptom signatures. The proposed algorithm's effectiveness relies on prior knowledge and definitions of the field's behavior regarding various types of defects and the designation of symptoms for each defect. Experimental validation is provided for specific faults.

**Keywords:** Faults · Diagnosis · Identification · Symptoms · Hybrid fault detection and localization algorithm · PV technologies · Tests

## 1 Introduction

The production of electrical energy based on photovoltaic conversion through PV fields can run into problems during the operating life of the project on various levels of the PV installation, due to the various faults encountered at various times inherent in the PV generator [1]. Photovoltaic field maintenance consists in diagnosing and visualizing the general behavior to guarantee the quality and continuity of the production service and consequently the protection of the PV devices and modules and thus protect the lifetime of the PV chain. Diagnostics is an essential step and an indispensable means in the maintenance of renewable energy systems and, of course, PV conversion systems. This method is designed to detect and locate faults in any renewable energy system. To maximize photovoltaic conversion efficiency and reduce maintenance costs, numerous

M. Hatti (Ed.): IC-AIRES 2023, LNNS 984, pp. 381–402, 2024.
https://doi.org/10.1007/978-3-031-60629-8_38

solutions have been proposed in the course of research efforts. Diagnostics is one of the most interesting solutions for operating PV modules at optimum power and competing with fossil fuels for economic reasons. Many studies have focused on assessing the impact of various faults by analyzing the resulting I-V characteristic. However, the appearance of these defects always has adverse consequences, leading to a drop in the relative performance of PV generators, such as: performance factor, maximum power output and photovoltaic conversion efficiency. In this study, a system for detecting and localizing seven faults in two photovoltaic modules of the same power output, but using two different technologies is designed. First, various diagnostic methods and symptoms frequently encountered in PV modules are discussed. Also, the measurement thresholds relevant to decisions regarding the presence of these symptoms by presenting the electrical circuit that describes the system implemented in this work are analyzed. Then the PV fault detection and localization algorithm are presented. The approach and implementation of the diagnostic algorithm are detailed, based on the fault detection and localization stages. Next, a hybrid diagnostic method using fuzzy logic with thresholding is adopted. The fault location flowchart is presented according to the type of module selected. The algorithm developed by an experimental study carried out to analyze the impact of defects on the performance of two (02) monocrystalline and polycrystalline photovoltaic modules is validated.

## 2  Presentation of the Diagnostic System

The aim of this study is to diagnose two monocrystalline and polycrystalline photovoltaic modules of the same power (100Wp), with the concept of detection, identification and localization of certain defects. The electrical circuit describing the system to be implemented and studied is shown in Fig. 1.

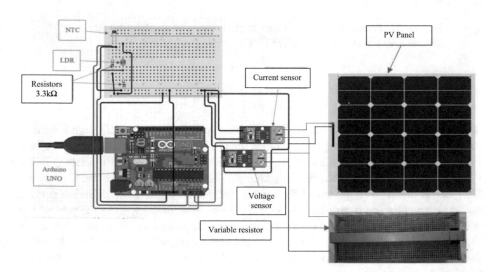

**Fig. 1.**  Presentation of electrical circuit.

Two PV modules of the DIMEL brand, monocrystalline and polycrystalline are used, of the same power (100 W peak). The tables below details their main characteristics and the two variables resistors used in this work.

**Table 1.** DP-100 et DM-100 module characteristics.

| Panel | Power (Wc) | $I_{sc}$ (A) | $V_{OC}$ (V) | Technology | $V_{MPP}$ (V) | $I_{MPP}$ (A) | N° cells | N° Diode Bypass |
|-------|-----------|--------------|--------------|------------|---------------|---------------|----------|-----------------|
| **DP-100** | 100 | 6.10 | 22.10 | Polycristallin | 17.90 | 5.60 | 36 | 2 |
| **DM-100** | 100 | 5.80 | 22.70 | Monocristallin | 18.35 | 5.45 | 36 | 2 |

**Table 2.** Characteristics of the two variables resistors.

| Variables resistors | Maximum current (A) | Maximum resistor (Ω) |
|---------------------|---------------------|----------------------|
| Resistor 1 | **5.7** | **10** |
| Resistor 2 | 8 | **10** |

For the acquisition board, we have chosen the Arduino board, as it is an open-source electronic prototyping project based on a flexible platform [2].In the data acquisition section, the use of various sensors is essential in order to collect information from the two photovoltaic modules (for current ACS712 module" and for voltage "HiLetgo sensor") as well as climatic data (for sunshine "LDR sensor" and for temperature "NTC sensor). For software support (Windows PC, Matlab), a various software packages used are presented to run the fault detection and localization algorithm and display the results. New versions of Matlab are compatible with Arduino boards. This has enabled us to write the code developed on Matlab instead of writing it in the Arduino IDE, simply by using direct communication blocks with Arduino called "Matlab Support Package for Arduino".

## 3   Diagnostic Approach

### 3.1   Threshold Method

The method adopted is based on the thresholding of the different electrical characteristics of the photovoltaic module (for monocrystalline or polycrystalline modules). This made it possible to classify a list of symptoms with which the diagnosis is established for the purpose of determining the operating state of the photovoltaic module and to instantly report the various defects that can be triggered.

### 3.2   Calcul et le Réglage des Seuils

A wide range of diagnostic methods have been described in previous work. However, they can be grouped into two broad categories according to the type of knowledge used to detect faults. Electrical and non-electrical diagnostic methods are finded, and each of these two types is itself the subject of several other methods [3].

### 3.3   PV Module Faults Signatures

Symptom selection is a critical step in the success and design of a diagnostic algorithm. In this study, seven symptoms frequently present in a PV module are studied, as shown in Fig. 2. The results of tests of the behaviour of a PV module during a considered failure show that potential symptoms can be identified and can be used to trace the nature of the failure. Any defect may produce one or more symptoms, but some symptoms are common to several defects. Thus, the combination of default signatures makes it possible to identify the defect or the set of defects responsible for the behaviour examined [3].

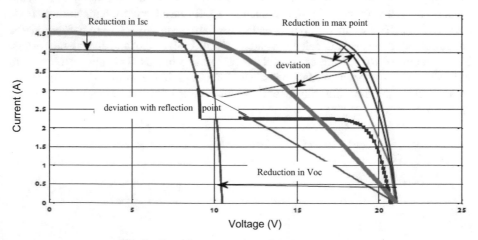

**Fig. 2.** Graphic presentation of the symptoms [3].

The table below shows the different symptoms used in this study and their characteristics.

**Table 3.** Characteristics of the various diagnostic parameters.

| Fault signature | Specified characteristic |
|---|---|
| S1: Power produced by photovoltaic module (W) | $\Delta P_{MPP} = P_{MPP(sain)} - P_{MPP(def)}$ |
| S2: Power factor PF (%) | $PF = \dfrac{P_{MPP(def)}}{P_{MPP(sain)}} \times 100$ |
| S3: Short-circuit current (A) | $\Delta Icc = I_{cc(sain)} - I_{cc(def)}$ |
| S4: Open-circuit voltage (V) | $\Delta V_{oc} = V_{oc(sain)} - V_{oc(def)}$ |
| S5: Current ratio CR | $\Delta C_R = \dfrac{I_{MPP(sain)}}{I_{cc(sain)}} - \dfrac{I_{MPP(def)}}{I_{cc(def)}}$ |
| S6: Voltage ratio VR | $\Delta C_R = \dfrac{V_{MPP(sain)}}{V_{oc(sain)}} - \dfrac{V_{MPP(def)}}{V_{oc(def)}}$ |
| S7: Presence of inflection point (%) | $Inflexion = 100\% - \dfrac{I_{Inflexion}}{I_{MPP(def)}} \times 100\%$ |

## 3.4 Threshold Setting

The decision as to whether a fault has occurred can be made using a fault catalog, in which the relationship between process errors and changes in coefficients is established. It can be based on simple threshold levels, or by using more sophisticated methods of statistical decision theory [4]. False alarms are caused by the variation of one or more symptoms within the tolerated range corresponding to normal system operation. This variation is due to different uncertainties in the generation of these symptoms [5]. In photovoltaic applications, the IEC 61724 standard [6] limits uncertainties. The following table shows the total error of the various uncertainties.

**Table 4.** Total error due to various uncertainties [7].

| Symptoms | Symptom name | Relative error (measurement) | Relative error (model) | Total error |
|---|---|---|---|---|
| S1 | Reduction of maximum power | 2% | 3% | 5% |
| S2 | Reduction in open-circuit voltage | 1% | 2% | 3% |
| S3 | Reduced short-circuit current | 1% | 5% | 6% |

### 3.5 Threshold Calcul

The table below shows the various thresholds for the diagnostic parameters (Table 5).

**Table 5.** Thresholds for selected diagnostic parameters.

| Symptom | Diagnostic parameters | Threshold | Detection threshold for DM-100 module | Detection threshold for DP-100 module |
|---------|----------------------|-----------|----------------------------------------|----------------------------------------|
| S1 | Reduction of maximum power | 5% | 3,1352 (W) | 3,1820 (W) |
| S2 | Performance factor reduction | 90 | 90 (%) | 90 (%) |
| S3 | Short-circuit current reduction | 6% | 0,31498 (A) | 0,3232 (A) |
| S4 | Open-circuit voltage reduction | 3% | 0,58431 (V) | 0,6019 (V) |
| S5 | Reduction of current ratio between normaly and faulty state | 3% | 0,0273 | 0,0262 |
| S6 | Reduction of voltage ratio between normaly and faulty state | 3% | 0,0203 | 0,0202 |
| S7 | Presence of inflection point | 40% | 40 (%) | 40 (%) |

# 4  Selection Faults

Seven faults are selected for diagnosis, as shown in Table 6.

**Table 6.** Overview of selected faults.

| Fault code | Fault name |
|------------|-----------|
| D1 | Shading (tree leaves) |
| D2 | Sand soiling (after normal operation) |
| D3 | Sand soiling (scattered) |

(*continued*)

**Table 6.** (*continued*)

| Fault code | Fault name |
|---|---|
| D4 | Soiling dust (after normal operation) |
| D5 | Dust-like dirt (scattered) |
| D6 | Shading with Bypass diode disconnected (Tree leaves) |
| D7 | Shading with Bypass diode (dust) |

## 5 Presentation and Process for Implementing the Fault Diagnosis Algorithm

The following flowchart illustrates the fault diagnosis process using the threshold method (Fig. 3).

Below, the detail how the algorithm works to acquire data from sensors, calculate diagnostic parameters and detect fault symptoms:

**Part 0: Importing the Healthy State Data of the Two Photovoltaic Modules**

The algorithm begins by initialising and importing the healthy state data of the two photovoltaic modules from an Excel file. This data is useful for comparison with the same data for a faulty module. Next, the algorithm asks the user to choose the type of Module in question, the choice being made by inserting the word "monocrystalline" to choose the DM-100 or "polycrystalline" to choose the DP-100 in the Matlab command window.

**Part 1: Data Acquisition From Sensors**

This part of the algorithm is designed to acquire data from the various quantities measured by the voltage, current, luminosity and temperature sensors. The Arduino is used to collect an analogue signal that can be used to measure the quantities in question. The characteristic of luminosity as a function of the resistance $L = f(R)$ of the photoresistor is in logarithmic form. To generate an equation for transferring the signal sent by the LDR into luminosity (LUX), an attribute measurement was taken, followed by injection of the natural logarithm for each value measured in order to obtain a linear curve. The equation is characterised by the following relationship:

$$luminosity\left(W \big/ m^2\right) = 10^{7,0152} \times R_{LDR}{}^{(-1.1291)} \times 0.0079 \tag{1}$$

The equation relating temperature to the variable resistance of the NTC is as follows [8]:

$$Temperature(C°) = \frac{TF_{NTC}}{\log\frac{R_{NTC}}{R_{int.NTC}} + \frac{1}{(25+273.15)}} - 273.15 \tag{2}$$

The voltage and current sensors were parameterised by means of practical tests to define the coefficients of the linear characteristic of these two sensors:

- Voltage sensor characteristic : $V_{C.voltage} = 5 \times V_{voltage.analog}$ (3)

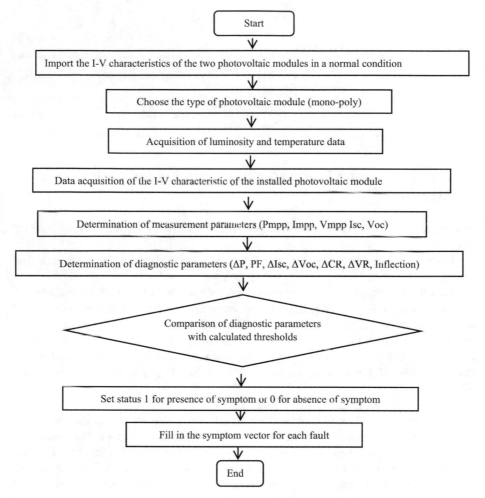

**Fig. 3.** Fault detection flowchart using the thresholding method.

• Current sensor characteristic : $I_{C.courant} = 14, 04 \times (2, 5 - V_{courant.analog})$    (4)

**Part 2: Fault Detection and Localization Algorithm (Thresholding Method)**

This part is the heart of the algorithm. It determines the calculation parameters for the reference healthy state and the faulty state, then compares them with the thresholds for each symptom. At the output, the algorithm constructs a signature matrix, which is then tuned as a fingerprint of the considered fault.

## 6   Test Results of Selected Faults

The following tables show the amplitudes of the diagnostic parameters for each fault expressed by its signature.

**For the DM-100 monocrystalline module.**

**Table 7.** Amplitudes of diagnostic parameters for each fault in the case of the DM-100 module.

| Faults | Symptoms | | | | | | |
|---|---|---|---|---|---|---|---|
| | S1 (W) | S2 (%) | S3 (A) | S4 (V) | S5 | S6 | S7 (%) |
| Normal state | 0 | 100 | 0 | 0 | 0 | 0 | 12.9498 |
| D1 | 2.4839 | 96.0386 | -0.0686 | 0.3665 | 0.0504 | -0.0155 | 46.6154 |
| D2 | 6.8269 | 89.1125 | -0.0686 | 0.4887 | 0.1407 | -0.0457 | 40.3355 |
| D3 | 54.8154 | 12.5801 | 4.2546 | 8.5044 | 0.1844 | -0.3227 | 28.5635 |
| D4 | 6.4580 | 89.7008 | 0.2059 | 1.1730 | 0.0309 | -0.0192 | 77.5197 |
| D5 | 31.7730 | 49.3282 | 3.1566 | 2.4438 | 0.0396 | -0.3235 | 11.3225 |
| D6 | 0.0887 | 99.8585 | -0.2744 | -0.1467 | 0.1197 | -0.0572 | 45.6693 |
| D7 | 2.1565 | 96.5609 | -0.2744 | 0.5865 | 0.0079 | 0.0308 | 57.9310 |

**For the DP-100 polycrystalline module.**

**Table 8.** Amplitudes of diagnostic parameters for each fault in the case of the DP-100 module.

| Faults | Symptoms | | | | | | |
|---|---|---|---|---|---|---|---|
| | S1 (W) | S2 (%) | S3 (A) | S4 (V) | S5 | S6 | S7 (%) |
| Normal state | 0 | 100 | 0 | 0 | 0 | 0 | 14.5985 |
| D1 | 0.4591 | 99.2786 | -0.0686 | 0.6598 | 0.0110 | -0.0179 | 49.6362 |
| D2 | 1.5984 | 102.5117 | 0.3431 | 0.1466 | -0.0321 | -0.0430 | 51.1275 |
| D3 | 23.9514 | 62.3639 | 2.6762 | 0.4643 | -0.0008 | -0.1806 | 8.6931 |
| D4 | 3.2680 | 94.8648 | 0.4117 | 1.1730 | 0.0105 | -0.0704 | 56.0004 |
| D5 | 35.2907 | 44.5459 | 3.3625 | 1.1974 | 0.0421 | -0.2190 | 4.0804 |
| D6 | 0.4083 | 99.3583 | -0.1372 | 0.3910 | -0.0404 | 0.0375 | 54.6195 |
| D7 | 1.0976 | 98.2753 | -0.1372 | 0.7820 | -0.0032 | 0.0043 | 12.7660 |

Values in red represent parameters above the thresholds set.
The binary representation of the signatures of the seven faults is given below:
**For the DM-100 monocrystalline module.**

**Table 9.** The binary code of the signatures for each fault in the case of the DM-100 module.

| Faults | Symptoms | | | | | | |
|---|---|---|---|---|---|---|---|
| | S1 (W) | S2 (%) | S3 (A) | S4 (V) | S5 | S6 | S7 (%) |
| Normal state | 0 | 0 | 0 | 0 | 0 | 0 | 0 |
| D1 | 0 | 0 | 0 | 0 | 1 | 0 | 1 |
| D2 | 1 | 1 | 0 | 0 | 1 | 0 | 1 |
| D3 | 1 | 1 | 1 | 1 | 1 | 0 | 0 |
| D4 | 1 | 1 | 0 | 1 | 1 | 0 | 1 |
| D5 | 1 | 1 | 1 | 1 | 1 | 0 | 0 |
| D6 | 0 | 0 | 0 | 0 | 1 | 0 | 1 |
| D7 | 0 | 0 | 0 | 1 | 0 | 1 | 1 |

This can be seen in Table 9, the presence of two (02) faults with the same symptom signature, faults D1 and D6 (faults diagnosed by the appearance of symptoms S5 and

S7) and faults D3 and D5 (S1, S2, S3, S4, S5). The similarity of the symptoms observed limits the thresholding method, hence the need to introduce another diagnostic method in order to obtain greater diagnostic accuracy and fault detection: D1 and D6, D3 and D5. The method introduced is based on fuzzy logic, which will be describe in more detail later.

**For the DP-100 polycrystalline module.**

**Table 10.** The binary code of the signatures for each fault in the case of the DP-100 module.

| Faults | Symptoms | | | | | | |
|--------|----------|----------|----------|----------|------|------|----------|
|        | S1 (W)   | S2 (%)   | S3 (A)   | S4 (V)   | S5   | S6   | S7 (%)   |
| Normal state | 0 | 0 | 0 | 0 | 0 | 0 | 0 |
| D1 | 0 | 0 | 0 | 1 | 0 | 0 | 1 |
| D2 | 0 | 0 | 1 | 0 | 0 | 0 | 1 |
| D3 | 1 | 1 | 1 | 0 | 0 | 0 | 0 |
| D4 | 1 | 0 | 1 | 1 | 0 | 0 | 1 |
| D5 | 1 | 1 | 1 | 1 | 1 | 0 | 0 |
| D6 | 0 | 0 | 0 | 0 | 0 | 1 | 1 |
| D7 | 0 | 0 | 0 | 1 | 0 | 0 | 0 |

The results of Table 10 show no similarity in signatures between the different defects. Therefore, in the case of the polycrystalline DP-100 module, the thresholding method used in the algorithm is sufficient to establish an adequate diagnosis for the seven faults selected.

# 7   Hybrid Diagnostic Method: Threshold-Fuzzy Logic

When using the DM-100 monocrystalline module, the thresholding method showed its limitations for few faults. It is therefore seen as an insufficient method for distinguishing between faults. The introduction of fuzzy logic as an intelligent diagnostic method offers a solution to optimize and correct problems related to the detectability of faults. The switch to fuzzy logic is conditioned by the appearance of signatures S = (0 0 0 0 1 0 1) and S = (1 1 1 1 1 0 0). According to the laws introduced in the fuzzy logic, it emits a new matrix of signatures perfectly adequate to distinguish between the defects: D1 and D6, D3 and D5. Four diagnostic parameter are introduced as inputs: **PF, CR, Voc and Inflexion** and seven symptom binary code as outputs: **S1, S2, S3, S4, S5, S6 and S7**.

The hybridization between the thresholding method and the fuzzy logic method is structured in three (03) steps as follows:

## 1. Inputs Fuzzification

The diagnostic parameters chosen to be introduced as inputs into the fuzzy logic are each guessed into two intervals (small and large) which are used to determine the degree of membership of each parameter. Each input has a membership function specified as follows (Figs. 4, 5, 6, and 7):

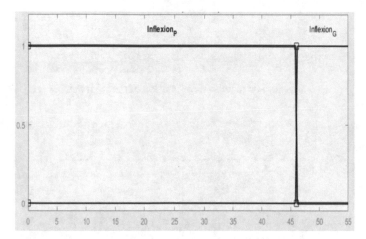

**Fig. 4.** The membership function of the inflection input.

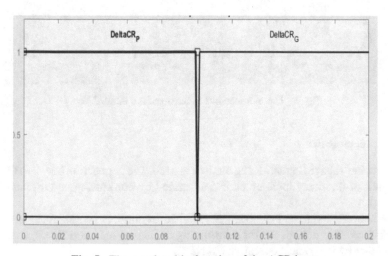

**Fig. 5.** The membership function of the $\Delta$CR input.

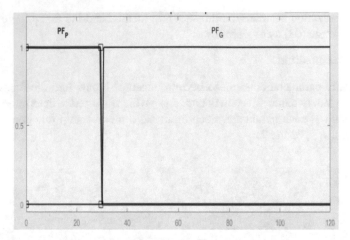

**Fig. 6.** The membership function of the PF input.

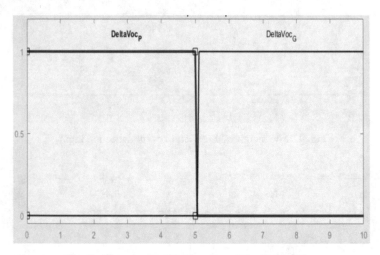

**Fig. 7.** The membership function of the input ΔVoc.

## 2. Interference Rules

In this step, the Takagi-Sugeno-Kang method is used. The connection between the inputs and outputs of the fuzzy logic block is determined by introducing certain interference rules.

Four interference rules have been established to eliminate cases of faults with the same signature. The parameters for determining degrees of membership are conditioned on each other by the logical operation ('AND').

Ex: If (Inflexion_G = 1) et (Δ CR_P = 1) and (PF_G = 1) and (Δ Voc_P = 1).
Then S = (0, 0, 0, 0, 0, 0, 0, 1).

All the rules are presented in Table 11 as follows:

**Table 11.** Fuzzy logic interference rules.

| Fuzzy logic inputs | | | | | | | | S1 | S2 | S3 | S4 | S5 | S6 | S7 |
|---|---|---|---|---|---|---|---|---|---|---|---|---|---|---|
| Inflexion | | Δ CR | | PF | | Δ Voc | | | | | | | | |
| P | G | P | G | P | G | P | G | | | | | | | |
| 0 | 1 | 1 | 0 | 0 | 1 | 1 | 0 | 0 | 0 | 0 | 0 | 0 | 0 | 1 |
| 1 | 0 | 0 | 1 | 1 | 0 | 0 | 1 | 1 | 0 | 1 | 1 | 1 | 0 | 0 |
| 1 | 0 | 0 | 1 | 0 | 1 | 1 | 0 | 0 | 0 | 0 | 0 | 1 | 0 | 0 |
| 1 | 0 | 1 | 0 | 0 | 1 | 1 | 0 | 1 | 1 | 1 | 0 | 0 | 0 | 0 |

## 3. Output defuzzification

The output of the fuzzy block is defuzzified using the zero-order Takagi-Sugeno-Kang method. This method produces a constant output signal. The zero-order Takagi-Sugeno-Kang method is used to construct adequate binary output signals to describe the presence of signatures for the faults: **D1 and D6, D3 and D5**. Figure 8 shows the value (0 or 1) of the output signal state (Fig. 9).

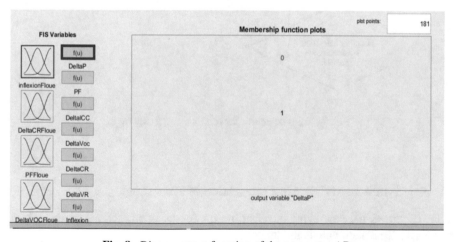

**Fig. 8.** Binary output function of the parameter $\Delta$P.

### 7.1  Fuzzy Logic Fault Detection Algorithm

The following flowchart describes the fault diagnosis algorithm using the fuzzy logic method.

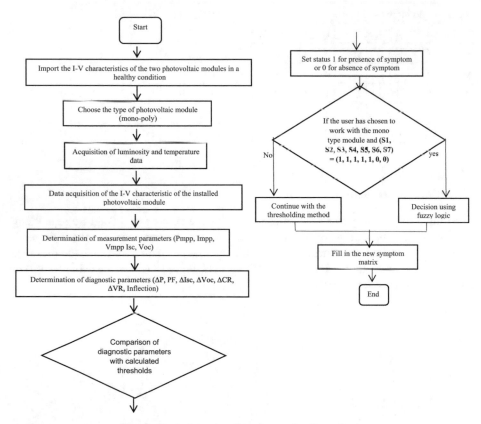

**Fig. 9.** Fault detection flowchart using fuzzy logic.

### 7.2  New Symptom Matrix After Integration of Fuzzy Logic

Table 12 shows the new symptom distribution for the malfunctions of the DM-100 monocrystalline photovoltaic module.

**Table 12.** Final signature matrix of the DM-100 monocrystalline module.

| Faults | Symptoms | | | | | | |
|---|---|---|---|---|---|---|---|
| | S1 (W) | S2 (%) | S3 (A) | S4 (V) | S5 | S6 | S7 (%) |
| Normal state | 0 | 0 | 0 | 0 | 0 | 0 | 0 |
| D1 | 0 | 0 | 0 | 0 | 0 | 0 | 1 |
| D2 | 1 | 1 | 0 | 0 | 1 | 0 | 1 |
| D3 | 1 | 0 | 1 | 1 | 1 | 0 | 0 |
| D4 | 1 | 1 | 0 | 1 | 1 | 0 | 1 |
| D5 | 1 | 1 | 1 | 0 | 0 | 0 | 0 |
| D6 | 0 | 0 | 0 | 0 | 1 | 0 | 0 |

# 8   Fault Location Algorithm for the DM-100 Module

The following flowchart describes the algorithm for locating faults in relation to their signatures for the.**DM-100 monocristallin** (Fig. 10).

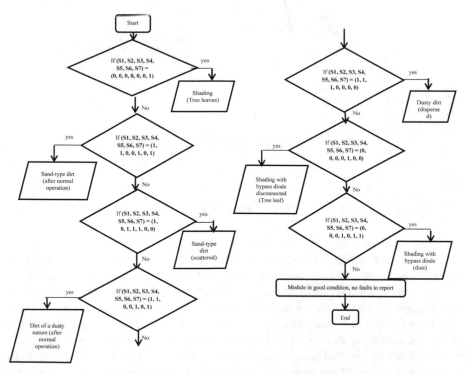

**Fig. 10.**  Fault location diagram for DM-100 module.

# 9 Fault Location Algorithm for the DM-100 Module

The following flowchart describes the algorithm for locating faults in relation to their signatures for the.**DP-100 polycristallin** (Fig. 11).

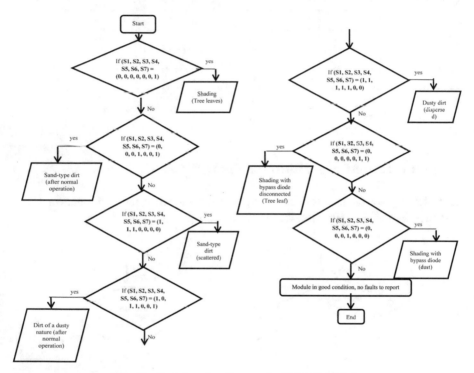

**Fig. 11.** Fault location diagram for DP-100 module.

# 10 Experimental Tests: Study Results

The study of the influence of faults on photovoltaic modules focuses on the analysis of electrical characteristics and diagnostic parameters such as: open-circuit voltage (Voc), short-circuit current (Isc), performance factor (PF), maximum power (Pmax), efficiency (η) and average power (Pmean). Our approach is to compare the effect of the degradation of the electrical characteristics and performance of PV modules caused by different defects on the two (02) DM-100 monocrystalline and DP-100 polycrystalline modules. To do this, number of tests are carried out, the first of which was a test carried out on the two modules under normal conditions to ensure that they were operating correctly and to determine the values of the parameters relating to normaly operation. Then a number of tests of the faults already determined were carried out. Table 13 shows the meteorological data collected and the tilt angle and azimuth of the PV module installation position.

**Table 13.**  Climatic conditions and system positioning.

| Azimuth | Inclination | Sunshine (W/m$^2$) | Temperature (C°) |
|---------|-------------|--------------------|------------------|
| Sud     | 30°         | 550                | 39               |

## 10.1  Test No. 01: Shading Tree Leaves)

In this test, the two PV modules are shaded by depositing tree leaves on their surfaces during initial healthy operation. Table 14 and Fig. 12 show the performance parameters and characteristics (I-V) of the two PV modules.

**Table 14.**  Parameters of the modules observed during test No.01.

|          | Technologies     | Reference | Parameters                         | Values   |
|----------|------------------|-----------|------------------------------------|----------|
| Module A | Monocrystalline  | DM-100    | Efficiency ($\eta$)                | 16,93%   |
|          |                  |           | Performance factor (PF)            | 96,03%   |
|          |                  |           | Average power ($P_{mean}$)         | 28,74 W  |
|          |                  |           | $\Delta\ \eta$ relative            | 3,92%    |
|          |                  |           | $\Delta$ PF relative               | 3,97%    |
|          |                  |           | $\Delta\ P_{mean}$ relative        | 17,88%   |
| Module B | Polycrystalline  | DP-100    | Efficiency ($\eta$)                | 15,80%   |
|          |                  |           | Performance factor (PF)            | 99,27%   |
|          |                  |           | Average power ($P_{mean}$)         | 27,74 W  |
|          |                  |           | $\Delta\ \eta$ relative            | 0,75%    |
|          |                  |           | $\Delta$ PF relative               | 0,73%    |
|          |                  |           | $\Delta\ P_{mean}$ relative        | 22,08%   |

Figure 12 and Table 14 show the results for the impact of partial shading on the (I-V) curves, form factor, average power and efficiency. We note that the (I-V) characteristic curves of the two PV modules are merged despite a slight difference between the short-circuit current (Isc) of the two PV modules, where the module (B) DP-100 has a slightly higher Isc current than module (A). However, the results shown in Table 14 show a decrease in relative efficiency $\Delta\eta$ more significant in the case of module (A) DM-100 despite the fact that its efficiency remains higher than that of module (B). This is due to the fact that monocrystalline technology has a greater overall efficiency compared to polycrystalline modules.

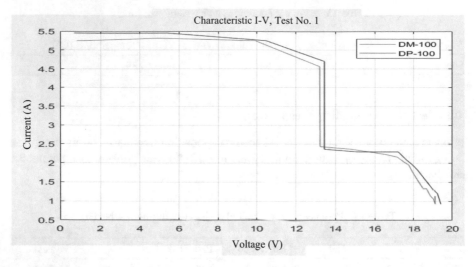

**Fig. 12.** Characteristics of the two modules in test No. 01.

## 10.2   Test N0. 02: Sand-Type Dirt After Normal Operation

Test No.02 is concerned with the deposition of sand on the surfaces of module (A) and (B) during initial healthy operation. Table 15 and Fig. 13 show the performance parameters and I-V characteristics of the two PV modules.

**Table 15.** Module parameters observed during test No.02.

|  | Technologies | Reference | Parameters | Values |
|---|---|---|---|---|
| Module A | Monocrystalline | DM-100 | Efficiency ($\eta$) | 15,71% |
|  |  |  | Performance factor (PF) | 89,11% |
|  |  |  | Average power ($P_{mean}$) | 27,86 W |
|  |  |  | $\Delta\ \eta$ relative | 10,89% |
|  |  |  | $\Delta$ PF relative | 10,89% |
|  |  |  | $\Delta\ P_{mean}$ relative | 20,40% |
| Module B | Polycrystalline | DP-100 | Efficiency ($\eta$) | 16,32% |
|  |  |  | Performance factor (PF) | 102,51% |
|  |  |  | Average power ($P_{mean}$) | 27,99 W |
|  |  |  | $\Delta\ \eta$ relative | -2,51% |
|  |  |  | $\Delta$ PF relative | -2,51% |
|  |  |  | $\Delta\ P_{mean}$ relative | 21,38% |

The study of the influence of sand deposits (after healthy operation) on the photovoltaic modules (A) and (B) involves the study of the (I-V) characteristic shown in

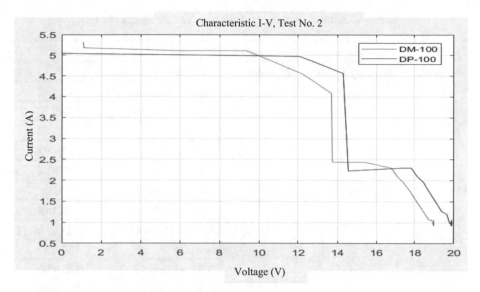

**Fig. 13.** Characteristics of the two modules in test No. 02.

Fig. 13. The shape of the two curves is almost identical, since there is an inflection point. However, the curve for module (B) DP-100 is shifted forward compared with that for module (A) DM-100. Open circuit voltage (Voc) and short-circuit current (Isc) are the main diagnostic parameters that can reflect the effect of degraded performance of the two (02) PV modules. Module (A) DM-100 has a slightly higher Isc than module (B) DP-100. However, its open circuit voltage (Voc) is lower than that of module (B) DP-100. Tthe parameters and performance of module (B) have higher values than those of the same module under normal conditions. This is indicated by the negative magnitudes of the parameters: $\Delta\eta$ and $\Delta PF$. Furthermore, the results of Table 15 state that the DM-100 monocrystalline type module (A) lost 10.89% of its efficiency and form factor compared to the healthy state due to the impact of sand deposition.

### 10.3   Test N0. 03: Sandy Dirt (Dispersed)

Test No. 03 involved the deposition of sand on the surfaces of modules (A) and (B) during the entire test period. Table 16 and Fig. 14 show the performance parameters and characteristics (I-V) of the two PV modules.

**Table 16.** Parameters of the modules observed during test No. 03.

|  | Technologies | Reference | Parameters | Values |
|---|---|---|---|---|
| Module A | Monocrystalline | DM-100 | Efficiency ($\eta$) | 2,21% |
|  |  |  | Performance factor (PF) | 12,58% |
|  |  |  | Average power ($P_{mean}$) | 5,03 W |
|  |  |  | $\Delta \eta$ relative | 87,46% |
|  |  |  | $\Delta$ PF relative | 87,42% |
|  |  |  | $\Delta P_{mean}$ relative | 85,63% |
| Module B | Polycrystalline | DP-100 | Efficiency ($\eta$) | 9,92% |
|  |  |  | Performance factor (PF) | 62,36% |
|  |  |  | Average power ($P_{mean}$) | 23,53 W |
|  |  |  | $\Delta \eta$ relative | 37,69% |
|  |  |  | $\Delta$ PF relative | 37,64% |
|  |  |  | $\Delta P_{mean}$ relative | 33,90% |

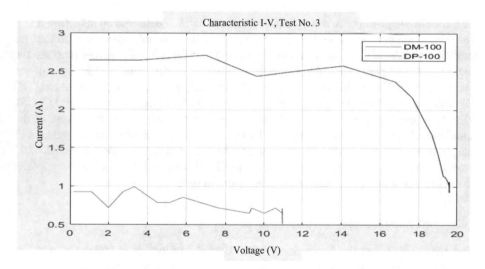

**Fig. 14.** Characteristics of the two modules in test No. 03.

The DM-100 and DP-100 PV modules were exposed to sand deposits throughout Test 03. Figure 14 shows the I-V curves of the two PV modules. The characteristic (I-V) of the monocrystalline type module (A) suffered a serious deterioration, the value of the short-circuit current (Isc) of this module is five (05) times lower than in normal operation and the open-circuit voltage (Voc) has lost 50% of its value in normal operation. The impact of this fault is not as great on the polycrystalline module (B), since the losses in efficiency and power factor are only 37.6%. On the other hand, the losses recorded for the same performance in the case of module (A) are much higher (around 87%). Similarly, the average power (Pmoy) of module (B) is much higher than that of module (A). This result confirms that polycrystalline modules are less sensitive to complete light blockage on small surfaces.

## 11  Conclusion

The approach to implementing the DLD algorithm is presented, and then introduce the diagnostic methods in detail. Next, The hybrid diagnostic method is adopted, which consists of using the intelligent fuzzy method with the thresholding method in order to offer an optimisation and correction solution to the problems associated with fault detectability. Two (02) signature matrices for each type of PV module are generated by the DLD algorithm in order to locate their relative faults. Experimental observations were used to analyse the relative performance of the two PV modules in the presence of faults:

- Dirt defect: sand or dust, and the manner of deposition: abrupt (after an initial healthy operating condition) or even dispersed (throughout the test) affect the sensitivity of both PV modules to defects.
- A fault involving the sudden deposition of dirt on the surface of the PV module causes a drop of more than 40% in the current produced. This forms an inflection point on the shape of the characteristic (I-V).
- Dispersed fouling deposits cause a deterioration in the short-circuit current Icc and a huge drop in performance.
- Shading defect caused by tree leaf deposition results in a drop in power output of up to 22.08%.
- A 36-cell PV module with only two (02) bypass diodes will not produce any power if 25% or more of its surface is shaded.

## References

1. https://www.mordorintelligence.com/fr/industry-reports/solar-photovoltaic-market
2. Bartmann, E.: Le grand livre d'Arduino 2eme édition
3. Bendel, C., Wagner, A.: Photovoltaic measurement relevant to the energy yield. In: The Proceeding of 2003 IEEE PEC (3rd World Conference on Photovoltaic Energy Conversion), Osaka, Japan, pp. 2227 – 2230 (2003)
4. Dhimish, M., Holmes, V., Mehrdadi, B., Dales, M., Mather, P.:.Photovoltaic fault detection algorithm based on theoretical curves modelling and fuzzy classification system. Energy 140(Part 1), 276–290 (2017), ISSN 0360–5442,
5. Haque, A., Bharath, K.V.S., Khan, M.A., Khan, I., Jaffery, Z.A.: Fault diagnosis of photovoltaic modules. Energy Sci. Eng. 7(3), 622–644 (2019 )
6. Hong, Y.-Y., Pula, R.A.: Methods of photovoltaic fault detection and classification: a review. Energy Rep. 8, 5898–5929 (2022)
7. Han, F., et al.: An intelligent fault diagnosis method for PV arrays based on an improved rotation forest algorithm. Energy Proc. 58, 6132–6138 (2019)

8. Naveen Venkatesh, S., Sugumaran, V.: Fault diagnosis of visual faults in photovoltaic modules: a Review. Int. J. Green Energy **18**(1), 37–50 (2021). https://doi.org/10.1080/15435075.2020.1825443
9. Bacha, M., Terki, A.: Diagnosis algorithm and detection faults based on fuzzy logic for PV panel. Mater. Today: Proc. **51**(Part 7), 2214–7853 (2022) ISSN 2131–2138
10. Dhimish, M., Holmes, V., Mehrdadi, B., Dales, M., Mather, P.: Photovoltaic fault detection algorithm based on theoretical curves modelling and fuzzy classification system J. Energy **140**, 276–290 (2017)

# Defect Detection in Photovoltaic Module Cell Using CNN Model

N. Drir[1(✉)], K. Kassa Baghdouche[1], A. Saadouni[1], and F. Chekired[2]

[1] Faculty of Electrical Engineering, University of Science and Technology Houari Boumediene (USTHB), BP 32, El Alia, 16111 Bab. Ezzouar, Algiers, Algeria
ndrir@usthb.dz
[2] Unité de Developpement des Equipements Solaires UDES/Centre de Developpement des Energies Renouvelables CDER, Tipaza, Algeria

**Abstract.** One way of examining surface defects on photovoltaic modules is the Electroluminescence (EL) imaging technique. The data set used in this work is an open data set for fault detection and classification of photovoltaic cells. In this article, we have used various deep learning (DL) techniques to ensure fault detection and diagnosis of photovoltaic modules. A binary classification model was developed that highlighted defective PV modules and normal modules. The subset of defective PV modules was used to design a multi class model of default detection (light, moderate, and severe). Evaluation results (confusion matrix, mean square error (mse)) showed that methods based on deep learning performed exceptionally well, making it possible to solve the problem of detecting and diagnosing faults in photovoltaic modules with good overall precision (mse = 0.060).

**Keywords:** Fault detection · photovoltaic module · deep learning · Electroluminescence images

## 1 Introduction

In the recent years governments and gents invested more in different sources of green energy in order to reduce the consumption of fossil fuels which causes more and more damage to the environment. Based on data provided by International Renewable Energy Agency (IRENA), By the end of 2022, global renewable energy capacity stood at 3,372 GW, a record increase of 295 GW, or 9.6% in one year [1].

Among renewable energy sources that have been the subject of great attention in recent years, is solar energy systems [2]. Solar power stations have been developed worldwide, leading to the activation of large-scale production facilities that create solar energy components [3]. To maintain long-term operational efficiency and reliability, it is imperative to implement monitoring and supervision protocols for photovoltaic (PV) installations. Solar cells can be damaged as a result of their environmental exposure such as hail, and the effect of falling tree branches which induces power losses in the system. Establishing a proficient methodology for identifying flaws in photovoltaic (PV) panels

© The Author(s), under exclusive license to Springer Nature Switzerland AG 2024
M. Hatti (Ed.): IC-AIRES 2023, LNNS 984, pp. 403–411, 2024.
https://doi.org/10.1007/978-3-031-60629-8_39

is paramount. This process is essential for the replacement or repair of faulty units, hence guaranteeing optimal operational efficiency of solar power plants.

To address the challenging issue of detecting surface imperfections in photovoltaic cells, several methods based on artificial intelligence have been developed; in reference [4] a supervised learning method using support vector machine (SVM) was applied, in [5] they proposed a end-to-end convolutional neural network (CNN). However, the rate of false positives from the CNN model is still high. In [6] they utilized a unique feature descriptor to propose a classification system for manufacturing faults in multicrystalline solar cells. Refence [7] they have used a optical CNN structure for identifying EL image defects, [8] proposed combination of deep-learning (CNN) and machine learning (SVM) approach to achieve automated identification of photovoltaic (PV) cell faults. The use of deep learning techniques has performed well in detecting defects in photovoltaic cells, however there are gaps in small unbalanced data sets our dataset contains electroluminescence images that provide us with information on a range of certain defects on the surface of PV modules. The process of manually reviewing EL pictures is a resource-intensive and time-consuming endeavor, necessitating a substantial understanding of the subject matter. Moreover, it is feasible to conduct such reviews only on a limited scale [9]. The detection of defects in photovoltaic modules in an intelligent and automatic way especially when working on a large scale is highly recommended for their current.

The primary aim of this study is to construct a novel supervised classification model utilizing modern deep learning methodologies for the purpose of accurately detecting and distinguishing various types of flaws in solar cells. In this context, the most important contributions of this study are:

The data volume was augmented and the weight class was implemented to address the challenges posed by limited data and imbalanced data models.

The performance of binary and multi-class classification CNN models is compared with the VGG 16 model for the.

fault detection and diagnosis of PV modules.

Three degrees of failure of photovoltaic modules were examined: light, moderate, and severe.

The present article is structured in the following manner. The concepts and methods employed for the experiments are outlined in Sect. 2. The results acquired are reported and analyzed in Sect. 3. Conclusions and potential areas for further research are discussed in Sect. 4.

## 2  Materials and Methods

The defect detection algorithm is depicted in Fig. 1. Initially, the system performs a binary classification on the input images, distinguishing between defective and normal photovoltaic (PV) cells. Subsequently, defective PV cells are classified by degree of degradation called multiple cells classifications. Finally, algorithm is then compared to the VGG16 deep learning algorithm.

The convolutional neural network (CNN) is a widely employed deep learning model that is frequently utilized for the purposes of picture categorization and recognition. Figure 2 depicts CNN's overall structure, it is composed of convolution layers, pooling

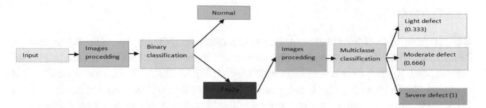

**Fig. 1.** CNN based fault detection algorithm

layers, fully connected (FC) layers, batch normalization, SoftMax, and classification layers.

Convolutional layer: The Convolutional layer is sometimes called the feature extractor layer due to its role in extracting visual properties. Then the input images are convoked using convolution nuclei of different sizes to produce the properties. After that aggregation process allows to combine of features extracted from various image parts [10].

The pooling layer is employed to decrease the spatial dimensionality of the input image subsequent to the convolution operation. This layer reduces the size of input images and makes the computation cost and duration much less.

The fully connected layer (FCL) establishes connections between neurons from one layer to neurons in another layer, using weights, biases, and neurons. The aim of this device is to employ a training process in order to classify photos into distinct categories [10].

The output layer represents the final layer of a Convolutional Neural Network (CNN). The entity's location is at the terminal point of the fully connected layer.

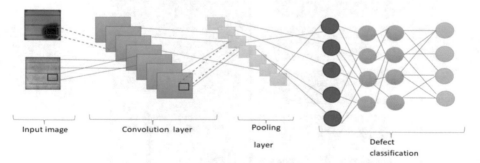

**Fig. 2.** CNN based classification.

Our proposed DeepCNN model will allow us to classification of defects based on collected images. This classifier can help users make informed decisions related to interventions in the photovoltaic installation, it will be compared to a deep CNN that uses VGG16 transfer learning approach.

## 2.1  VGG16

VGG16 is a convolutional neural network model designed by K. Simonyan and A. Zisserman. The VGG-16 model demonstrates a test accuracy of 92.7% on the ImageNet dataset, comprising over 14 million images distributed across 1000 distinct classes, it is called VGG-16 and well simply because this network of neurons includes 16 deep layers. Although we can create this network ourselves, then find the best hyperparameters to finally train it but it will consume a lot of time and resources. With the idea of transfer learning, make enables the utilization of pre-existing shape recognition methods, specifically the feature maps generated by the VGG16 model, to detect different shapes. The weight of the convolutive layers of models learned previously are used, only the final layers require training.

## 2.2  Proposed Model

Convolutional neural networks (CNNs) have garnered significant utilization within the image processing domain since they represent a strong deep learning model, which has proven its effectiveness in image processing and classification with excellent performance. Figures 3 and 4 show us the architecture of Deep CNN proposed in this work.

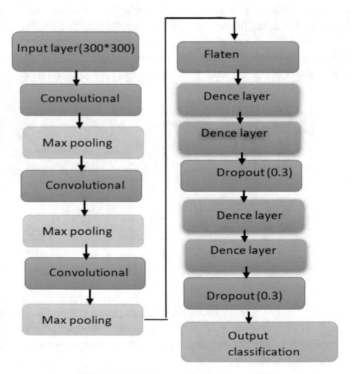

**Fig. 3.** CNN binary classification model

CNN for binairy classification it composed of 13 layers (Fig. 3): the first six layers (three conv2D layers, and three MaxPooling layers) were used for feature extraction; the second seven layers (one flatten layer, four dense layers, two dropout layer, and a final dense layer containing the Softmax function).

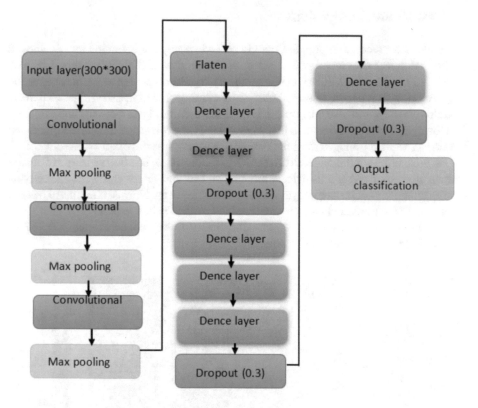

**Fig. 4.** CNN multi classes classification model

CNN for multiclass classification it composed of 17 layers (Fig. 4): the first six layers (three conv2D layers, and three MaxPooling layers) were used for feature extraction; the second eleven layers (one flatten layer, six dense layers, three dropout layer, and a final dense layer containing output classification).

## 2.3 Data Set and Metrics

We used a database containing 2624 images of solar cells taken from 44 distinct photo-voltaic modules, encompassing monocrystalline and polycrystalline. The images were of high resolution, with dimensions of 300 by 300 pixels. Expert evaluators divided the images into four distinct categories based on the severity of the identified issues [11]. The normal module that does not present defects (1508 images), modules with a 33% chance of operation (295 images), the modules with percentage of possibility of faulty operation is 66% (106 images), the modules defected a 100% (715 images).

This collection encompasses many forms of flaws, such as microcracks, cells having electrically isolated and damaged parts, shorted cells, open interconnections, and weld defects.

## 3   Results and Discussions

Our code was executed in google Colab in order to apply the detection and diagnostic method of defect developed, we selected an image size of 300*300 pixels. Figure 5 shows the results of the binary classification, it represents a confusion matrix for the proposed model, it can be observed that 82 image defect are classified right as defects module (class 2), and also 110 image are classified correctly as normal module (class 1). However, a total of 36 photos that were actually defective were incorrectly classed as normal, while 37 images from the normal class were misclassified as defective.

Figure 6 is the confusion matrix for VGG 16 model, it can be observed that 100 image defects are classified right as defects module; and 67 image are classified correctly as normal module. 87 defective images were misclassified in the Normal class, and 9 images belonging to the normal class were misclassified in the Defective class.

Our model generated 192 good classifications through contributions, and 167 of them were realized by the VGG model.

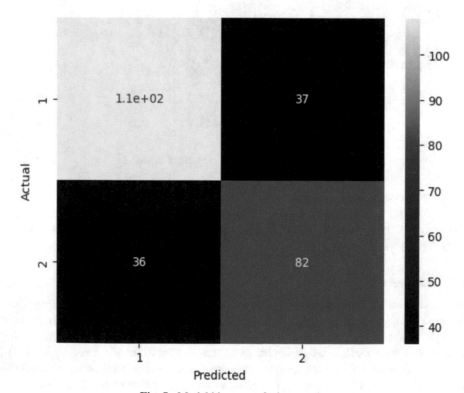

**Fig. 5.** Model binary confusion matrix

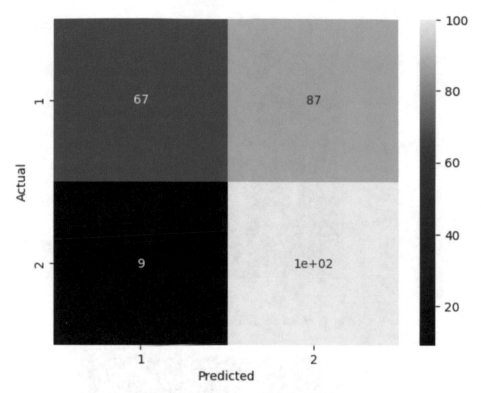

**Fig. 6.** VGG16 binary confusion matrix

After that we develop the second model to classe the defects images into three classes: (0.333 for light defect; 0.66666 for moderate defects, and 1 for several defects). Since we are only going to work on the images of the defective cells, the database is therefore reduced, which is why we opted for an increase of data and an assignment of class weight to the images; simulation results are presented in Figs. 7 and 8

As can be seen in Fig. 7 the developed model is able to detect multi-class anomalies with a very good precision that the VGG16 model. Indeed 60 images was well classified as images that has severe defects (class 1), 19 images as cells with light class defaults (class 0.33333), and 3 images with medium class defaults (class 0.66666), making a total of 82 images well classified by the model against 75 classified by the model VGG 16 (see Fig. 8).

The mean squad error for our model have achieve 0.061, and VGG 16 have achieve 0.068. we can say that our detection model with adoption of data and class weight increase gave us a good precision of classification of defaults of solar cells in electroluminescent images.

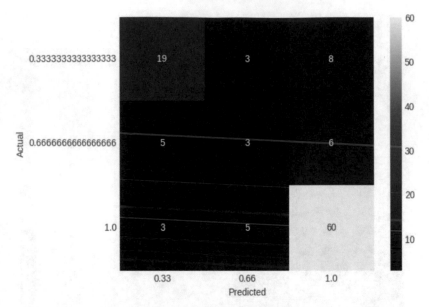

**Fig. 7.** Model mutli classes confusion matrix

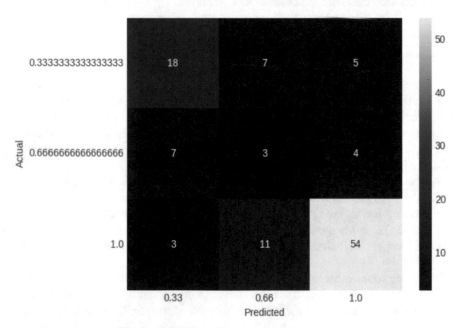

**Fig. 8.** VGG16 mutli classes confusion matrix

# 4 Conclusion

This research work presents a study of photovoltaic cell defect classification in electroluminescence images. First, we proposed a CNN model that performs binary classification between good and defective solar cells. After that we proposed a multiclass classification model using the image subset of cells with defects: cells with slight defects, cells with moderate defects, and defective cells. For this opted to increase the data and assign weights to the different classes, which enabled us to have a balanced database. The obtained results shows that the model was well performed and the classification results were very satisfactory. In the continuity of this work, our next work aims to integrate the new hybrid model between deep leaning and machine leaning in order to make a better classification of defects.

# References

1. « HOME VERT », Homevert. https://www.homevert21.fr
2. Benjamin, E., Misra, S., Damaševičius, R., Maskeliunas, R.: Hybrid microgrid for microfinance institutions in rural areas – a field demonstration in West Africa. Sustain. Energy Technol. Assess. **35**, 89–97 (2019)
3. Demirci, M., Besli, N., Gümüşçü, A.: Efficient deep feature extraction and classification for identifying defective photovoltaic module cells in electroluminescence images. Expert Syst. App. **175**, 114810 (2021). https://doi.org/10.1016/j.eswa.2021.114810
4. Bedrich, K.G., Bokalic, M., Bliss, M., Topic, M., Betts, T., Gottschalg, R.: Electroluminescence imaging of PV devices: advanced vignetting calibration. IEEE J. Photovoltaic. **8**, 1297–1304 (2018). https://doi.org/10.1109/JPHOTOV.2018.2848722
5. Hwang, S.-K., Kim, W.-Y.: Fast and efficient method for computing ART. IEEE Trans. Image Process. **15**(1), 112–117 (2005)
6. Su, B., Chen, H., Zhu, Y., Liu, W., Liu, K.: Classification of manufacturing defects in multicrystalline solar cells with novel feature descriptor. IEEE Trans. Instrum. Measure. **68**, 1–14 (2019). https://doi.org/10.1109/TIM.2019.2900961
7. Akram, M.W., et al.: CNN based automatic detection of photovoltaic cell defects in electroluminescence images. Energy **189**(C), 116319 (2019)
8. Deitsch, S., et al.: Segmentation of photovoltaic module cells in uncalibrated electroluminescence images. Mach. Vision Appl. **32**(4), 84 (2021)
9. Deitsch, S., et al.: Automatic classification of defective photovoltaic module cells in electroluminescence images. Solar Energy. **185**, 455–468 (2019). https://doi.org/10.1016/j.solener.2019.02.067
10. Espinosa, A.R., Bressan, M., Giraldo, L.F.: Failure signature classification in solar photovoltaic plants using RGB images and convolutional neural networks. Renew. Energy. **162**, 249–256 (2020)
11. A Benchmark for Visual Identification of Defective Solar Cells in Electroluminescence Imagery. ZAE Bayern. https://github.com/zae-bayern/elpv-dataset

# Photovoltaic System with Fuzzy-Direct Torque Control Technique for Standalone Application

Z. Mokrani(✉), D. Rekioua, T. Rekioua, and K. Kakouche

Université de Bejaia, Faculté de Technologie, Laboratoire de Technologie Industrielle et de l'Information, 06000 Bejaia, Algérie
zahra.mokrani@uni-bejaia.dz

**Abstract.** This paper's main goal is to apply the Direct Torque Control (DTC) method with fuzzy logic control to a solar system. In order to accomplish this, the study also compares the performance properties of two alternative control approaches, Indirect Field Oriented Control Technique (IFROC) and Closed Loop Scalar Control (V/f). The photovoltaic system is intended primarily for standalone applications. However, the power production of solar panels might vary during the day depending on the various climatic conditions, making it unpredictable. For this, Maximum Power Point Tracking (MPPT) strategies are being developed. Three different solar irradiance profiles were used for the examined system and collected using a data acquisition system used in our laboratory. To evaluate and present the efficacy of the proposed system for standalone applications, simulation modeling with Matlab/Simulink is used.

**Keywords:** MPPT · Direct torque control · IFROC · V/f Control

## 1 Introduction

The numerous advantages of photovoltaic energy make it an attractive and environmentally friendly option for meeting the increasing global energy demand and transitioning towards a more sustainable energy future [1]. However, the power production of solar panels might vary during the day depending on the various climatic conditions, making it unpredictable. To address this problem, MPPT strategies are being developed [2–8], and [9, 10]. The fuzzy logic control (FLC) methodology is widely used [2, 6–8]. It's adaptability, robustness, and performance advantages contribute to its growing popularity as an efficient and reliable MPPT technique, particularly in scenarios with dynamic and challenging operating conditions.

The use of advanced control approaches in induction motor drive systems for photovoltaic applications has emerged as a promising trend, enabling higher efficiency, improved performance, and enhanced adaptability to changing solar energy inputs [11–13]. As research and technological advancements continue, we can expect these control strategies are important in the development of sustainable and efficient photovoltaic energy systems. The first control presented in the market was the scalar control because it is simple. It entails keeping an eye on the induction motor speed and changing the

© The Author(s), under exclusive license to Springer Nature Switzerland AG 2024
M. Hatti (Ed.): IC-AIRES 2023, LNNS 984, pp. 412–422, 2024.
https://doi.org/10.1007/978-3-031-60629-8_40

stator voltages' magnitude and frequency. However, this method does not make use of the motor's field orientation.

One of the vector control approaches called Field Oriented Control (FOC) is currently a very intriguing research topic [3, 6]. This method involves adjusting the flux components that produce torque and magnetize matter individually. As a result, a vector is used to represent the stator currents. PWM drives rely on input frequency and voltage. The DTC approach has various advantages, including torque response, and accuracy in dynamic speed.

The current paper describes the DTC method's application and argument for using them. Then, a thorough overview of electrical machine control methods is suggested. The purpose is to defend the DTC method's selection. The DTC method modeling and its application to a solo photovoltaic system utilizing experimental data of solar irradiance are the topics of the fourth section of this research. Finally, a comparison of the model outputs and conclusions is suggested.

## 2  Studied System

The studied photovoltaic application is given in Fig. 1.

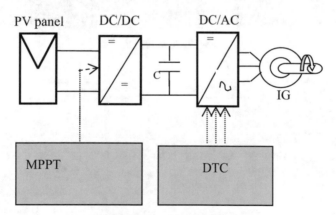

**Fig. 1.** Photovoltaic application.

### A. Photovoltaic Generator Modeling

A multitude of models are used to compute PV panel electrical curves. Their level of accuracy and complexity varies. We applied the one-diode model in our research [3, 5, 10]. It is simple but useful for characterizing a solar generator's electrical properties.

The single-diode model's electrical current is [3, 13]:

$$I_{pv} = I_{ph} - I_d - I_{sh} \tag{1}$$

The following bench (Fig. 2) is used to obtain the following curves (Fig. 3).

**Fig. 2.** Test bench.

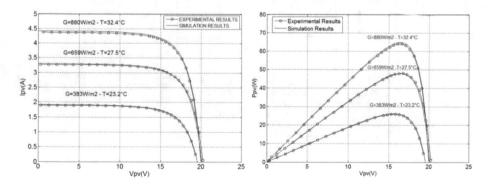

**Fig. 3.** PV electrical characteristic.

### B. MPPT Algorithms

The P&O approach can sometimes cause oscillations around the MPP, particularly if the perturbations (changes in the operating point) are too large. This oscillation can lead to inefficient operation and power losses. An FLC is a control system based on fuzzy logic [10]. It provides an output reference, which is a desired set point for the system. The FLC takes two inputs to make decisions and converge to the optimal operating point. These inputs are power fluctuation and voltage variation (Fig. 4). The FLC is expected to provide a more refined and efficient control strategy compared to the simple P&O approach, helping the solar system achieve better [8].

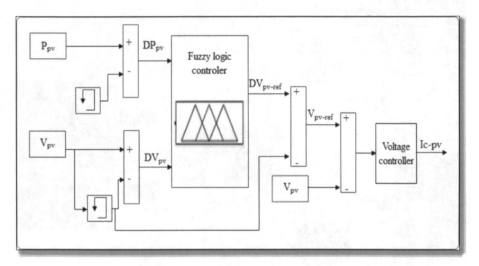

**Fig. 4.** Structural scheme of the FLC.

# 3 Control Method Used in Electrical Machines

The literature provides in-depth descriptions of a wide range of variable speed control techniques. In this section of the study, the advantages and disadvantages of these control systems are addressed in Fig. 5. The best space vector must be chosen for this using a hysteresis controller (Fig. 6).

The stator flux $(\alpha, \beta)$ are given as [2, 5]:

$$
\begin{cases}
\Phi_{s\alpha}(t) = \int_0^t (V_{s\alpha} - R_s.i_{s\alpha})dt \\
\Phi_{s\beta}(t) = \int_0^t (V_{s\beta} - R_s.i_{s\beta})dt
\end{cases}
\tag{2}
$$

$$
\Phi_s = \sqrt{\Phi_{s\alpha}^2 + \Phi_{s\beta}^2}
\tag{3}
$$

The electromagnetic torque is written as:

$$
\Gamma_e = \frac{3}{2}P.\left(\Phi_{s\beta}\, i_{s\alpha} - \Phi_{s\alpha}\, i_{s\beta}\right)
\tag{4}
$$

The regulator keeps the hysteresis bands' torque and flux errors within them [2]. Six equally sized sectors make up the flux, as suggested in [5]. The stator flux and torque can be changed in each sector by choosing different vectors (Table 1). Figure 6 shows the DTC structure.

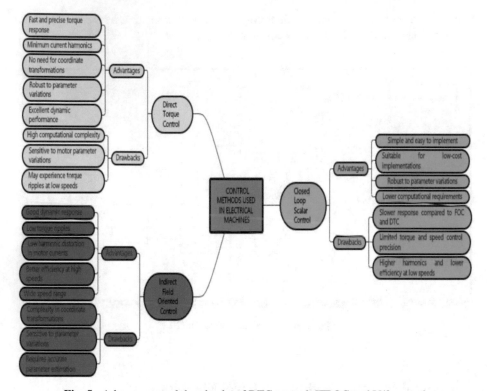

**Fig. 5.** Advantages and drawbacks of DTC control; IFROC and V/f control.

**Table. 1.** Switching table

|         | N          | 1   | 2   | 3   | 4   | 5   | 6   |
|---------|------------|-----|-----|-----|-----|-----|-----|
| $b_\Phi = 1$ | $b_\Gamma = 1$ | 110 | 010 | 011 | 001 | 101 | 100 |
|         | $b_\Gamma = 0$ | 111 | 000 | 111 | 000 | 011 | 001 |
|         | $b_\Gamma = -1$ | 101 | 100 | 110 | 010 | 011 | 001 |
| $b_\Phi = 0$ | $b_\Gamma = 1$ | 010 | 010 | 011 | 001 | 101 | 100 |
|         | $b_\Gamma = 0$ | 000 | 000 | 111 | 000 | 111 | 000 |
|         | $b_\Gamma = -1$ | 001 | 001 | 101 | 100 | 110 | 010 |

## 4   Simulation Study

Solar irradiance of two different days were used (Fig. 7) for the examined system. The different PV powers are given in Fig. 8 and the SOC of battery in Fig. 9.

Around the load torque value, the electromagnetic torque waveform oscillates (Fig. 10). Figure 11 shows the stator currents. The stator flux is kept constant (Figs. 12 and 13).The rotor speed is represented in Fig. 14.

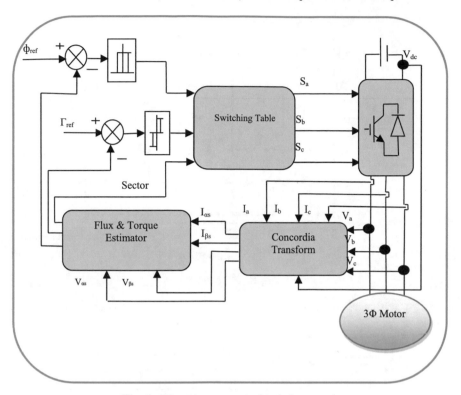

**Fig. 6.** Direct torque control technique used.

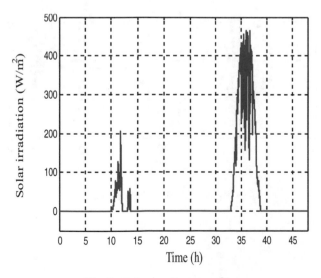

**Fig. 7.** Solar irradiance profile.

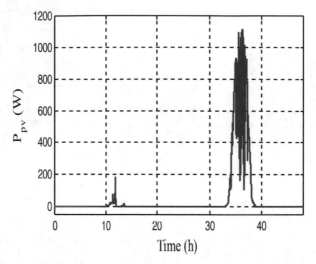

**Fig. 8.** PV powers.

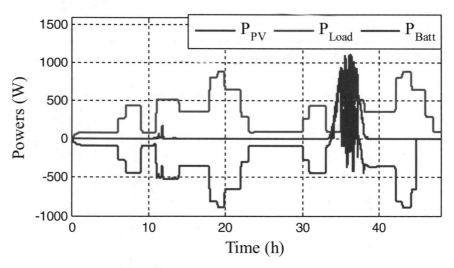

**Fig. 9.** The different powers.

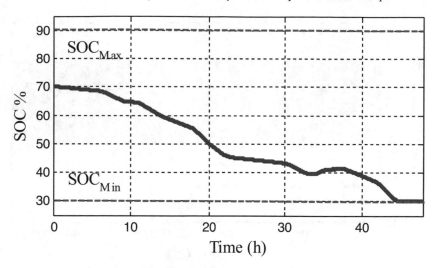

**Fig. 10.** Battery state of charge.

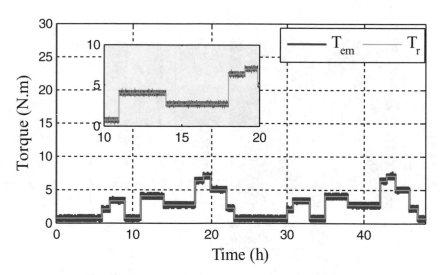

**Fig. 11.** Electromagnetic torque waveform and its zoom.

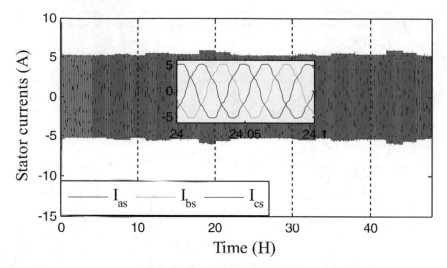

**Fig. 12.** Stator current waveform.

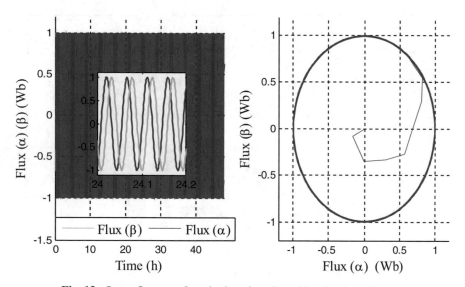

**Fig. 13.** Stator flux waveform in time domain and its circular trajectory

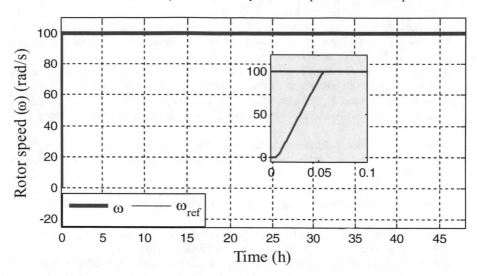

**Fig. 14.** Rotor speed waveform.

## 5   Conclusion

The research demonstrates the effectiveness of using DTC in an induction generator powered by a photovoltaic system. By integrating MPPT and employing a modulated hysteresis controller, the system's performance is optimized, minimizing torque ripples and maximizing power generation. This improved tracking facilitates the optimization of power generation, leading to enhanced system effectiveness. It will be interesting to use in future the combination of MPPT-DTC with the modulated hysteresis controller which will results in a more efficient and reliable photovoltaic system.

## References

1. Mohapatra, B., Sahu, B.K., Pati, S., Ghamry, N.A., Ghoneim, S.S.M.: Application of a novel metaheuristic algorithm based two-fold hysteresis current controller for a grid connected PV system using real time OPAL-RT based simulator. Energy Rep. **9**, 6149–6173 (2023)
2. Rekioua, D., Rekioua, T., Idjdarene, K., Tounzi, A.: An approach for the modeling of an autonomous induction generator taking into account the saturation effect. Int. J. Emerg. Electr. Power Syst. **4**(1) (2005)
3. Rekioua, D., Matagne, E.: Modeling of solar irradiance and cells. In: Optimization of Photovoltaic Power Systems, Green Energy and Technology. Springer, London (2012). https://doi.org/10.1007/978-1-4471-2403-0_2
4. Okedu, K., Kingsley-Amaehule, M., Uhunmwangho, R., Nwazor, N.: Investigation of the impact of soot on the efficiency of solar panels using a smart intelligent monitoring system. Int. J. Smart Grid – ijSmartGrid **7**(1), 1–14 (2023)
5. Casadei, D., Profumo, F., Serra, G., Tani, A.: FOC and DTC: two viable schemes for induction motors torque control. IEEE Trans. Power Electronics **17**(5), 779–787 (2002)
6. Venu Gopal, G.T.: Comparison between direct and indirect field oriented control of induction motor. Int. J. Eng. Trends Technol. (IJETT) **43**(6), 364–359 (2017)

7. Mohammedi, A., Rekioua, D., Rekioua, T., Bacha, S.: Valve Regulated Lead Acid battery behavior in a renewable energy system under an ideal Mediterranean climate. Int. J. Hydrogen Energy **41**(45), 20928–20938 (2016)

8. Kakouche, K., et al.: Model predictive direct torque control and fuzzy logic energy management for multi power source electric vehicles. Sensors **22**, 5669 (2022)

9. Hassani, H., Zaouche, F., Rekioua, D., Belaid, S., Rekioua, T., Bacha, S.: Feasibility of a standalone photovoltaic/battery system with hydrogen production. J. Energy Storage **31**, 101644 (2020)

10. Mebarki, N., Rekioua, T., Mokrani, Z., Rekioua, D.: Supervisor control for stand-alone photo-voltaic/hydrogen/battery bank system to supply energy to an electric vehicle. Int. J. Hydrogen Energy **40**(39), 13777–13788 (2015)

11. Boglou, V., Karavas, C.-S., Arvanitis, K., Karlis, A.: A fuzzy energy management strategy for the coordination of electric vehicle charging in low voltage distribution grids. Energies **13**(14), art. no. 3709, 1–34 (2020)

12. Akhtar, I., Kirmani, S., Hasan, S., Khan, S.: Advanced control approach for solar assisted electric vehicle drive. Mater. Today Proc. **46**(20), 10806–10810 (2021)

13. Aissou, R., Rekioua, T., Rekioua, D., Tounzi, A.: Robust nonlinear predictive control of permanent magnet synchronous generator turbine using Dspace hardware. Int. J. Hydrog. Energy **41**(45), 21047–21056 (2016)

14. Belaid, S., Rekioua, D., Oubelaid, A., Ziane, D., Rekioua, T.: A power management control and optimization of a wind turbine with battery storage system. J. Energy Storage **45**, 103613 (2022)

# Small UWB Antenna for GPR Applications in VHF/UHF Bands

El-Hadi Khoumeri[1](✉), Henia Fraoucene[2], and Aissa Khoumeri[3]

[1] LTI Lab Ecole Nationale Supérieure des Technologies Avancées, Algiers, Algeria
elhadi.khoumeri@ensta.edu.dz
[2] Laboratory of Advanced Technology, UMM Tizi-Ouzou, Tizi-Ouzou, Algeria
henia.fraoucene@ummto.dz
[3] Thales Air System S.A.S, Le Mont Jarret, 76520 Ymare, France

**Abstract.** This work presents a new miniaturized UWB antenna for UHF/VHF GPR. The antenna is a folded exponential tapered slot antenna (ETSA) miniaturized using load impedances, which ensures adaptation in lower frequencies. The antenna is designed on FR4 substrate and has an area of $30 \times 30$ cm$^2$, simulations are done on CST environment, results show that the antenna is matched, at $-6$ dB, in the frequency bandwidth 140–2000 MHz, within that it's gain is higher than 2 dBi. The antenna is fabricated and tested, measurement results are in good concordance with simulation.

**Keywords:** Antenna · folded · GPR · load impedance · miniaturization · small · UHF · UWB · VHF

## 1 Introduction

Ultra-wideband (UWB) technology is gaining more attention in recent years, it has a wide range of applications, including radar, communications, and geolocation [1]. As a form of UWB systems, impulse-radio (IR) UWB technology plays an important role in system localization and detection, it uses very short pulses in time covering a very wide frequency spectrum.

Ground Penetration Radar (GPR) is one of the major applications of IR-UWB technology, it's an advanced geophysical system that is used quickly and widely in deep detection application for civil engineering [2] and archaeology applications [3]. The principle is based on measuring the time it takes a pulse to travel to and from a target, indicating its depth and position. For a deeper penetration, the system needs to work in lower frequency like VHF band, which unfortunately requires an antenna with a larger size. In the other side, a detection with a high resolution requires an UWB system with low temporal distortion. It is then very important to design an antenna with good quality properties and compact size [4].

The tapered slot antenna is largely used in GPR systems, it has a planar structure, directional radiation pattern and quite low impulse distortion in contrast to other UWB antennas. The first exponential tapered slot antenna (ETSA) was proposed by Gibson

© The Author(s), under exclusive license to Springer Nature Switzerland AG 2024
M. Hatti (Ed.): IC-AIRES 2023, LNNS 984, pp. 423–437, 2024.
https://doi.org/10.1007/978-3-031-60629-8_41

[5] in 1979, it consists of two adjacent metal chips with conical exponential shapes supported by a dielectric substrate. The antenna size is usually limited by the places where they operate, such as those on vehicles, aircraft and armored vehicles. However, size is one of the most important factors affecting the electrical performance of the antenna. The characteristics of radiation and impedance deteriorate as the electrical magnitude shrinks. Different antenna miniaturization techniques are therefore used to balance both sizes and performances. A type of coplanar waveguide (CPW) to coplanar strip line (CPS) transition was used to feed the Vivaldi antenna [6], as a result, the low frequencies of matching bandwidth is decreased, but the antenna had a bad radiation performances. A tapered slot antenna loaded by a resistor was proposed for ground penetrating radar in [7], the shape of flare arms of slotline was modified and loaded in order to miniaturize the antenna. Another effective method, investigated in [8], uses the electromagnetic band gap (EBG) structure as defected ground structure to reduce the antenna size, increase the bandwidth and improve the antenna matching while keeping the other performance within the operating band almost unchanged. In [9] an artificial materials lens (AML) and sidelobe suppressor (SSR) are used to load the antenna to improves the gain. In [10], slots are added on the metal patch to improve and optimize antenna but keeping a large size. Corrugated structures and a direct connection to a microstrip feeder are used in [11], this structure has ultra-wideband performance and small area.

The goal of this work is to design a miniature antenna in the UHF/VHF bandwidth with good performances, working in the VHF bandwidth allows a deeper penetration of electromagnetic (EM) wave underground as the frequency is lower. The antenna miniaturization is done by folding firstly the ETSA and secondly by using loaded charges. For the last one, calculation and optimization is carried out with MATLAB. In All the work, EM simulations are done with CST software.

## 2   Antenna Modeling and Design

### 2.1   Basic Shape of the Antenna

Firstly, the ETSA is designed in an area of $30 \times 30$ cm$^2$ (Fig. 1), this size allows us to use this antenna in many application (installed in Unmanned Aerial Vehicle, mobile vehicles GPR). As a printed antenna, an FR4 substrate is used, it has a thickness of 1.6 mm, an electrical permittivity of 4.8 and a tangent loss of 0.02. In the Fig. 1, the blue color presents the copper and the yellow color presents the opening of the antenna (FR4).

The exponential profile curve employed to define the ETSA antenna are described with the next equations:

$$
\begin{cases}
Y(x) = A_0 e^{rx} \\
0 \le x \le X_{max} \\
A_0 \le y \le \frac{Y_{max}}{2}
\end{cases}
\tag{1}
$$

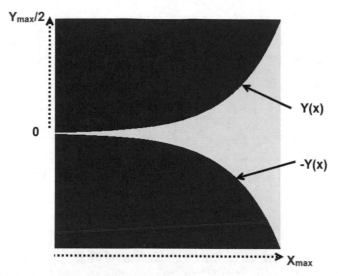

**Fig. 1.** Basic ETSA with size of 30X30 cm2

The ETSA is printed in a substrate size of $30 \times 30$ cm$^2$ (Fig. 1), the calculation of the $r$ constant is done according to the Eq. (1) as flow:

With the condition:

$$\begin{cases} x = x_{max} \to Y(x_{max}) = \frac{Y_{max}}{2} \\ x = 0 \to Y(0) = A_0 \end{cases}$$

We have:

$$\frac{Y_{max}}{2 * A_0} = e^{r(X_{max})} \to ln\left(\frac{Y_{max}}{2 * A_0}\right) = r(X_{max})$$

Then:

$$r = \frac{ln\left(\frac{Y_{max}}{2*A_0}\right)}{X_{max}} \tag{2}$$

To design an antenna with dimension of $X_{max}=$ 300 mm, $Y_{max}=$ 300 mm and $A_0=$ 1.2 mm, the value of $A_0$ is chosen in order to have an input characteristic impedance of 100 $\Omega$. From Eq. (2) the calculated $= r$ 0.016.

The result of the simulated reflection coefficient magnitude (S11) of the antenna is shown in Fig. 2.

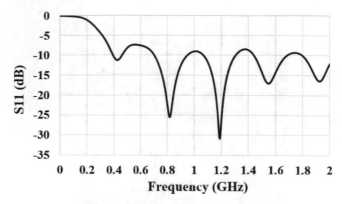

**Fig. 2.** S11 parameter of the ETSA

From Fig. 2, the reflection coefficient is less than −6 dB in the frequency range from 330 MHz to 2000 MHz. The goal of this work is to enhance the matching bandwidth in lower frequency without changing the size.

To match the antenna in lower frequency, a new shape is presented in the following paragraphs:

## 2.2 Enhancement of the Matching in the Lower Frequency Band by Folding the ETSA

To improve the matching bandwidth of the antenna in lower frequency, slots are done in the original shape of the ETSA. The goal is to lengthen the current path along the metallic surface. Figure 3 shows the new design where the blue color shows the copper region and yellow color shows the FR4 region without copper.

This new shape of antenna contains two exponentials, internal and external, each one is defined using the ETSA curve profile given in Eq. (1). The first exponential is the external one which defines the boundary limit of the copper antenna part (bleu). The second exponential is the internal formed by etching copper obtained from the first shape antenna (Fig. 1).

The equation below describes the profile of the external exponential:

*1) External exponential:*

The design is based on the equation below:

$$\begin{cases} Y_1(x) = A_0 e^{r_1(x-x_0)} \\ x_0 < x < X_{max} \\ A_0 < y < \frac{Y_{max}}{2} \end{cases} \tag{3}$$

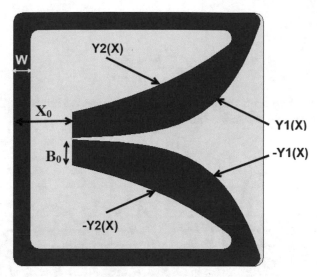

**Fig. 3.** Folding ETSA.

An important parameter to define the profile of the antenna is the value of $r_1$, this parameter is calculated as following.

For the condition:

$$\begin{cases} x = x_0 \rightarrow Y(x_0) = A_0 \\ x = X_{max} \rightarrow Y_1(X_{max}) = \frac{Y_{max}}{2} \end{cases} \tag{4}$$

where $r_1$ is given by:

$$r_1 = \frac{ln(\frac{Y_{max}}{2*A_0})}{X_{max} - x_0} \tag{5}$$

The calculated value $r_1 = 0.02099$ knowing that $= x_0$ 70 mm fixed to get best impedance matching.

2) *Internal exponential:*

As in the external exponential, the calculation of the internal exponential is given by:

$$\begin{cases} Y_2(x) = (A_0 + B_0)e^{r_2(x-x_0)} \\ x_0 < x < X_{max} - w \\ A_0 + B_0 < y < \frac{Y_{max}}{2} - w \end{cases} \tag{6}$$

For the condition:

$$\begin{cases} x = X_{max} - w \rightarrow Y(X_{max} - w) = \frac{Y_{max}}{2} - w \\ x = x_0 \rightarrow Y(x_0) = A_0 + B_0 \end{cases}$$

where $r_2$ is given by:

$$r_2 = \frac{\ln(\frac{Y_{max}-2*w}{2*(A_0+B_0)})}{X_{max} - x_0 - w} \tag{7}$$

The structure based on the calculation made before is simulated. The reflection coefficient (S11) of the antenna is shown in Fig. 4.

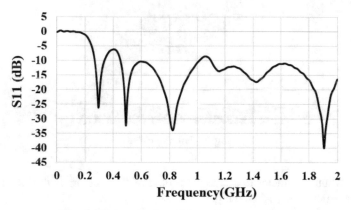

**Fig. 4.** S11 parameter of the new shape antenna

The simulation results ($Z_c = 100\ \Omega$) show that the S11 is less than $-6$ dB for frequencies higher than 251 MHz. Comparing the results with the original ETSA (Fig. 1), the lower frequency of matching bandwidth is decreased from 330 MHz to 250 MHz. Folding the antenna has permitted a gain of 80 MHz matching bandwidth in lower frequency.

## 2.3  Enhancement of the Matching in the Lower Frequency Band by Loading the ETSA

The To decrease more the lower frequency of matching bandwidth, a new technique is applied here, which consists on adding discrete loads in a specific place of the folded ETSA.

### 1)  Placement of the load impedance

In order to get the optimal places to add the load elements, current distribution is plotted in the bandwidth of interest. When the magnitude of current is maximum, means that the characteristic modes can be manipulated, adding a specific load allows to match the input impedance of the antenna. Figure 5 presents surface current distribution for the frequency bandwidth 150–300 MHz.

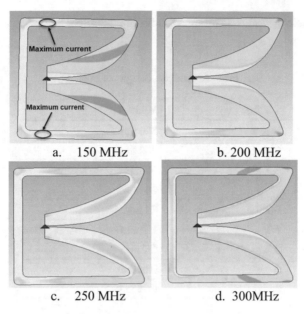

Fig. 5. Surface current of the ETSA

Surface currents show the presence of maxima at certain area of the antenna, at this area slots are added. Figure 6 shows the location of the two slots where the load will be added (Slot 1, Slot 2) with a width S of 2.7 mm.

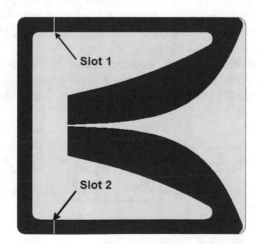

Fig. 6. Cutting the boarder of the folding ETSA with a slot

Optimal load can be calculated to match the antenna in the frequency bandwidth 150–250 MHz.

The table below (Table1) summarize all dimension parameters used to design the ETSA:

**Table 1.** Antenna design parameters

| Variable | Value |
|----------|-------|
| Xmax | 300 mm |
| Ymax | 300 mm |
| X0 | 50 mm |
| W | 20 mm |
| B0 | 30 mm |
| A0 | 1.2 mm |
| S | 2.7 mm |
| $r_1$ | 0.02099 |
| $r_2$ | $6.795 \times 10^{-3}$ |

2) *Calculating optimal loads*

As the Vivaldi antenna theory indicated, the aperture width of the Vivaldi antenna is about half the wavelength corresponding to the cutoff low frequency. Folding the antenna has allowed decreasing the lower frequency of matching bandwidth to 250 MHz. To reduce more to the lower operating frequency while kipping the same size of antenna $(30 \times 30 \text{ cm}^2)$, three loads are added in the cutting places (Slot 1, Slot 2) of the design shown in Fig. 6.

Loads permit to match the characteristics mode crossing the antenna in the frequency bandwidth 150–250 MHz. This matching is done in the price of decreasing the antenna efficiency [12]. The new ETSA can be modeled by three chains matrix as shown in Fig. 7 [13].

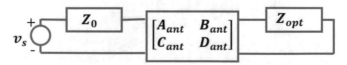

**Fig. 7.** Electrical diagram of the ETSA

The input bloc is related to the characteristic impedance of the generator $Z_0$, its ABCD matrix is:

$$ABCD_{gen} = \begin{bmatrix} 1 & Z_0 \\ 0 & 1 \end{bmatrix}$$

The radiated bloc of the ETSA is modeled by ABCD matrix, which is extracted from [S] parameters got from CST software (Fig. 8), and is given by:

$$ABCD_{ant} = \begin{bmatrix} A_{ant} & B_{ant} \\ C_{ant} & D_{ant} \end{bmatrix}$$

The output bloc is the optimal load, it is considered to be a serial impedance $Z_{load}$:

$$ABCD_{load} = \begin{bmatrix} 1 & Z_{load} \\ 0 & 1 \end{bmatrix}$$

$Z_{opt}$ is the value of $Z_{load}$ Which allows the input of the antenna to be matched, for that the following equation will be confirmed:

$$\begin{bmatrix} 1 & Z_0 \\ 0 & 1 \end{bmatrix} = \begin{bmatrix} A_{ant} & B_{ant} \\ C_{ant} & D_{ant} \end{bmatrix} \begin{bmatrix} 1 & Z_{load} \\ 0 & 1 \end{bmatrix}$$

According to [13] the optimal impedance will be:

$$Z_{opt} = \frac{B_{ant} - D_{ant} \times Z_0}{C_{ant} \times Z_0 - A_{ant}}$$

*Simulation result with optimal loads*

Due to the symmetric shape of the antenna, simulation and optimized loads are done only between the antenna input (port 1) and slot 1 (port 2). Figure 8 presents S parameters obtained by EM simulations:

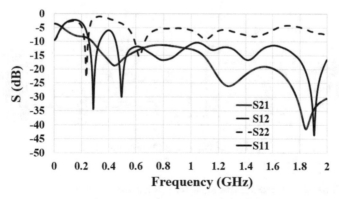

**Fig. 8.** S parameters of the ETSA

The optimal load is computed using MATLAB, Fig. 9 presents its values in the frequency band 100–250 MHz:

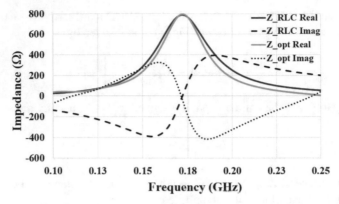

**Fig. 9.** Optimum load and calculated component RLC

The optimal load has a real behavior similar to RLC serial circuit but an inverse imaginary behavior where it increases beside the resonance frequency. A serial RLC with discrete components circuit is used to model the optimal impedance, it is calculated to have the same real impedance and the same resonance frequency.

The value of the discrete elements RLC serial circuit are: R = 780 $\Omega$, C = 10 pF and L = 150 nH. Figure 9 presents the optimal load and the impedance of serial RLC circuits used to model it, near the resonance frequency 172 MHz, the two impedances are equal, and when it takes away, the real impedance stays near and the imaginary part diverge.

Using the value of the calculated optimal load, a co-simulation using CST as shown in Fig. 10 an optimization is done to get a suitable S11.

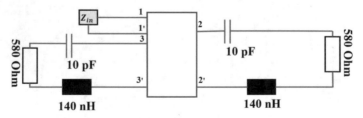

**Fig. 10.** Circuit load optimization in CST

Values of the discrete components after optimization are R = 580 $\Omega$, C = 10 pF and L = 140 nH, simulation result shows an S11 less than −6 dB from 140 MHz to 2 GHz (Fig. 11).

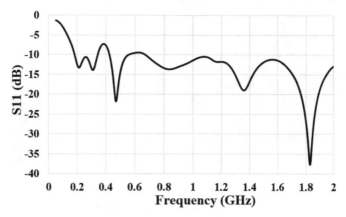

**Fig. 11.** S11 of the new ETSA loaded.

The antenna group delay simulation is shown in Fig. 12, in the frequency bandwidth 140–2000 MHz, it's value is less than 3.5 ns, except a pics located around 500 MHz and 1900 MHz. A lower group delay allows a high detection resolution.

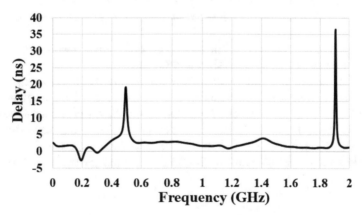

**Fig. 12.** Group delay of the ETSA loaded

The antenna gain is an important parameter in the equation of propagation (pathloss), higher gain means a deeper detection of GPR system. Figure 13 shows the simulated realized gain, it's value is higher than 2 dBi, in the frequency upper than 140 MHz, the gain tends to increase with frequency, as with the same structure, the electrical length of antenna is bigger when frequency increases.

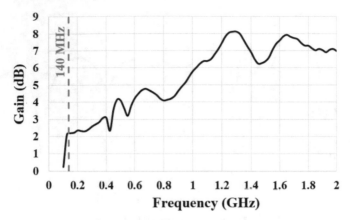

**Fig. 13.** Gain of the ETSA loaded

## 3 Experimental Study and Results

With objectives to demonstrate simulation results of the proposed antenna, the ETSA loaded antenna have been fabricated using a FR4 substrate which have a permittivity of 4.8, a loss tangent of 0.02 and a thickness of 1.6 mm. Figure 14 shows the fabricated antenna.

Load impedances are soldered. The reflection coefficient is measured by the Agilent E5071C vector network analyzer as shown in Fig. 15.

The measures are carried out using differential method. The Fig. 16 shows the reflection coefficient, it's noticed that the antenna is matched, at −6 dB from 140 MHz to 2 GHz. There is also a good concordance between simulation and measurement results.

Figure 17 presents simulated and measured group delay, they have similar shape except a high pic obtained in simulation at 1900 MHz. Measure group delay is less than 3 ns over the matching bandwidth 140–2000 MHz except a pic of 5 ns observed around 500 MHz.

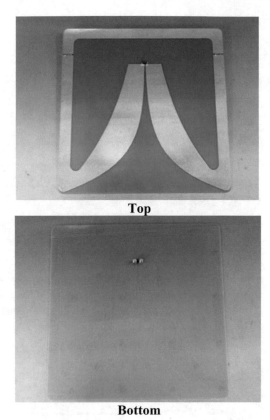

**Top**

**Bottom**

**Fig. 14.** Top and bottom view of the fabricated ETSA

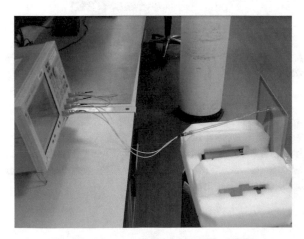

**Fig. 15.** Measurement of S11 of the ETSA

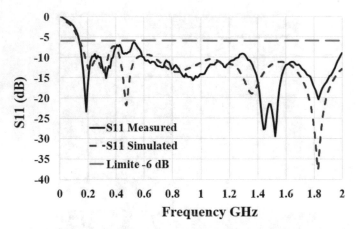

**Fig. 16.** Simulated and measured S11 of the ETSA

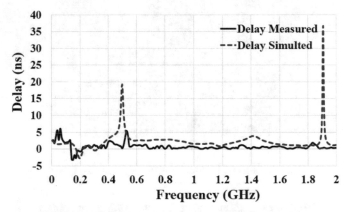

**Fig. 17.** Simulation and measurement of group delay

## 4 Conclusion

This paper presents a new ETSA for UWB GPR applications, two miniaturizations techniques are applied to make the antenna small, a folding technique to lengthen the current path in addition to loaded impedances. With a surface of only 30x30 cm2, the antenna is matched in the frequency bandwidth 140–2000 MHz, has a gain of more than 2 dBi, a group delay less than 3.5 ns, which make this antenna a good candidate for GPR application in the frequency bandwidth UHF/VHF.

## References

1. Porcino, D., Hirt, W.: Ultra-wideband radio technology: potential and challenges ahead. IEEE Commun. Mag. **41**(7), 66–74 (2003)

2. Manacorda, G., Persico, R., Scott, H.F.: Design of Advanced GPR equipment for civil engineering applications. In: Civil Engineering Applications of Ground Penetrating Radar, pp. 3–39. Springer, Cham (2015)
3. Verdonck, L., Launaro, A., Vermeulen, F., Millett, M.: Ground-penetrating radar survey at Falerii Novi: a new approach to the study of Roman cities. Antiquity **94**(375), 705–723 (2020)
4. Travassos, X.L., Pantoja, M.F.: Ground Penetrating Radar, pp. 1–38. Springer, Cham (2018)
5. Gibson, P.J.: The vivaldi aerial. In: 1979 9th European Microwave Conference, pp. 101–105. IEEE, September 1979
6. Zhang, F., Fang, G.Y., Ji, Y.C., Ju, H.J., Shao, J.J.: A novel compact double exponentially tapered slot antenna (DETSA) for GPR applications. IEEE Antennas Wirel. Propag. Lett.Wirel. Propag. Lett. **10**, 195–198 (2011)
7. Li, Y., Meng, S., Sheng, Y.-Z., Dong, L.: Ultra-Wideband dual polarized probe for measurement application. Antenna Propagation (ISAP) **02**, 1025–1028 (2013)
8. Wang, N.B., Jiao, Y.C., Song, Y., Zhang, L., Zhang, F.S.: A microstrip-fed logarithmically tapered slot antenna for wideband applications," 1. Electromagn. Waves Appl. **23**(10), 1335–1344 (2009)
9. Cheng, H., Yang, H., Li, Y., Chen, Y.: A compact Vivaldi antenna with artificial material lens and Sidelobe suppressor for GPR applications. IEEE Access **8**, 64056–64063 (2020)
10. Li, H., Li, P.: Modified Vivaldi antenna. Electronic Sci. & Tech, pp. 15–17 (2011)
11. Abbosh, A.M.: Miniaturized microstrip-fed tapered-slot antenna with ultrawideband performance. IEEE Antennas Wireless Propag. Lett. **8**, 690–692 (2009)
12. Wu, Q., Huang, Y.: A miniaturized tapered slot antenna with ultra-wide bandwidth. In: 2016 IEEE International Conference on Ubiquitous Wireless Broadband (ICUWB), pp. 1–3. IEEE, October 2016
13. Hamouda, C., Pintos, J.F.: Ultra-miniature loaded loop antenna for VHF pager. In: 2018 IEEE International Symposium on Antennas and Propagation & USNC/URSI National Radio Science Meeting, pp. 297–298. IEEE, July 2018

# Energy Monitoring on Stand-Alone Solar PV Driven Cold Storage in the Rural Desert Area

M. Tizzaoui[1,2(✉)], H. Soualmi[1], and F. Mguellati[1]

[1] Centre de Développement des Energies Renouvelables, Unité de Recherche en Energies Renouvelables en Milieu Saharien, URERMS, CDER, 01000 Adrar, Algeria
m.tizaoui@urerms.dz
[2] Laboratory of Energy on Dry-Zones (ENERGARID), Faculty of Technology, University of Tahri Mohamed Bechar, 417, Bechar, BP, Algeria
tizaoui-miloud2@univ-bechar.dz

**Abstract.** Rural populations either have no reliable electricity from the grid or have it in very distant locations. Small farmers lack on-farm cold storage facilities, leading to post-harvest losses. In addition, conventional refrigeration options have frequently not been accessible enough. Therefore, a decentralized, autonomous and intelligent solar-powered refrigeration system can contribute to the direct preservation of crops. The main objective of this study is to monitor and control decentralized refrigerated warehouses for fresh vegetables using the option of remote console control. This study examines aspects of automatic alerts used to warn of warehouse failure in remote rural areas, in particular the limitations imposed by desert environmental conditions. Data visualization via remotely connected Victron software displays low battery voltage, overload alerts and other warnings over the IP network. The results obtained describe the synthesis of the real-time remote monitoring electrical energy management system based on the Ethernet communication protocol. As a result, it simplifies intervention by farmers and enhances growth in rural off-grid areas. Nevertheless, the socio-economic status of end customers and the lack of financing options for small farmers are the main operating constraints.

**Keywords:** Cold Storage · Remote Monitoring · Ethernet Protocol · Off Grid Area · Saharan Environment · Solar PV Energy

## 1 Introduction

Smart energy control refers to the utilization of modern technology to enhance the efficiency, cost-effectiveness, and sustainability of energy production, distribution, and consumption. The farming sector has undergone structural improvements and adaptations in response to the rapid growth of the modern agricultural sector, the digital information revolution, and the advent of cutting-edge technology. The covid-19 pandemic in 2020 will have led to considerable worldwide destruction in various areas of human life. As a result, demand for cold storage facilities has increased due to health restrictions [1]. Moreover, the need to maintain perishable foodstuffs, essential medicines, and vaccines for consumers living in isolated areas or in developing countries is progressively increasing to improve their quality of life. Refrigeration appliances consume an increasing amount of energy, accounting for a significant proportion of total energy consumption. Recently, scientists have become more involved in decentralized autonomous refrigeration systems powered by photovoltaic solar energy [2, 3, 4].Cold stores are generally built close to manufacturing areas in rural regions. However, this proximity often leads to substantial operational constraints. For this reason, the use of a computer-based monitoring system is of enormous practical value in terms of improving the running effectiveness of these facilities. In addition, the implementation of remote monitoring techniques can significantly reduce the time spent on repairs, maintenance, and enhancements in faraway and isolated places. Many technical papers and standards provide a framework for remote monitoring that, when followed, ensures the reliability, objectivity, and representativeness of the recorded equipment data. Continual real-time monitoring can assist in the discovery of power supply failures and the deployment of remedial strategies [5, 6]. The automated control chain for cold stores is an important way of saving energy and minimizing consumption. For example, it is still possible to upgrade the automated control system for cold stores [7]. Ethernet is a widely used communication protocol in modern electronic systems. The main advantages of this technology are its high reliability, fast transmission capabilities, and high resistance to noise and electromagnetic interference. It is essential to note that *modbus* is a communication protocol that has been widely applied in the industrial field to connect electronic devices [8]. The system has the possibility of establishing a connection with various devices thanks to the implementation of the widely used *Modbustcp/ip protocol*. It should be noted that this standard protocol is available for use at no extra cost. The objective of our approach is to integrate a real-time Ethernet-based remote monitoring system into an autonomous solar-refrigerated warehouse located in a Saharan environment, famous for having the longest and most intense hot and dry seasons globally. The intention of the system is to continuously monitor and assess the electrical effectiveness of the device under examination. Furthermore, authorized operators can check the information in real-time on the Internet. The previously mentioned operation minimizes failures in the specific cold storage facility.

## 2   Apparatus Overview Setup

The Consideration parameters that used for the realization of the different experiments of refrigerated room were described in our previous works [9]. Figure 1 describes the sketch of the pilot scale cold storage remote monitoring process, tested under constraints imposed by desert climatic.

The prototype has a well-insulated cold storage facility that uses vapor compression cycles and electricity produced by a solar photovoltaic system. During periods of limited sunlight or at night, power is supplied from a battery bank through an electrochemical conversion process. The construction of the cold store was specifically designed to ensure capability in remote locations, near farms in the province of Adrar. The evaluation of efficiencies and achievements at the pilot scale utilized control strategies that developed Victron's real-time remote monitoring of the energy process. The establishment of the

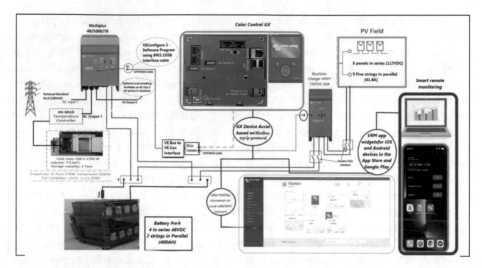

**Fig. 1.** Sketch of cold storage remote monitoring process.

connection between a Multiples inverter, the blue MPPT solar controller, the battery pack, and the AC output can be assisted through the use of the Color Control GX device. The control panel enhances the performance of the arrangement process and provides immediate alerts for any specific failures occurring in the cold store. The efficiency of the gadget data is systematically stored in an online database for convenient subsequent access and analysis. The VRM portal provides continuous monitoring and surveillance of your system, utilizing either Ethernet local LAN or WIFI network connectivity. Additionally, it has the capability to notify you via E-mail. The VRM app widgets, which can be downloaded from the app store for iOS devices and Google play for android devices, enable users to remotely monitor, control, and manage their Victron energy use system(s) from anywhere in the world.

## 3 Data System Visiualization

### 3.1 VRM Portal

The VRM Portal constantly monitors and watches over your system and can also inform you by email if something is amiss. There are four categories of monitoring:

- The no data alarm monitors the connection between the Portal and the Victron installation.
- Automatic alarm monitoring monitors a predefined list of parameters on all connected products.
- Geofence: monitors location (requires a Color Control GX with a USB-GPS)
- User-configurable alarms

The remote console is the central element for making or changing settings on the GX device. In GX devices with a display this remote console feature may be disabled by default and need to be enabled. GX devices without a display have remote console enabled by default. There are several ways to get access described in the following table.

**Table 1.** Different connexion access to GX device [10].

| Access type | Collor Control GX |
|---|---|
| Victron Connect via Bluetooth | - This feature can be easily added by connecting a USB Bluetooth dongle<br>- Bluetooth functionality is limited to assist with initial connection and networking configuration. You cannot use Bluetooth to connect to the Remote Console or other Victron products (e.g., Smart Solar charge controllers) |
| Built-in WiFi Access Point | This feature can be easily added by attaching a USB Wi-Fi dongle |
| Local LAN/Wi-Fi network | The CCGX does not have built-in Wi-Fi. This feature can be easily added by attaching a USB Wi-Fi dongle |
| VRM Portal | Requires the GX device connected to the internet |

One of the primary benefits of the application is its capability to visually represent stored data through a graphical display. The ability to observe logged data in real time is provided. The summarized data will undergo processing. The task involves the generation of periodic diagram reports and their subsequent delivery via email. This will be accomplished by utilizing an Excel spreadsheet. Besides email alerts and alarms, the system also provides remote access to the database through a web browser. The charts are updated in real time by continuously fetching data from the server in the background, eliminating the need to manually refresh the browser window.

### 3.2 ESS Baterry Life Controls

The ESS system offers users the option to switch between various settings, including battery life optimization, non-battery life optimization, battery charging maintenance, and external control [11]. The cancellation of any modifications made to the settings before their transmission to the device in question will be enabled for an extra 5 s. In addition, you will also be able to change the minimum charge. Figure 2 shows how the screen was set up afterward.

### 3.3 Victron Connect Widget

The Victron Connect app is utilized for the purpose of configuring, monitoring, updating, and diagnosing your Victron items. The application is compatible with Android, iOS, Windows, and Mac OS X operating systems. The device has the capability to establish hyperlinks with Victron products through different ways, including Bluetooth, USB, and Wi-Fi/LAN/Internet [12]. The setting IP configuration steps on Victron connect Widget shown in Fig. 3.

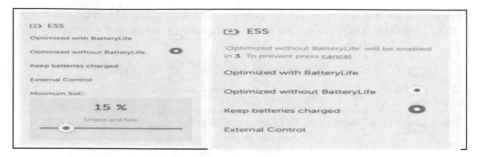

**Fig. 2.** Screenshot of the ESS remote mode on the VRM Portal

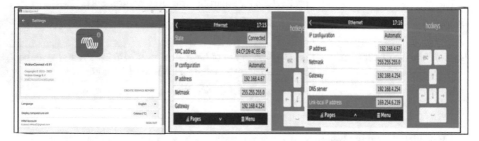

**Fig. 3.** Victron Connect app-based Internet IP configuration screenshot.

### 3.4 Warnings and Alarms

The device checks a multitude of established criteria for all the products now connected. This function makes it easier to set alarm trigger parameters for a large proportion of data without having to do this manually. If any of the parameters listed below become alarming, a message will be sent to your email address [13]:

- VE.Bus products (Multi, Inverter, and Quattro)
- VE.Bus state
- VE.Bus Error
- Temperature alarm
- Low battery alarm
- Overload alarm
- AC Input phase rotation (for three-phase systems)

An email notification is sent when the warning's status switches. When the parameter returns to its normal value, a recovery email is sent too. The default setting for new installations is "Alarms only".

The figure below showed the setting alarms and warnings for our installation.

**Fig. 4.** Alarm and Warning setup screenshot

# 4 Results and Data Analysis

The Full data (hourly and daily) on system performance is available on the web. Data recorded in June were used for this analysis. To carry out a full performance test, some data was also recorded on software from suitable instruments and control systems, which include temperature and pressure sensors and relative humidity sensors. Examples of the handling of alarms and warnings were shown Screenshot graphical display.

## 4.1 Adrar Climate Constraints

The province of Adrar is located to the south-west of Algiers. This particular region is located in the vast North African desert, which has an extreme continental climate. There are two distinct seasons in this climate: winter and summer. Winter covers a relatively short period, usually from December to March. By contrast, summer, which runs from early May to October, seems to be very long. Compared with the other periods, July is recognized as a peak hot season with extremely high temperatures. The province of Adrar is renowned for its excellent ability to harness solar energy, making it one of the world's most favorable locations for solar power production. The region benefits from sufficient sunshine, minimal humidity, and large unoccupied areas close to roads and grid lines. As selected sample test day on June 22 from 14:15 to 15:21 afternoon, Fig. 5 shows the climatic constraints of humidity, outdoor temperature, and solar irradiance recorded at the frame time transient regime running of the compressor. As shown in this period, the external temperature profile reach to 45 °C with low humidity value, thus coincide with high intensity sunlight and cooling demand. This situation reported as challenge for cooling devices appliance.

## 4.2 Alarms Warnings Analysis

The case studied for the alarm and warning responses during the daily experiment with ice from June 22 to 23, 2023, is described in the following figures: Fig. 6 describes the refrigeration test, performed with a water load of 500 L from June 22 at 14:15 to June 23 at 14:16. The phase 1 AC output and solar irradiation profiles were observed

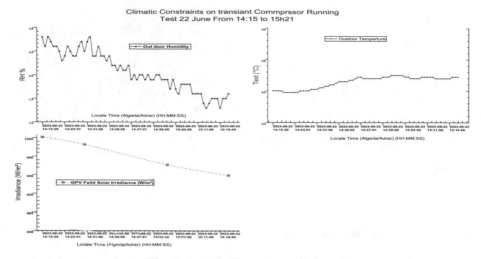

**Fig. 5.** Climatic Constraint on 22 June.

during the 24 h of refrigeration. During the trial, it should be noted that the intensity of solar radiation can reach a maximum value of around 1025 w/m2. This coincides with a higher requirement for cooling capacity. At the same time, a peak in current and power was also recorded. Subsequently, an overload warning was communicated directly to the remote monitoring panel. Overnight, the battery pack energy source alarm and remote warning were triggered in the event of a high overload and low VDC battery voltage. A shutdown time was observed from 03:31a.m to 10:17 a.m. due to a failure of the Battery voltage (VDC) drain. In response, the Victron remote management solution switches the Multiplus UPS to off mode until the VDC battery voltage configuration is re-established within the system. The Fig. 7 displays the daily profile of the DC source throughout the designated test period. The control mechanism employed in the conversion coupling of the Victron (PV-battery) power system depends on monitoring the states of charge and discharge. The battery voltage, current, and power profiles associated with the operation of the cold store display a peak demand during each compressor start-up. As previously stated, the electrical stop period indicates a failure in the electrochemical battery conversion process. The data obtained via the scan proved a battery voltage of roughly 33 VDC, which falls below the minimum limit configuration of 37 V. In this scenario, the battery configuration was continuously monitored by management in real-time using the well-designed ESS Life mode. The alert is transmitted via the internet, which allows the user to make adjustments as needed. This strategy validates the efficacy of the remote Victron application widget. The cold store was operated automatically without any human involvement as long as the battery voltage was fully set and the state of charge status was satisfactory.

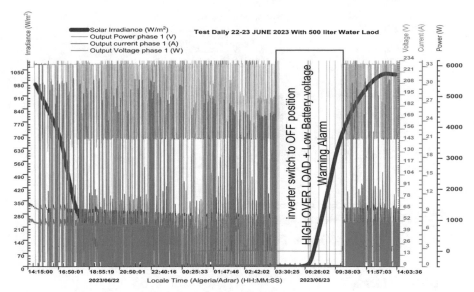

**Fig. 6.** Profile Evolution of AC Output Phase 1 and Solar Irradiance on a Daily Test, June 22–23

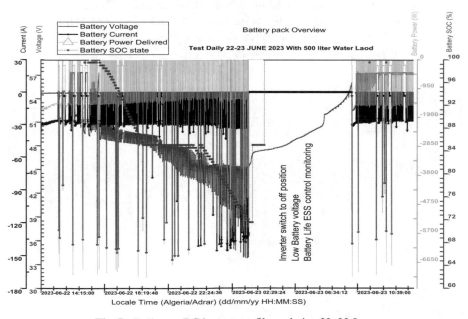

**Fig. 7.** Daily test DC battery profile evolution 22–23 June

The Fig. 8 displays the message notifications (OK, Alarm, and Warning) that are presented during the chosen daily test. The experiment recorded alerts for overload and low battery voltage. The subsequent text messages also communicate the knowledge found in the aforementioned graphs.

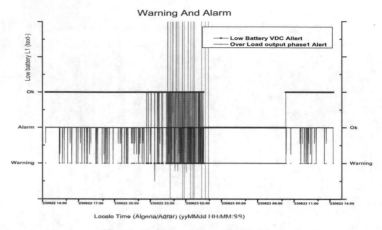

**Fig. 8.** Daily test alerts, 22–23 June.

## 4.3 Historical Data

In accordance with the provided data, this component will show a bar graph illustrating the production and consumption of kilowatt-hours (kWh). Also, it will include a blue line indicating the state of charge. The following screenshot in Fig. 9 presents the historical data of selected day. Throughout the typical test, the VE Bus State, as depicted in Fig. 10, includes the inverting, AES mode, and fault state.

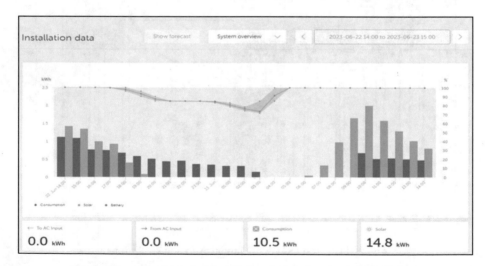

**Fig. 9.** Screenshot historical installation data during day test 22–23 June.

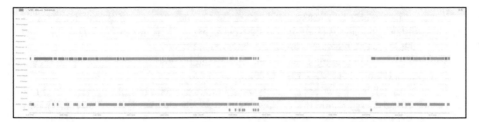

**Fig. 10.** Screenshot of historical VE bus states on the daily test, June 22–23

## 5 Conclusion

In this paper, we explore the thrilling potential of Victron energy in a data acquisition system for remote monitoring the functioning of a stand-alone PV cold storage facility in the Adrar province, recognized for its distinctive Saharan environment. Monitoring solar PV power source systems is crucial for ensuring a reliable and consistent process. In fact, the early identification of failures is a great opportunity to save money and improve the reliability of the sustainability concept. The system in question encompasses a web-based VRM portal, allowing the easy shipping of data over the Internet to remote users. The main services provided to the user are management, observation, notification, submitting reports, and data export, which offer great opportunities for controlling and optimizing various tasks. During the experiment, the online monitoring has been highly useful in spotting various operation causes such as high overload at compressor startup, high temperature of the batteries, low battery voltage, and more. On the bright side, historical recorded data has been thoroughly analyzed to assess the effectiveness of small-scale cold storage, regardless of its harsh desert climate. As an example, the remote mode of the EES life battery, along with the management of the minimum VDC and SOC battery states, has the potential to significantly improve the life cycle of the pack battery. Moreover, the remote-based internet enables farmers to provide real-time assessments without causing any displacement in isolated places. One potential area for improvement in this modern technology is exploring more financing options.

## References

1. Takeshima, H., Yamauchi, F., Edeh, H., Hernandez, M.: Solar-powered cold-storage and agrifood market modernization in Nigeria. Agric. Econ. **1–22** (2023). https://doi.org/10.1111/agec.12771
2. Munir, A., Ashraf, T., Amjad, W., et al.: Solar-hybrid cold energy storage system coupled with cooling pads backup: a step towards decentralized storage of perishables. Energies **14**, 7633 (2021). https://doi.org/10.3390/en14227633
3. Rutta, E.W.: Understanding barriers impeding the deployment of solar-powered cold storage technologies for post-harvest tomato losses reduction: insights from small-scale farmers in Tanzania. Front. Sustainable Food Syst. **6**, 990528 (2022). https://doi.org/10.3389/fsufs.2022.990528
4. Sadi, M., Arabkoohsar, A.: Techno-economic analysis of off-grid solar-driven cold storage systems for preventing the waste of agricultural products in hot and humid climates. J. Clean. Prod. **275**, 124143 (2020). https://doi.org/10.1016/j.jclepro.2020.124143

5. Torres, M., Muñoz, F.J., Muñoz, J.V., Rus, C.: Online monitoring system for standalone photovoltaic applications-analysis of system performance from monitored data. J. Sol. Energy Eng. **134**(3) (2012). https://doi.org/10.1115/1.4005448
6. Guidelines for Monitoring Stand-Alone Photovoltaic Power Systems Methodology and Equipment Report IEA PVPS T3–13 (2003)
7. Rydh, C.J., Sandén, B.A.: Energy analysis of batteries in photovoltaic systems. Part I: Performance and energy requirements. Energy Convers. Manage. **46**, 1957–1979 (2005). https://doi.org/10.1016/j.enconman.2004.10.003
8. http://www.modbus.org/; (Accessed 21 July 2023)
9. Tizzaoui, M., et al.: Solar PV power-driven cold room storage for Saharan rural area. E3S Web Conf. **152**, 01004 (2020). https://doi.org/10.1051/e3sconf/202015201004
10. https://www.victronenergy.com/media/pg/Cerbo GX/en/connecting-victron-products.html; (Accessed 24 July 2023)
11. https://www.victronenergy.com/upload/documents/VRM_app_Widgets/123168-VRM_App_Widgets-pdf-en.pdf; (Accessed 24 July 2023)
12. https://www.victronenergy.com/upload/documents/Energy_Storage_System/6292ESS_design_and_installation_manual-pdf-en.pdf (Accessed 24 July 2023)
13. https://www.victronenergy.com/live/vrm_portal:alarms, [Accessed 25 July 2023)

# Design and Implementation of a Solar-Powered Irrigation Pivot System with Remote Control via Android Application for Sustainable Agriculture

M. Ferroukhi[✉], I. Drouiche, H. Saadi, M. A. Boucheneb, and I. Otsmane

Electronics Deptartment, Faculty of Electrical Engineering, University of Science and Technology Houari Boumediene (USTHB), P.O. Box. 32, 16111 Bab-Ezzouar, Algeria
{merzak.ferroukhi,imane.drouiche,hyem.saadi}@usthb.edu.dz

**Abstract.** Irrigation plays a vital role in modern agriculture, ensuring optimal crop growth and efficient water usage. However, traditional irrigation methods often lack automation and require frequent manual intervention. To address these challenges, this work focuses on the design and implementation of a remotely controlled photovoltaic irrigation pivot.

The objective of this work is to develop an intelligent and automated irrigation system using solar energy to power the pivot and controlled remotely via a user-friendly Android application. By integrating photovoltaic panels into the irrigation pivot system, the reliance on external power sources can be significantly reduced, making it more sustainable and cost-effective.

**Keywords:** Irrigation pivot · solar panel · sensors · android · IoT

## 1   Introduction

The development of remote monitoring and control systems for irrigation mechanization and automated irrigation in large-scale farmland, commercial, and non-commercial platforms have gained significant attention in modern agriculture. Extensive research has been conducted worldwide on automated and remote irrigation monitoring and control, utilizing various platforms to facilitate task execution and equipment management from any location and at any time, while also reducing the need for human intervention through the linkage of devices and sensors [1–5].

To address the cost and environmental concerns associated with traditional power sources for pivot irrigation systems, several studies have explored the feasibility and efficiency of using photovoltaic (PV) energy. For example, El-Agamy et al. (2021) found that a PV-powered pivot irrigation system had higher efficiency and lower operating costs compared to a grid-powered system. Similarly, Oyedepo et al. (2014) found that a PV-powered pivot irrigation system was more cost-effective than a diesel-powered system. Other studies have focused on optimizing the design and operation of PV-powered

M. Hatti (Ed.): IC-AIRES 2023, LNNS 984, pp. 449–459, 2024.
https://doi.org/10.1007/978-3-031-60629-8_43

pivot irrigation systems. For example, Zhang et al. (2019) developed a model to optimize the sizing of the PV array and battery storage system for a pivot irrigation system in China, taking into account factors such as crop water demand, solar radiation, and battery capacity. In addition to PV-powered pivot irrigation systems, Muyambo et al. (2019) presented a smart irrigation system that utilizes Android mobile technology to remotely monitor and control irrigation systems. The system improves water-use efficiency and reduces water waste by enabling farmers to irrigate their crops based on real-time environmental conditions such as soil moisture, temperature, and humidity.

This work focuses on the study and realization of a remotely controlled photovoltaic irrigation pivot, aiming to revolutionize irrigation practices in agriculture. One of the key features of this work is the integration of remote-control functionality through an Android application. This mobile application will provide farmers with the ability to control and monitor the irrigation pivot from any location, using their Android smartphones. By leveraging the convenience and ubiquity of smartphones, farmers will be able to adjust essential irrigation parameters such as duration, frequency, and water flow rate according to their specific crop and environmental requirements. Additionally, real-time data collected from sensors embedded within the system will be displayed on the application, allowing farmers to monitor crucial information such as the pivot's status, water levels, and soil moisture content.

## 2 Existing Pivot Systems

Existing pivot irrigation systems are composed of both hardware and software components. The hardware includes the pivot structure, pumps, sprinklers, and sensors, while the software includes control systems, data analysis tools, and remote monitoring systems. The pivot structure is the backbone of the system, providing the physical support for the rotating arm that carries the sprinklers. The sprinklers distribute water evenly across the crops, and pumps are used to move water from the source to the sprinklers. Sensors are used to monitor soil moisture, weather conditions, and other factors that affect crop growth and water requirements. The control system is responsible for managing the irrigation schedule and adjusting water delivery based on the sensor data. Data analysis tools are used to analyze the sensor data and optimize irrigation schedules, while remote monitoring systems enable farmers to manage their irrigation systems from a distance. Several studies have investigated the hardware and software components of pivot irrigation systems, including their performance and effectiveness. The study by Zhang et al. (2021) demonstrated the use of precision irrigation software to optimize water delivery in pivot systems, resulting in improved crop yields and water use efficiency. Another study by Kandelous and Grismer (2018) developed a low-cost microcontroller-based system for remote monitoring and control of pivot irrigation systems. Overall, the hardware and software components of pivot irrigation systems play a critical role in their performance and efficiency, with ongoing research focused on further advancements in technology and management practices [11, 12].

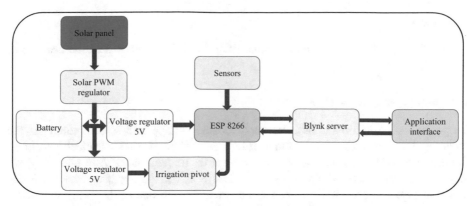

**Fig. 1.** Comprehensive synoptic diagram of the irrigation pivot system

## 3 Pivot Proposed Solution

The pivot irrigation system presented in Fig. 1 is composed of various parts, including both hardware and software components. The primary aim of this proposed solution is to achieve an autonomous pivot irrigation system that operates using solar energy. The hardware components of the system include the pivot structure, pumps, sprinklers, and sensors, while the software components consist of control systems, data analysis tools, and remote monitoring systems. The system's autonomous operation is made possible by the integration of solar panels, which provide the necessary power to run the system without relying on external energy sources.

## 4 Realization of the Pivot Irrigation System

### 4.1 Photovoltaic Installation

The photovoltaic installation typically consists of a solar panel, solar regulator, and battery.

– Solar Panel

   A 150 W 12 V solar panel has been used. This type of panel has a voltage that varies between 15 to 18 V under standard conditions. However, almost all components in the system operate at a 12 V voltage, a 12 V battery and a solar regulator are needed.

– PWM Solar Regulator

   The role of a PWM solar regulator is to:

- Convert the output voltage of our solar panel into a voltage suitable for the battery charge (often 13.6 V for 12 V batteries);
- Prevent reverse current and protect the solar panel;
- Protect the batteries by monitoring their charge level. Once the batteries are fully charged, the regulator will cut off the charging process. At the same time, it will stop the power consumption by connected devices if the battery charge level falls below a certain safety threshold (deep discharge limit).

– Battery

We have used a 12 V battery to store solar energy and reuse it in the absence of sunlight.

## 4.2 Voltage Regulator Circuit

To power the ESP8266 and other sensors, a stable voltage of 5 V is required. As the solar regulator provides a voltage range of 12 V to 13 V, an intermediate electrical circuit is necessary to convert and stabilize the voltage to the required 5 V. In this study, we implemented an electrical circuit using the LM7805 voltage regulator to achieve this functionality. The LM7805 voltage regulator was chosen due to its ability to convert higher voltages to a stable 5 V output. The performance of this circuit was evaluated in terms of its ability to provide a stable and reliable power supply to the ESP8266 and other sensors (Fig. 2).

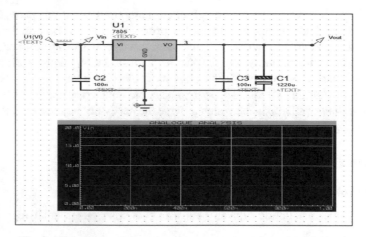

**Fig. 2.** Simulation of the voltage regulator circuit based on LM7805

## 4.3 Sensors

– Ultrasonic sensors HC-SR04: Ultrasonic sensors are commonly used to measure distances, but they can also be utilized to detect the water level in a tank or reservoir. The basic principle involves emitting high-frequency sound waves and measuring the time it takes for the waves to bounce back after hitting a surface. This time measurement can be used to calculate the distance to the water surface.
– FC-28 (YL-69) soil moisture: The soil moisture sensor uses a two-electrode probe that acts as a variable resistor. The resistance of this probe varies depending on the moisture content of the soil. When the soil contains more water, its conductivity increases, resulting in a lower resistance. Conversely, when the soil contains less water, its conductivity decreases, leading to higher resistance. The sensor measures

this resistance and converts this information into an analog voltage, which is then delivered to the A0 output. Thus, the voltage at the A0 output of the sensor varies depending on the soil resistance, allowing for an estimation of the soil moisture level.
– DHT11: The DHT11 is a relatively basic and affordable humidity and temperature sensor. It is widely used in electronic projects and home automation systems. The DHT11 sensor is capable of measuring the relative humidity of the air as well as the ambient temperature. It uses a capacitive humidity sensor and a thermistor to perform these measurements.

### 4.4 Controlling the Rotational Speed of a Motor

To manage the rotational speed of the motor in the rotating irrigation system, it is necessary to use an electrical circuit that allows us to control and adjust its speed according to our specific requirements (Fig. 3).

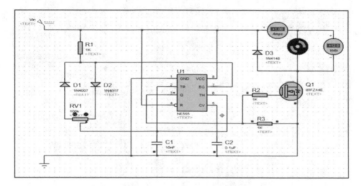

**Fig. 3.** Simulation of the motor speed controller circuit

### 4.5 Connection of Microcontroller with Sensors and Relays

This circuit includes a DHT11 sensor for measuring air temperature and humidity, as well as a soil moisture sensor to monitor soil moisture. We have also used an ultrasonic sensor to determine the water level inside the reservoir, and relays to control the pumps and motor as shown in the following (Fig. 4).

### 4.6 The Final Circuit

This circuit integrates the three circuits mentioned earlier and serves as the overall and final circuit of the system (Fig. 5).

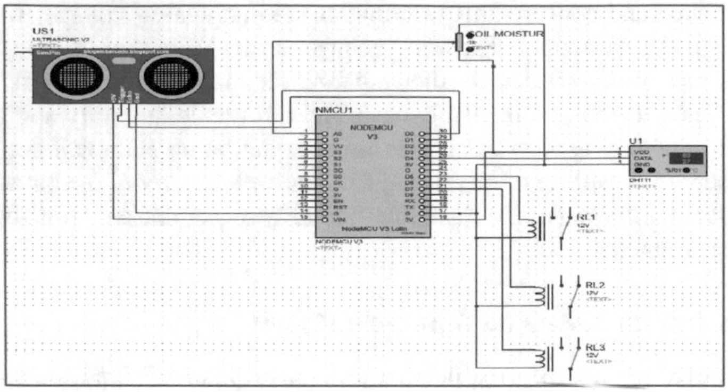

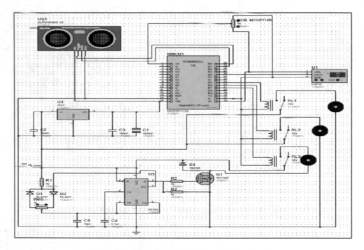

**Fig. 4.** Simulation of the microcontroller with sensors and relays

**Fig. 5.** Simulation of the final circuit

## 4.7  Android Application

To create the application, we used the Blynk platform. This latter allows users to easily create mobile applications to control and monitor various devices and projects remotely [13] (Fig. 6).

In the application we have developed, the user has the choice between two modes:

- Manual Mode: In this mode, the user can manually activate or deactivate the pumps based on the humidity sensor data. If the humidity is low, the user can activate irrigation. Additionally, if the water level in the reservoir is low, the user can activate the reservoir pump (Fig. 7).

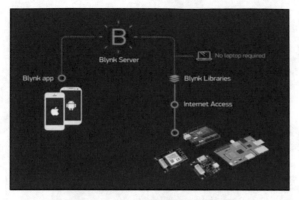

**Fig. 6.** Blink's functionality [13]

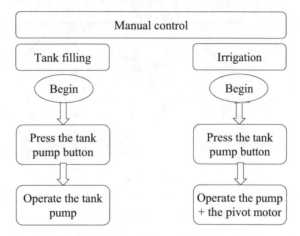

**Fig. 7.** Organizational chart of the manual mode

– Automatic Mode: In this mode, the user simply presses the automatic mode button, and watering is done automatically based on the soil's needs. Moreover, the reservoir pump opens automatically when the water level is low (Fig. 8).

Furthermore, the application displays sensor data in graphs to facilitate the monitoring of the crop (Fig. 9).

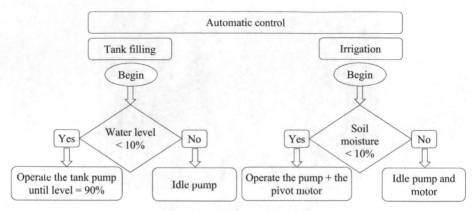

**Fig. 8.** Organizational chart of the automatic mode

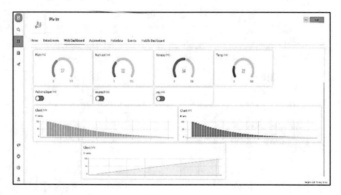

**Fig. 9.** Interface of the application

## 5   Tests and Results

### 5.1   PWM Solar Regulator Test

In Fig. 10, we can observe that the voltage generated by the solar panel is measured before being connected to the PWM regulator. However, in Fig. 11, it can be seen that the voltage of the load and the battery is reduced to a suitable level for battery charging, enabling a direct connection to our irrigation system.

### 5.2   Test of the Motor Speed Controller Circuit

During the test of this circuit, we noticed that the current drawn by the motor decreases as the resistance of the potentiometer increases. As a result, the motor speed also decreases. Similarly, when the resistance of the potentiometer decreases, the current drawn by the motor increases, leading to an increase in the motor speed.

**Fig. 10.** Voltage test of the solar panel without the PWM

**Fig. 11.** Voltage test of the solar panel with the PWM

## 5.3 Application Testing

After testing the application, we can see that the changes in sensor data and the control of actuators occur almost in real-time (depending on the internet connection state) and with very good accuracy (Fig. 12).

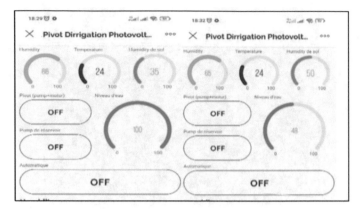

**Fig. 12.** Test of the application

## 6 Conclusion

In this study, we have successfully developed and evaluated a remotely controlled photovoltaic irrigation pivot system that offers efficient water management while utilizing solar energy as a power source. The system comprises a solar panel and battery that captures and stores solar energy, making the irrigation pivot self-sufficient and independent of the electrical grid. The development of a user-friendly Android application has enabled remote control of the irrigation pivot, allowing farmers to adjust irrigation parameters, monitor real-time data, and receive crop information from a distance. Through our study, we have observed that the remotely controlled photovoltaic irrigation pivot system offers numerous advantages, including precise irrigation management that adjusts water needs based on weather conditions and soil characteristics.

This study represents a significant contribution to sustainable agriculture, providing an efficient and cost-effective solution for crop irrigation that combines the use of solar energy with remote control technology. The system allows for efficient water utilization, reduces energy costs, and promotes intelligent resource management, which can benefit farmers and the environment in the long term.

## References

1. Wang, S., Chen, J., Liu, X., Li, X., Zhang, Y.: IoT-based smart agriculture: a review. Comput. Electron. Agric. **175** (2020)
2. Shahzad, M.K., Hussain, A., Hussain, M.: Internet of Things (IoT) based smart irrigation systems: a review. Comput. Electron. Agric. **181** (2021)
3. Chen, Y., Cheng, Y., Zhang, Y., Li, Y.: Research progress of irrigation automation technology based on the Internet of Things. J. Agric. Inf. **31**(1), 1–10 (2019)
4. Zhang, D., Li, J., Li, Y., Wang, J.: A review of smart irrigation systems: from traditional irrigation to intelligent irrigation. Agric. Water Manage. **253** (2021)
5. Bhattarai, R., Pandey, R.: IoT-based smart irrigation systems: a comprehensive review. Comput. Electron. Agric. **182** (2021)
6. El-Agamy, M.S., et al.: Performance and economic analysis of photovoltaic-powered center pivot irrigation systems. Energies **14**(12) (2021)

7. Bhattarai, U., et al.: A review of solar photovoltaic irrigation systems. Renew. Sustain. Energy Rev. **145** (2021)
8. Oyedepo, S.O., Mekhilef, S., Ohunakin, O.S.: Comparative study of centralized and decentralized solar PV-diesel hybrid power supply systems for remote homes in Nigeria. Renew. Sustain. Energy Rev. **32**, 736–749 (2014)
9. Zhang, Y., et al.: Optimization of photovoltaic-battery irrigation system design: a case study of center pivot irrigation in China. Energy Convers. Manage. **183**, 8–18 (2019)
10. Muyambo, F., van der Waals, J., van der Zaag, P.: Smart irrigation system using android mobile technology. Comput. Electron. Agric. **164** (2019)
11. Zhang, Y., Zhang, R., Li, X., Xu, X., Wang, J., Li, Y.: Effects of precision irrigation on yield, water use efficiency, and soil water content in maize (Zea Mays L.) under pivot irrigation. Agric. Water Manage. **244** (2021)
12. Kandelous, M.M., Grismer, M.E.: Remote sensing and control of an irrigation pivot using a low-cost microcontroller-based system. Comput. Electron. Agric. **151**, 292–302 (2018)
13. https://booteille.github.io/blynk-docs-fr/

# Shoot-Through Duty Ratio in Modulated Model Predictive Control for Quasi-Z-Source Based on Fuzzy Logic

Abdelouahad May, Fateh Krim$^{(\boxtimes)}$, and Hamza Feroura

Department of Electronics, University of Setif-1, Setif, Algeria
{abdelouahad.may,h_feroura}@univ-setif.dz, krim_f@ieee.org

**Abstract.** This paper presents a novel approach to control a grid-connected Quasi-Z-Source Inverter (qZSI), employing a combination of Modulated Model Predictive Control (M$^2$PC) and Fuzzy Logic Control (FLC). The objective was to achieve continuous operation of the PV system at its maximum power point while ensuring efficient tracking and delivery of power to the grid. To facilitate this, a current-based Perturb and Observe (P&O) Maximum Power Point Tracking (MPPT) algorithm was utilized to generate reference current for the FL controller. The system exhibited remarkable tracking dynamics and demonstrated autonomous grid current injection capabilities. Extensive simulations confirmed the feasibility and effectiveness of the proposed control techniques, showcasing their strong performance in regulating the grid-connected PV system.

**Keywords:** Power quality · qZSI · FLC · M$^2$PC · Photovoltaic (PV)

## 1 Introduction

In recent years, there has been a remarkable surge in research and development efforts in the field of power electronics and renewable energy systems. As the demand for more efficient, reliable, and environmentally friendly power conversion solutions grows, inverters have emerged as critical components in various applications, ranging from renewable energy integration to motor drives and industrial power supplies.

In this context, one noteworthy and innovative inverter topology that has garnered significant attention is the qZSI. The qZSI is a variation of the traditional Z-Source Inverter (ZSI), characterized by its unique impedance network. This specialized inverter has demonstrated promising capabilities to enhance power conversion performance, offer better efficiency, and address certain operational challenges [1–4].

Control methods in the qZSI have been a subject of extensive research and development, aiming to enhance its efficiency and performance across various applications [5]. One of the conventional control techniques commonly used with the qZSI is the Model Predictive Control (MPC), which offers simple implementation and satisfactory regulation of the output voltage and current. MPC is gaining traction due to its ability to predict future system behavior and optimize control signals accordingly [6, 7]. M$^2$PC combines

© The Author(s), under exclusive license to Springer Nature Switzerland AG 2024
M. Hatti (Ed.): IC-AIRES 2023, LNNS 984, pp. 460–469, 2024.
https://doi.org/10.1007/978-3-031-60629-8_44

MPC and modulation techniques, resulting in a flexible control strategy with improved transient response. FLC employs linguistic variables to handle complex nonlinearities, enhancing the qZSI's robustness and adaptability to varying operating conditions [8–10].

This paper proposes a combination of M2PC and FLC for a qZSI connected with a grid, to enhance the control quality and reduce the harmonic distortion of the injected current into the grid.

## 2 Quasi-Z-Source Inverter Structure

Figure 1 illustrates a photovoltaic system consisting of solar panels and a qZSI converter connected to the grid. A qZSI is a type of power electronic converter that plays a crucial role in energy conversion and control systems. This innovative inverter topology is designed to overcome the limitations of traditional voltage-source (VSI) and current-source inverters (CSI). The qZSI operates by utilizing an impedance network, typically implemented using a coupled inductor and a capacitor, to provide a unique feature of voltage buck-boost capability. The qZSI consists of several key components that work together to enable its functionality. These components include two capacitors $C_1$ and $C_2$, and two inductors $L_1$ and $L_2$.

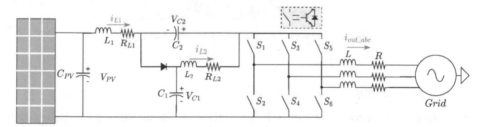

**Fig. 1.** Structure for PV grid-connected qZSI.

The non-shoot-through state (NST) and the shoot-through state (ST) are the two operating states of the qZSI, as shown in Fig. 2. When in active mode, the inverter functions similarly to a VSI, as shown in Fig. 4(b). And in the ST mode, both switches in the same leg are switched on simultaneously at the same time.

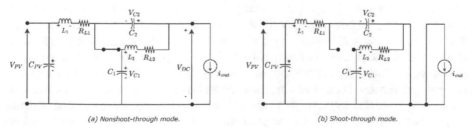

(a) Nonshoot-through mode.          (b) Shoot-through mode.

**Fig. 2.** qZSI equivalent circuit.

## 3  Proposed Controller

Considering the distinctive characteristics of qZSI, an appropriate control technique must be implemented to ensure maximum utilization of these features. This paper proposes $M^2PC$ with an FLC of the ST. This combination of control has been proposed to enhance the control quality of the direct current (DC) and reduce the harmonic distortion of the injected current into the grid. The suggested system's general controller concept is shown in Fig. 3.

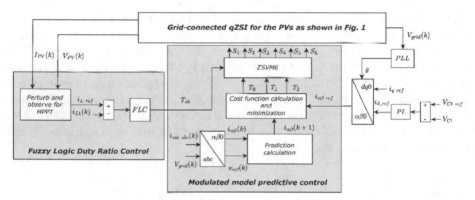

**Fig. 3.** The proposed control system's block diagram.

### 3.1  Modulated Model Predictive Control

$M^2PC$ integrates an appropriate modulation scheme into minimizing the cost function within the MPC algorithm. In this study, a modulation scheme designed explicitly for qZSI control is implemented within the framework of $M^2PC$ with space vector modulation (SVM) for the qZSI, called the ZSVM6. The ZSVM6 controller concept is shown in Fig. 4(a). To achieve the concept of ZSVM6, the firing times were derived by proposing an FLC, as explained in the following section. In this section, we present the method of extracting traditional time intervals for SVM by controlling the output current of the converter with an $M^2PC$.

The mathematical model utilized in this study employs the inverter's current output voltage vector ($V$) to inject power into the grid voltage ($V_{grid\_\alpha\beta}$) through an RL filter, resulting in the predictive equation for the output current. The continuous-time equation for the inverter grid interface is represented by Eq. (1). To make the computation more tractable, the equation is then transformed into its equivalent discrete-time expression using the forward Euler method, as provided by Eq. (2). To further streamline the computation process, each three-phase parameter is defined within a complex frame of reference represented by ($\alpha\beta$). This approach effectively reduces the number of equations from three to one complex equation, significantly reducing the critical computation time for algorithms such as MPC.

$$V = Ri + L\frac{di}{dt} V_{grid\_\alpha\beta} \tag{1}$$

$$i_{\alpha\beta}(k+1) = i_{\alpha\beta}(k)\left[1 - \frac{R}{L}T_S\right] + \frac{T_S}{L}\left[V(k) - V_{grid\_\alpha\beta}(k)\right] \tag{2}$$

A cost function (Eq. 3) is formulated to achieve the control objective, which incorporates the output current of the qZSI.

$$g = \left\| i_{\alpha\beta}(k+1) - i_{\alpha\beta}(k+1)^* \right\| \tag{3}$$

To determine the duty cycles in M²PC, the cost function is evaluated for each prediction, considering the optimal two vectors and the ST. This evaluation results in three dus named $g_1$, $g_2$, and $g_0$, with Ts representing the sample time. The respective duty cycles are defined by Eq. 4:

$$\begin{cases} d_1 = T_s g_0 g_2 / (g_1 g_2 + g_0 g_1 + g_0 g_2) \\ d_2 = T_s g_0 g_1 / (g_1 g_2 + g_0 g_1 + g_0 g_2) \\ d_0 = T_s g_1 g_2 / (g_1 g_2 + g_0 g_1 + g_0 g_2) \end{cases} \tag{4}$$

### 3.2 Fuzzy Logic Duty Ratio Control

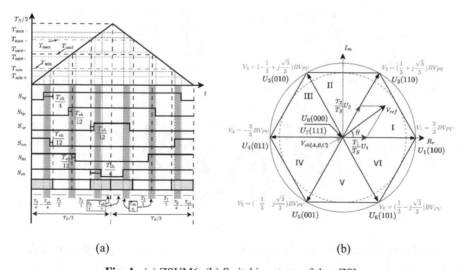

(a)                 (b)

**Fig. 4.** (a) ZSVM6, (b) Switching states of the qZSI

In this work, a new strategy for controlling the shoot-through time of qZSI has been developed using FLC, and this is without the need for a detailed mathematical model of the system. Fuzzification, a fuzzy rule base, and defuzzification are the three components that make up the fuzzy controller. FLC has emerged as one of the most successful applications of fuzzy set theory. Its key feature is the use of linguistic variables instead of numerical variables [10, 11]. The FL control technique leverages human intuition to comprehend the system's behavior and is built on qualitative control rules. This approach allows for a more intuitive and human-like control strategy, making it particularly suitable for systems with complex dynamics or uncertain environments.

The translation of the input physical variables into fuzzy sets is possible through fuzzification. The error "$E$" and the variation of the error "$\Delta E$" are our two inputs in this instance, and they are defined as follows:

$$E = i_{L\_ref}(n) - i_L(n) \tag{5}$$

$$\Delta E = E(n) - E(n-1) \tag{6}$$

In the inference step, logical relationships are established between the inputs and the output, represented by their respective membership functions as depicted in Fig. 5. These membership functions are utilized to determine the inference rules. Subsequently, a table of inference rules is constructed, as presented in Table 1.

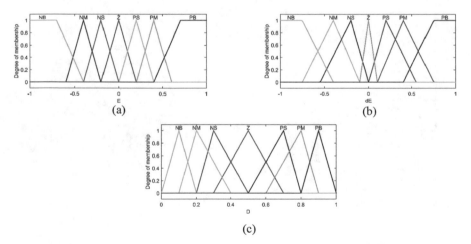

**Fig. 5.** Membership function: (a) The error, (b) The variation of the error, (c) Shoot through duty ratio.

**Table 1.** Fuzzy Rule Base.

| CE/E | NB | NM | NS | Z | PS | PM | PB |
|------|----|----|----|----|----|----|----|
| NB | Z | NS | NM | NB | PS | PM | PB |
| NM | NS | Z | NS | NM | PS | PM | PM |
| NS | NM | NS | Z | NS | PS | PS | PS |
| Z | NB | NM | NS | Z | PS | PM | PB |
| PS | NS | NS | NS | PS | Z | PS | PM |
| PM | NM | NM | NS | PM | PS | Z | PS |
| PB | NB | NM | NS | PB | PM | PS | Z |

## 4 Simulation and Results

The proposed control technique for the studied system has been validated through computer simulations using 'Simpower systems' in MATLAB/Simulink®. The system is analyzed separately to track the MPP of the system, while also independently controlling the active power injection into the grid and observing the dynamic response of the controlled parameters. The results of these analyses are verified based on the specified criteria and are presented in Figs. 6, 7, 8 and 9.

In Fig. 6(b), it can be observed that the power delivered by the PV system remains consistently at its maximum with minimal oscillations, even under varying irradiance levels. This achievement is attributed to the combined implementation of the FL controller and the $M^2PC$. The FL controller and $M^2PC$ effectively maintain the inductor current by accurately tracking its reference, which is generated by the P&O-MPPT algorithm. As a result, the PV system operates efficiently, ensuring optimal power extraction from the varying environmental conditions.

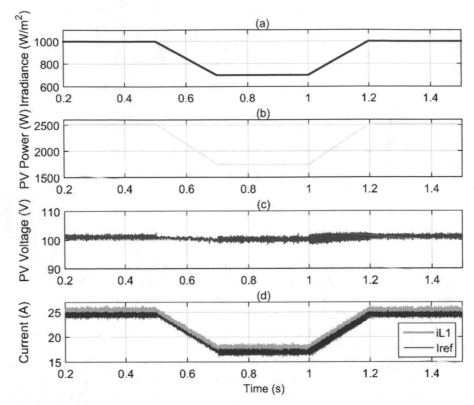

**Fig. 6.** PV side simulation results: (a) Solar irradiance levels, (b) PV power, (c) PV Voltage, (d) Inductance current and inductance reference current.

Figure 7 illustrates the three-phase grid current, which appears to be practically sinusoidal with low distortion, adhering to the required standards. The grid current's Total Harmonic Distortion (THD) remains below 5%, fluctuating between 0.90% and 1% as shown in Fig. 9. Additionally, Fig. 8 demonstrates the attainment of unity power factor (PF) operation, indicating that the PF is close to 1. However, if desired, non-unity PF operation can be achieved by modifying the reference value of the reactive current ($Iq^*$). This flexibility allows the system to adapt to different PF requirements based on specific needs or grid conditions.

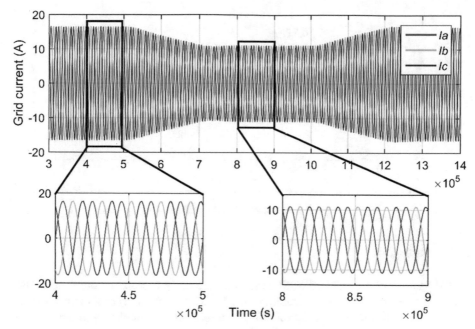

**Fig. 7.** Response of three-phase grid currents.

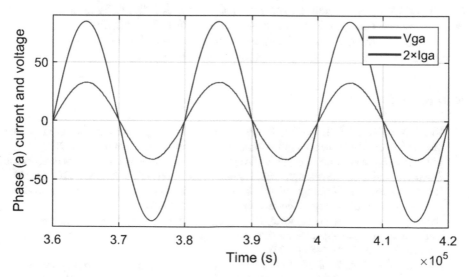

**Fig. 8.** Phase (a) grid current and voltage.

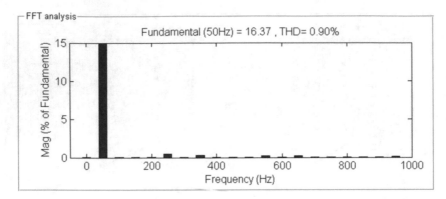

**Fig. 9.** Phase (a) grid current THD and harmonics spectra.

## 5 Conclusion

In this paper, a grid-connected qZSI was controlled using a combination of $M^2PC$ and FLC. To achieve continuous operation of the PV system at its maximum power point, a straightforward current-based P&O-MPPT algorithm was employed to generate the reference current for the FL controller. The system exhibited excellent capabilities in tracking and delivering maximum power from the PV source, characterized by rapid and efficient tracking dynamics. Moreover, it demonstrated the ability to independently inject current into the grid. The simulation results verified the feasibility of the proposed system and the effectiveness of the implemented control techniques, showcasing their strong performance in regulating the grid-connected PV system.

## References

1. Nayak, M.R., Tulasi, V., Teja, K.D., Koushic, K., Naik, B.S.: Implementation of quasi Z-source inverter for renewable energy applications. Mater. Today Proc. **80**, 2458–2463 (2023)
2. Liu, Y., Ge, B., Abu-Rub, H., Peng, F.Z.: Overview of space vector modulations for three-phase Z-source/quasi-Z-source inverters. IEEE Trans. Power Electron. **29**(4), 2098–2108 (2013)
3. Ho, A.-V., Chun, T.-W.: Single-phase modified quasi-Z-source cascaded hybrid five-level inverter. IEEE Trans. Industr. Electron. **65**(6), 5125–5134 (2017)
4. Bubalo, M., Bašić, M., Vukadinović, D., Grgić, I.: Hybrid wind-solar power system with a battery-assisted quasi-Z-source inverter: optimal power generation by deploying minimum sensors. Energies **16**(3), 1488 (2023)
5. Ahangarkolaei, J.M., Izadi, M., Nouri, T.: Applying sliding mode control to suppress double frequency voltage ripples in single-phase Quasi-Z-source inverters. CSEE J. Power Energy Syst. **9**(2), 671–681 (2023)
6. Ahmed, A.A., Bakeer, A., Alhelou, H.H., Siano, P., Mossa, M.A.: A New modulated finite control set-model predictive control of Quasi-Z-Source inverter for PMSM drives. Electronics **10**(22), 2814 (2021)
7. Saavedra, J.L., et al.: Comparison of FCS-MPC strategies in a grid-connected single-phase Quasi-Z Source inverter. Electronics **12**(9), 2052 (2023)

8. Devaraj, U., Ramalingam, S., Sambasivan, D.: Comparative evaluation of PI and fuzzy logic controller for PV grid-tie quasi Z-source multilevel inverter. Mehran Univ. Res. J. Eng. Technol. **40**(3), 465–473 (2021)
9. Say, G., Hosseini, S.H., Esmaili, P.: Hybrid source multi-port Quasi-Z-Source converter with fuzzy-logic-based energy management. Energies **16**(12), 4801 (2023)
10. Mazouz, F., Belkacem, S., Ouchen, S., Harbouche, Y., Abdessemed, R.: Fuzzy control of a wind system based on the DFIG. In: Hatti, M. (ed.) ICAIRES 2017. LNNS, vol. 35, pp. 173–181. Springer, Cham (2018). https://doi.org/10.1007/978-3-319-73192-6_18
11. Ouchen, S., Betka, A., Gaubert, J.P., Abdeddaim, S., Mazouz, F.: Fuzzy-direct power control of a grid connected photovoltaic system associate with shunt active power filter. In: Hatti, M. (ed.) ICAIRES 2017. LNNS, vol. 35, pp. 164–172. Springer, Cham (2018). https://doi.org/10.1007/978-3-319-73192-6_17

# The NDT-FC Simulation Using the 3D FEM

S. Khelfi[1,2(✉)], B. Helifa[2], L. Hachani[2], and I. K. Lefkaier[2]

[1] ENS of Laghouat, Laghouat, Algeria
s.khelfi@ens-lagh.dz
[2] LPM Laboratory, University Amar Telidji, Laghouat, Algeria
l.hachani@lagh-univ.dz

**Abstract.** During the realization of a non-destructive control tool (NDC), the experimental parametric study can be difficult for several reasons, including the multiplication of the number of tests or the manufacturing cost of a test prototype. The design, development, and optimization of eddy current NDT processes are made possible through the modeling and numerical simulation of electromagnetic systems. The purpose of the present study is to identify the non-destructive control (NDC) of eddy currents (FC) in its various modeling and experimental aspects. The TEAM Workshop 15-1 benchmark issues have been considered to validate a 3D electromagnetic system model using FEM. The results of our calculations are extremely accurate and consistent with the experimental data. This obviously enables us to examine many more examples in a perfect manner. Without the need to carry out experiments, we can now rely only on using the simulation model to study more phenomena related to our model and control its variables.

**Keywords:** non-destructive control · eddy currents · electromagnetic system · Finite Element Method FEM

## 1 Introduction

In all industrial sectors, materials must generally be checked regularly during production, during operation or during maintenance to determine whether they have defects or not. The techniques used to perform this check must obviously not damage the structure. Many methods have been developed in recent decades and have all been grouped under the term Non-destructive Testing (NDT) [1]. The term defect is relative and not very precise, but its negative connotation evokes the role played by non-destructive control in the search for quality [2]. A defect (defect) detect in a part is physically, highlight a heterogeneity of material, a local variation of physical or chemical property detrimental to the proper use of it. That said, it is common to classify defects into two broad categories related to their location: surface defects, internal defects [2–4].

NDT methods are used either to assess characteristic quantities of the product (thickness, conductivity . . .) or to detect and characterize defects. Some of the most widely used methods are ultrasound, methods using ionizing radiation (radioscopy) and electromagnetic methods (magnetoscopy, eddy currents . . .). The choice of a method depends on a large number of factors such as the nature of the materials constituting the parts to be checked, the nature of the information sought (defect opening or buried . . .), the conditions of implementation . . . [5]. This work focuses on Eddy Current Testing (ECT), which involves inducing currents in the electrically conducting material being inspected. The interaction between currents and materials is realized and a density image of the magnetic field is obtained.

## 2  Experimental Setup

The experience consists of an air coil that moves over a metal plate with an outlet crack oriented according to the ox axis. The crack and the coil-to-part distance (lift-off) are both imposed (see Fig. 1 and Table 1) [6].

Table 1.  Geometric and electromagnetic characteristics of the system

| Conductive plate | Length: 120 mm |
|---|---|
| | Width: 103 mm |
| | Thickness: 20 mm |
| | Electrical Conductivity: $3.85 \ 10^7$ |
| | Magnetic Permeability: $4\pi \ 10^{-7}$ |
| Defect | Length: 14 mm |
| | Width: 1 mm |
| | Depth: 5 mm |
| Coil | Coil inner radius |
| | Coil outer radius |
| | Coil length |
| | Lift off |
| | Resonance frequency: fc = 717k |

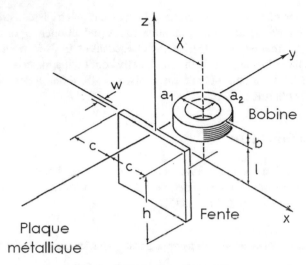

**Fig. 1.** Sensor-crack system [6].

## 3  Use of Software

*A. Formulation*

The magnetic vector and electric scalar potential equations that need to be solved are as follows:

$$\iiint_D [\nabla \times \left(v\nabla \times \vec{A}\right) - \nabla\left(v\nabla.\vec{A}\right)]dxdydZ + j\omega \iiint_D [\sigma\left(\vec{A} + \nabla V\right)dxdydz =$$

$$\iiint_D \vec{J}_s dxdydz \tag{1}$$

$$\iiint_D \nabla.[-j\omega\sigma\left(\vec{A} + \nabla V\right)dxdydz = 0$$

where $\vec{A}$ and $V$ are respectively the magnetic vector potential and electric scalar potential, $v$ is the magnetic reluctivity and $\sigma$ the electrical conductivity of the conductive plate [6–8].

*B. Mesh*

The quality of the mesh is crucial for the quality of the results. Tetrahedral elements are used for forming the mesh. A fine mesh is required in the region where field variation is important. As for example, we need to refine the mesh around the coil or in the depth of the skin in the plate since the magnetic induction field varies significantly in these regions [9–12] (Fig. 2).

*C. Boundary condition*

The magnetic induction B is supposed tangent and then the magnetic insulation condition (nxA = 0) is imposed (default condition in COMSOLMultiphysics) [6].

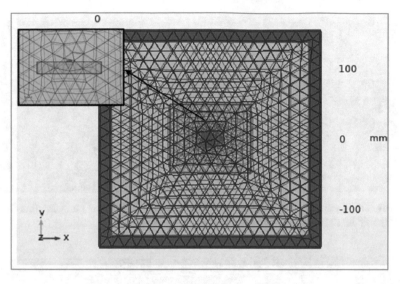

**Fig. 2.** Mesh

## 4 Validation

The benchmark result gives as a function of the position, the variations of $\Delta R$ and $\Delta L$ of the measured coil. In this part, we compared the numerical results of 3D simulations with the experimental findings (Figs. 3 and 4).

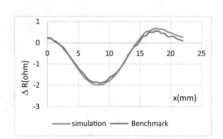

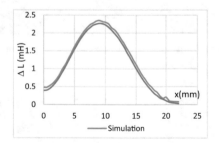

**Fig. 3.** Variation of the resistance according to the displacement of the sensor [6]

**Fig. 4.** Variation of inductance as a function of sensor displacement [6]

## 5 Simulation

A. The effect of electrical conductivity

Eddy currents are electrical currents created in a conductive mass, either by the variation over time of an external magnetic field through this medium (the flux of the field through the medium) by a displacement of this mass in a magnetic field. They are a consequence of electromagnetic induction (Figs. 5, 6, 7 and 8).

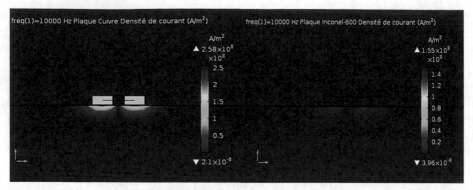

**Fig. 5.** Module of current density of a copper plate ($\sigma = 5.91 \times 10^7$) and Inconel-600 plate ($\sigma = 10^6$) without defect.

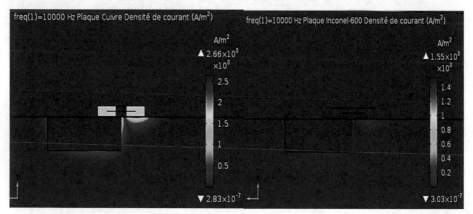

**Fig. 6.** Module of current density of a copper plate ($\sigma = 5.91 \times 10^7$) and Inconel-600 plate ($\sigma = 10^6$) with a defect.

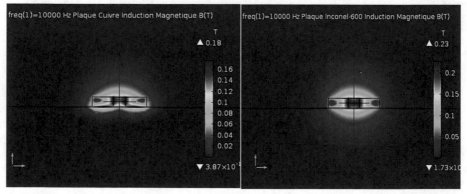

**Fig. 7.** Module of magnetic flux density of a copper plate ($\sigma = 5.91 \times 10^7$) and Inconel-600 plate ($\sigma = 10^6$) without defect.

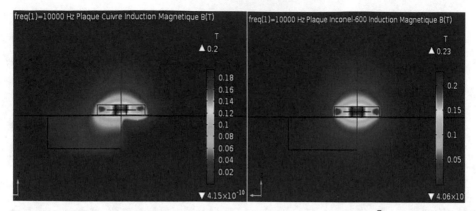

**Fig. 8.** Module of magnetic flux density of a copper plate ($\sigma = 5.91 \times 10^7$) and Inconel-600 plate ($\sigma = 10^6$) with defect.

## 6 Conclusion

In this study, we presented the modeling and simulation software for a 3D eddy current control problem, COMSOL Multiphysics 6. All simulation steps to solve a type of NDT-FC problem have been detailed and explained.

Our findings from simulations using COMSOL Multiphysics are in excellent agreement with the experimental data, which undoubtedly enables us to study many additional examples in an optimal way.

From the results of the simulations obtained, we notice that the amplitude of the induced currents increases when the electrical conductivity of the plate rises, and that the magnetic induction is more and more absorbed in the conductive plate when the electrical conductivity of the plate reduces.

## References

1. Kuhn, É.: Contrôle non destructif d'un matériau excité par une onde acoustique ou thermique. Observation par thermographie, Thèse de doctorat, Université Paris Ouest Nanterre la defense, 6 Décembre 2013
2. Zemouri, N.: Etude et Réalisation d'un Système Multicapteurs Destiné au Diagnostic des Matériaux Conducteurs, Mémoire de Magister, Université de Mouloud Mammeri de Tizi-Ouzou, 21 Juillet 2016
3. Lakhdari, A.-E.: Etude de modélisation de capteurs en CND par Courant de Foucault: Application à la détection des fissures, Mémoire de Magister, Université Mohamed Khider – Biskra, 22 Mai 2011
4. Ramdane, B.: Contribution à la modélisation tridimensionnelle de la technique thermo inductive de contrôle non destructif: Développement d'un outil de conception, d'analyse et d'aide à la décision, Thèse de Doctorat, l'Université de Nantes, 16 Novembre 2009
5. Choua, Y.: Application de la méthode des éléments finis pour la modélisation de configurations de contrôle non destructif par courants de Foucault, Thèse de doctorat, Université Paris-sud11 (2009)

6. Khelfi, S., Helifa, B., Lefkaier, I.K., Hachani, L.: Simulation of electromagnetic systems by COMSOL Multiphysics. In: 3th International Conference on Artificial Intelligence in Renewable Energetic Systems, IC-AIRES2019 (2019)

7. Nam, T., Pardo, T.A.: Conceptualizing smart city with dimensions of technology, people, and institutions. In: Proceedings of the 12th Annual International Digital Government Research Conference: Digital Government Innovation in Challenging Times, pp. 282–291, June 2011

8. Helifa, B., Zaoui, A., Feliachi, M., Lefkaier, I.K., Boubenider, F., Cheriet, A.: Simulation du CND par courants de Foucault en vue de la caractérisation des fissures débouchantes dans les aciers austénitiques

9. Bensetti, M., et al.: Adaptive mesh refinement and probe signal calculation in Eddy current NDT by complementary formulations. IEEE Trans. Magn. **44**(6), 1646–1649 (2008)

10. Santandrea, L., Le Bihan, Y.: Using Comsol-Multiphysics in an Eddy current non-destructive testing context. In: Proceedings of the Comsol Conference (2010)

11. Keciba, A.: Modelisation D'un Systeme De Controle Par Methode Electromagnetique, Memoire Master Academique, Universite Amar Telid (2019)

12. Cardoso, J.R.: Electromagnetics Through the Finite Element Method. Taylor & Francis Group, New York (2017). ISBN 9781498783576

# Magnetic Field Modeling and Study Produces of High Voltage Overhead Power Line

Salah-Eddine Houicher[✉], Rabah Djekidel, and Sid Ahmed Bessedik

Electrical Engineering Department, LACoSERE Laboratory, Ammar Telidji University of Laghouat, BP 37G Route of Ghardaïa, 03000 Laghouat, Algeria
{s.houicher,r.djekidel,s.bessedik}@lagh-univ.dz

**Abstract.** With the expansion development of electric power plants systems is raised the continuous consumption of energy resources, this caused more need of the production of high amount power. Consequently, it was an increasing concern about the possible potential dangers resulting from the influence of the power lines on the objects and facilities located nearby. In this presented paper, a simple approach and a two-dimensional (2D) quasi-static numerical modelling of the magnetic field using image method to determine and examine the magnetic induction produced by high voltage overhead transmission lines and its lateral profile is discussed. Several parameters affecting the magnetic induction strength have been studied; it is observed that taking into account the effect of the exact catenary curve of the power line conductors is much more interesting particularly at the mid-span level in the center distance where the magnetic field becomes very significant. According to these values, we note that the exposure limits set by the International guidelines for magnetic induction strength are respected for occupational and public exposure. The simulation results obtained by this method are also verified then are available applied in the literature, a good calculation is achieved and satisfied full enough and it sufficient that confirms the validity and accuracy of the proposed method.

**Keywords:** Magnetic field · Overhead transmission line · Circuit line · High Voltage · Impact factors · Lateral profile

## 1 Introduction

The high growth in the population and great increasing demand for production and consumption in electric power at many places is in liaison with expansion of various transmission lines design with high voltage operating. These power lines by high voltage (HV) and current intensities produce huge levels of electric and magnetic fields that impact in the public health [1, 2]. Electricity is necessary aspect to promote the economic expansion available to improve and simplify the people's everyday lives of all human beings, the continuous world population increase and with the tendency for people to live in big cities. This has provided many opportunities for electrical energy consumption to previously unknown levels and has accelerated the construction of new

© The Author(s), under exclusive license to Springer Nature Switzerland AG 2024
M. Hatti (Ed.): IC-AIRES 2023, LNNS 984, pp. 477–489, 2024.
https://doi.org/10.1007/978-3-031-60629-8_46

high voltage power transmission lines close and in the vicinity of the most densest populations areas. Since more than thirty years, people are more worried about the potential consequences of continual exposure to extremely low frequency electromagnetic fields, If there is no question of challenging the huge benefits made by electricity [3, 4]. With the development of technology and the rise of novel electrical applications, this exposure has grown ever more important. However, the better comprehension of exposure impacts to electromagnetic fields produced by high-voltage overhead power lines has become a significant worry; The human population must comply to the safety exposure limits set by international norms and regulations like those of the International Commission on Non-Ionizing Radiation Protection (ICNIRP).The exposure limits established by this international standard at extremely low frequency of 50 Hz for the general public are 200 µT about the magnetic field and 5 kV/m for electric field, respectively. In other hand, this levels for occupational exposure are 1000 µT and 10 kV/m [5–7]. In the interest of improving protection of human health and the environment, it is important to accurately evaluate and calculate the distribution of electric and magnetic fields profile near high voltage overhead transmission lines [8–10]. The electric and magnetic fields from power lines have become a significant issue in recent years; this aspect has led growing community concern over possible negative impacts on human health and the environment. These caused effects have inspired a variety of research activities focused at exactly examining the very low frequency electric and magnetic fields generated by power lines. Several publications studies have been made for the calculation and measurement of very low frequency electric and magnetic fields (ELF) created by overhead power lines. Most assume that the power lines are horizontal, straight, long, infinity and parallel to the flat ground plane. The sag due to the power line weight is neglected or modeled by introducing an average height for the horizontal line [11–15]. In view of the above, the main objective of this paper is to analytical simulation and simplified 2D quasi-static numerical model, which makes it possible to determine approximately and obtain a precise computational result of the lateral profile of magnetic field distribution under high voltage overhead power transmission lines in any region of space at height 1m above the ground level, using the GARSON method, in order show the factors affect in the magnetic induction around the power lines. However, this technique has been mostly used in the literature and the results reported with this method are almost very precisely which allows us to validate it.

## 2 Magnetic Field Evaluation Under Three-Phase High Voltage Overhead Power Line

The overhead power lines with high voltage (HV) and current levels generate large values of electric fields intensity and magnetic fields density that possibly affect the public health. Moreover, create serious concerns about its impact in the human and environment. Many research publication works and experimental simulation studies have been made to calculate and measure precisely the electromagnetic fields around the power lines. Based on the results obtained and recommendations reported by these research studies, a numerous number of international standards and guidelines have recommended for occupational and public exposure limits for electric and magnetic fields values induce from overhead transmission lines.

## 2.1 Magnetic Flux Density Calculation

In the modeling quasi-static regime consider that the low frequency magnetic and electric fields are analyzed separately at independently manner, suppose that the electric field is created according the rated voltage of overhead power line, while the magnetic field is based greatly in the current flowing through the phase conductors [16]. In this work, the static technique will also be used analytically to compute the magnetic flux density under power line, in 2D profile using Ampere's law and Biot-Savart theorem [17]. To simplify the integration of magnetic field components around the charged lines, the -component of the magnetic flux density will be decomposed into two x- and y- components in a Cartesian coordinate system, and then used the superposition theorem to get the resultant field generates from high voltage overhead power line at observation point.

## 2.2 Description Technique of the Magnetic Field Calculation

In the present work, the proposed geometry model of single-circuit 275 kV overhead power line, with a symmetrical horizontal phase conductors configuration and earth wire is considered, carries sagging conductors is illustrated in Fig. 1, described the simplified schematic diagram of this transmission line structure used in this performed study, with the arrangement and geometric caracterestiques details in the vicinity of the suspension pylon of the power line. On observe in this figure we have three phases positioned horizontally, each phase contains a bundle of two sub conductors separated by 40 cm. The phase conductors are located at a constant height of 12 m above ground, and the phase spacing between the conductors that is to say the horizontal separation between the phase conductors is 10 m, we show two guard cables in this power line, are situated at distance of 7 m horizontally, and located at height 20 m. The radius of conductors is 14.31 mm, and for the guard cables is 11.2 mm. In the simulation computation of the magnetic field intensity around this power line, we used the resistivity of homogeneous ground is 100 $\Omega$.m, with operating power source frequency $f = 50$ Hz, and the current magnitude flows in the phase conductors of the power line is 500A, that is consider in symmetric system, and take in positive sequence, with angle phase rotation of 120. The effect of earth wires currents are ignored because are very low. The intensity of the magnetic induction due to the currents carrying in supposed infinitely long straight and parallel horizontal conductors along the span length of this power line is obtained by the direct application of the Ampere's law and superposition principle of the partial results. A set of image conductors located symmetrically below the ground level at a depth equal to the conductor's height were also considered to model the ground plane and regularly describe a catenary shape, then explain the real curvature of the power line. In this 2 D quasi-static numerical modelling for the magnetic field computation, the proximity effect of the grounded towers and neighbouring spans which influences the magnetic field intensity distribution under overhead transmission line can be neglected.

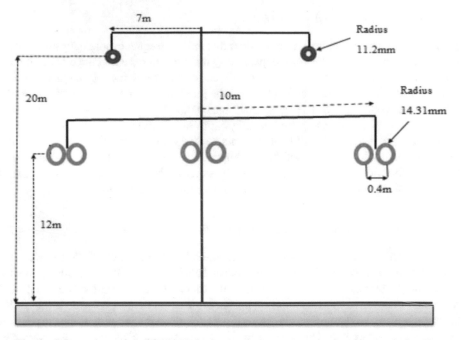

**Fig. 1.** Schematic model of 275 kV single circuit three-phase overhead transmission line.

For precise magnetic field calculation, the image of the filamentary current in a given sub-conductor is located at penetration depth $D_e$ different from the real sub-conductor height above ground, can be expressed as follows [18–20]:

$$D_e = \sqrt{2}\sqrt{\frac{\rho_s}{\pi.\mu_0.f}}e^{-j\frac{\pi}{4}} \tag{1}$$

Where, $\rho_s$ is the electrical resistivity of earth expressed as $\Omega.m$; $f$ is the frequency of power source current in Hz.

The GARSON method that as to say the image theorem is based in the notion of current returns over the ground. The image conductors are located in the distance equivalent to the height of reel conductors above ground $y_i$ plus the depth of penetration $D_e$ defines by this expression: $y_i + D_e$. The current carry conductors of the line generates magnetic field that is composed and reversed the magnetic field induces of source line (see Fig. 2).

In this accurate assessment of the magnetic induction produced by high voltage transmission line, the catenary form of overhead power line conductor that to say sagging conductor is taken into account, the horizontal and vertical components of the magnetic flux density ($B_x$ and $B_y$) caused by transmission line conductors, located at $(x_i, y_i)$ above a uniform earth surface at any point in the space $(x, y)$ around the power line can be

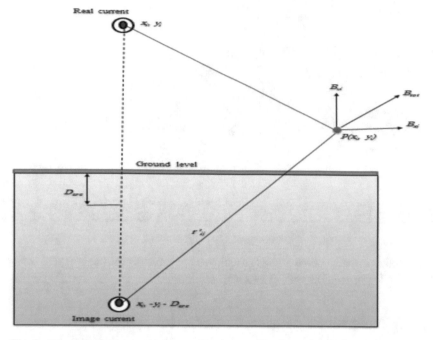

**Fig. 2.** Magnetic field generated by real current and its image in an observation point.

easily calculated by [21–24]:

$$B_x = -\frac{\mu_0}{2 \cdot \pi} \sum_{j=1}^{n} I_j \cdot \left[ \frac{y_i - y_j}{r_{ij}^2} - \frac{y_i + y_j + D_e}{r'^2_{ij}} \right] \tag{2}$$

$$B_y = \frac{\mu_0}{2 \cdot \pi} \sum_{j=1}^{n} I_i \cdot \left[ \frac{x_i - x_j}{r_{ij}^2} - \frac{x_i - x_j}{r'^2_{ij}} \right] \tag{3}$$

Where, $\mu_0$ is the permeability of free space; $I_i$ denotes the current flowing through the conductors; $n$ is the number of the conductors; $(x_i, y_i)$ and $(x_j, y_j)$ are the coordinates of observation point and location of simulation line current, respectively; $r_{ij}$ is the distance between each conductor and observation point above ground, and $r'_{ij}$ is the distance between each image conductor and observation point. The resultant magnetic induction at any desired point $P$ is the summation of horizontal and vertical components in each phase of three-phase overhead power line is calculated by:

$$B_t = \sqrt{B_x^2 + B_y^2} \tag{4}$$

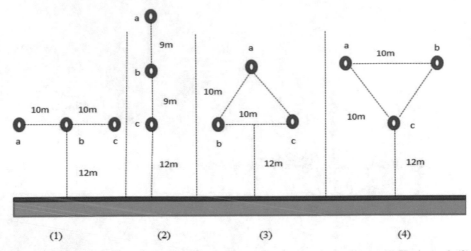

**Fig. 3.** Different configurations of single circuit overhead transmission lines - (1) Horizontal, (2) Vertical, (3) Triangular, (4) Inverted triangular.

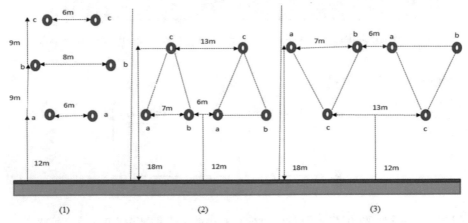

**Fig. 4.** Different configurations of double circuit overhead transmission lines (1) Vertical, (2) Triangular (3), Inverted triangular.

## 3   Simulation Results and Discussion

Figure 5 shows the lateral profile of the resultant magnetic field at 1m above the ground level under high voltage overhead transmission line as a function of the axis of the lateral distance from the power line center. This magnetic field strength depends on the electric currents values and the coordinates of desired calculation point. It can be observed that taking into account the effect of sagging conductor of the power line. It may be interesting to note that the strongest value of the magnetic field gradually increases to a maximum value 7.24 µT register in the side phase conductor x = 7 m, then reduces little to lower value 0.67 μT register below the symmetry center point x = 0 m at mid-span length, is given by the middle phase and produced close to the conductors surface, and then increases again to the same maximum value 7.24 μT at the other side phase conductor, after these levels start decrease laterally to the smaller value 0.27 μT when one moves away from both edges directions of the right of way ROW transmission line, with increasing the lateral distance at x = 80 m a very far from the center line, because the conductors are situated at their highest point.

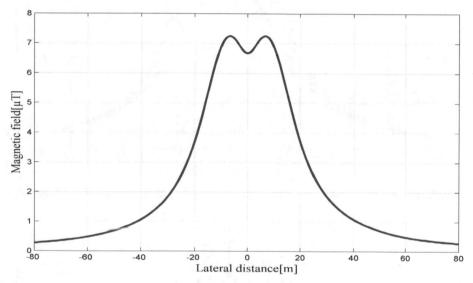

**Fig. 5.** Lateral profile of the magnetic field distribution calculated at 1 m above the ground under high voltage overhead power line.

The lateral profile variation of magnetic field under overhead transmission line at 1 m above ground as a function of different conductor heights directly is shown in Fig. 6. It can be seen from this figure, increasing the height of phase conductors from the ground level leads to a decrease rapidly in the maximum magnetic field intensity of power line.

Figure 7 presents effect of varying the phase spacing between conductors on the magnetic field. As we can be seen from this figure, an increase of the spacing between phases indicates a significant increase in the magnetic field intensity at full overhead transmission line corridor. Consequently, placing the three phases as close together as possible that is to say phases compaction creates some field cancellation, then caused the magnetic field is reduced.

Figure 8 shows the calculation of magnetic field intensity executed at different observation points above the ground surface. As it can be clear observed, increasing the height of observation point from the ground level leads to a significant increase in the peak value of magnetic field under the power line. As we approach the phase conductors with increase the observation point height above the ground level, the magnetic induction strength tends to increase.

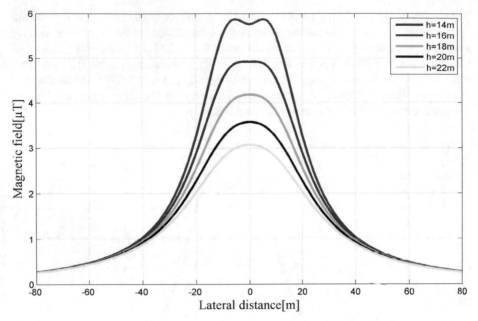

**Fig. 6.** Magnetic field profile at 1 m above the ground as a function of the conductors height.

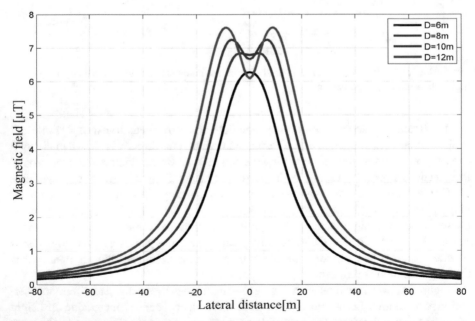

**Fig. 7.** Magnetic field profile at 1 m above the ground as a function of different spacing between phase conductors.

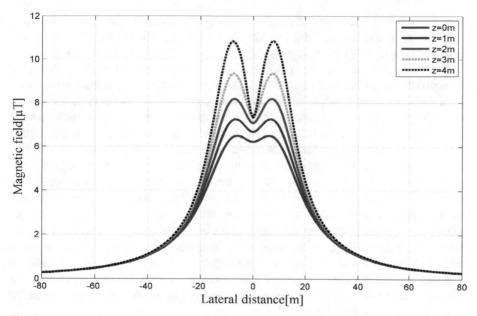

**Fig. 8.** Magnetic field lateral profile for different heights of observation point above ground level.

Figure 9 shows the effect of the soil relative permeability at 1m above the ground level on the lateral profile of the magnetic induction underneath overhead power line. An increase in the relative permeability of the soil in the range accuired in both directions along the right-of-way corridor will result in a slight increase of the magnetic induction.

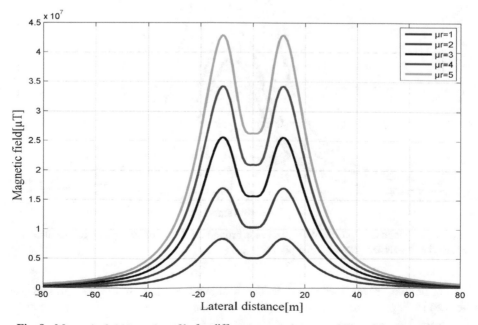

**Fig. 9.** Magnetic field lateral profile for different magnetic permeability of the ground plane.

In order to know what type of circuit line configuration creates a maximum magnetic induction, A comparison of the lateral distribution of magnetic field profiles for all the four single circuit three phase configurations of power line (horizontal, vertical, delta and inverted delta) (see Fig. 3), it can be seen and is clearly shown from Fig. 10 that the horizontal configuration produces the highest magnetic field in all corridor transmission line, at both edges directions along the lateral distance around the power line centre than the other circuit line configurations, with a maximum value of 7.23 μT since all three phase conductors are close to the ground level, while the inverted triangular configuration produces the lowest magnetic field with a maximum value of 3.88 μT in the center lateral distance of the circuit line configuration for the same loading current amplitude of 500A due to the better cancellation effect by 120° three phase shift of the line currents. The magnetic field values calculated at 1m above ground for the triangular and vertical circuit line are 5.36 μT and 5.05 μT, respectively. This observed deviation and variation of the magnetic field profiles is due to difference in the height of phase conductors above the ground in each line configuration. As a result, an appropriate arrangement of the phase conductors can creates some limitation and reduction of the magnetic field density in the power line. Consequently, placing the three phases as close together as possible that is compaction creates some field cancellation, and the magnetic field is reduced under the power line along span length.

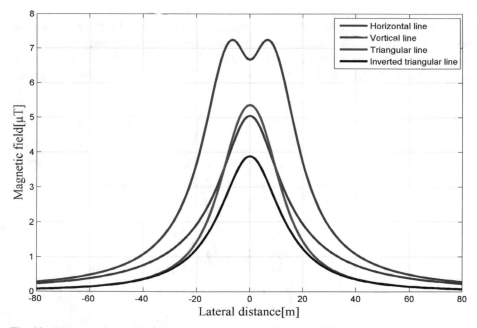

**Fig. 10.** Magnetic field profile at 1m above the ground level for different phase arrangements of single-circuit overhead transmission line.

For double circuit lines shown in Fig. 4 above the same phasing abc-abc, as can be seen in Fig. 11 illustrates the lateral distribution of the magnetic induction profile for different circuit configurations vertical, triangular and inverted triangular line at 1m above ground. On observe in this below curve that the maximum magnetic field values corresponding to this double circuit high voltage overhead transmission lines are 6.83 $\mu$T, 5.97 $\mu$T and 3.30 $\mu$T, respectively. Has a lower value under the power line center in middle phase conductor at mid-span level, and then increases to a maximum value nearly under the lateral conductor, near the side phase conductor. From this point, it decreases rapidly with the lateral distance from the power line increases is lower as one moves away from the line. On show that the vertical line configuration causes higher magnetic induction values at all points along the right of way corridor of transmission line range over the lateral distance at either side in both directions of the power line center than all other arrangements of phase conductors. On the other hand, the inverted triangular configuration produces smaller values of the magnetic induction nearby to center power line. In double circuit lines, the phase arrangements has an important influence in the magnetic induction evaluation around the power line; it is very possible to organize the phase conductors to improve the cancellation then obtain the maximum percent of the reduction factor that give permissible diminution. As an example, we study the lateral profile of the magnetic induction distribution for different phase arrangements in the same circuit line configuration with the transposition of the phase conductors, this technique is almost used for high reduce the magnetic field intensity. In this case study, on note obvious that the inverted triangular line is the best solution for significative reduction of the magnetic field intensity under high voltage overhead power lines, when change the basic geometry of the circuit line configuration.

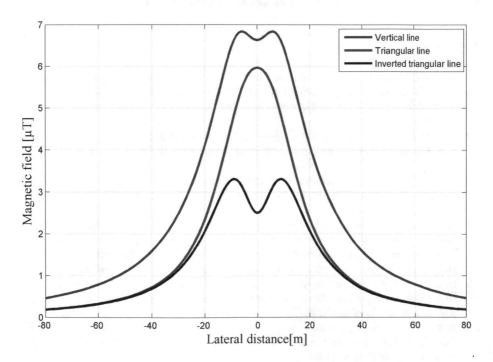

**Fig. 11.** Magnetic field profile at 1 m above the ground for different phase arrangements of double-circuit overhead transmission line.

# 4 Conclusion

This paper presents a quasi-static modeling and assessment to evaluate and calculate the magnetic induction generates an 275 kV overhead power transmission line. From these results it is very clear that using a 2D quasi-static numerical simulation for analytically computation of the magnetic field produced by high voltage overhead power lines at height 1m above the ground level is presented, utilizing GARSON method. According to the results obtained, it is evident that the magnetic field strength is less intense and minimum under the middle phase conductors at center distance of the power line and increases to a maximum intensity near under the side phase conductors in the immediate vicinity of close conductor, then decreases progressively of the same lower value at the other end of the span length with increasing the lateral distance from the center power line in both edges directions of the right of way ROW, to reach the smaller values as one moves away from the conductors, a very far from the power line center, in which it becomes very negligible, which confirm and verify the accuracy of the proposed method. According to these computation values, on represent that the magnetic field strength under the power line at observation point is strongly affected by several factors such as the height of conductors above ground, the spacing between phase conductors, the height of calculation point, and the phases arrangement of three-phase line conductors. These simulation results obtained show that the maximum values of magnetic induction were well below the threshold limits levels set by the ICNIRP recommendations for the general public and occupational exposure The performance and effectiveness of the proposed method are confirmed, the results obtained are also sure verified.

# References

1. C95.1–2019. IEEE standard for safety levels with respect to human exposure to electric, magnetic, and electromagnetic fields, 0 Hz to 300 GHz. In: IEEE Std C95.1–2019 (Revision of IEEE Std C95.1–2005/ Incorporates IEEE Std C95.1–2019/Cor 1–2019), C95, pp. 1–312 (2019)
2. Alihodzic, A., Mujezinovic, A., Turajlic, E.: Electric and magnetic field estimation under overhead transmission lines using artificial neural networks. IEEE Access. **9**, 105876–105891 (2021)
3. Portier, C.J., Wolfe, M.S.: Assessment of health effects from exposure to power line frequency electric and magnetic fields. In: Working Group Report, NIEHS and EMFRAPID, pp. 1–523 (1998)
4. Olden, K.: Health effects from exposure to power-line frequency electric and magnetic fields. National Institute of Environmental Health Sciences National Institutes of Health, NIEHS Report, Prepared in Response to the 1992 Energy Policy Act (PL 102-486, Section 2118), NIH Publication No. 99-4493, Supported by the NIEHS/DOE, pp. 1-80 (1999).
5. Havas, M.: Biological effects of low frequency electromagnetic fields. *Chapter 10*. In: Clements-Croome, D. (ed.) Electromagnetic Environments and Health in Buildings, pp. 207–232. Spon Press, London (2004)
6. Samaras, T.: Preliminary opinion on potential health effects of exposure to electromagnetic fields. In: Scientific Committee on Emerging and Newly Identified Health Risks SCENIHR, Health effects of EMF. pp.1–288 (2013)

7. ICNIRP Standard: International Commission on Non-Ionizing Radiation Protection, guidelines for limiting exposure to time-varying electric and magnetic fields (1Hz to 100 kHz). Health Phys. **99**(6), 818–836 (2010)
8. Al-Bassam, E., Elumalai, A., Khan, A., Al-Awadi, L.: Assessment of electromagnetic field levels from surrounding high-tension overhead power lines for proposed land use. Environ. Monit. Assess. **188**(5), 1–12 (2016)
9. Petrović, G., Kilić, T., Garma, T.: Measurements and estimation of the extremely low frequency magnetic field of the overhead power lines. Elektronika IR Elektrotechnika. **19**(7), 33–36 (2013)
10. Bürgi, A., Sagar, S., Struchen, B., Joss, S., Röösli, M.: Exposure modelling of extremely low-frequency magnetic fields from overhead power lines and its validation by measurements. Int. J. Environ. Res. Public Health **14**(9), 949 (2017). https://doi.org/10.3390/ijerph14090949
11. Ztoupis, I.N., Gonos, I.F., Stathopulos, I.A.: Calculation of power frequency fields from high voltage overhead lines in residential areas. In: 18th International Symposium on High Voltage Engineering, ISH 2013, PA- 01. Seoul, Korea, pp. 61–66 (2013)
12. Sheppard, A.R., Kavet, R., Renew, D.C.: Exposure guidelines for low –frequency electric and magnetic fields: report from the Brussels workshop. Health Phys. Soc. J. **83**(3), 324–332 (2002)
13. Rifai, A.B., Hakami, M.A.: Health hazards of electromagnetic radiation. J. Biosci. Med. **2**(8), 1–12 (2014)
14. CIGRE: Description of phenomena Practical Guide for calculation. Working Group 01 (Interference and Fields) of Study Committee 36. Paris (1980)
15. James, G., Paolo, R., Elisabeth, C.: Potential health impacts of residential exposures to extremely low frequency magnetic fields in Europe. In Environ. Int. J. **62**, 55–63 (2014)
16. Filho, M.L.P., Cardoso, J.R., Sartori, C.A.F., et al.: Upgrading urban high voltage transmission line: impact on electric and magnetic fields in the environment. In: 2004 IEEE/PES, Transmission and Distribution Conference and Exposition: Latin America (IEEE Cat. No. 04EX956). Sao Paulo, Brazil, vol. 8, pp. 788–793 (2004)
17. Dahab, A.A., Abu-Elhaija, W.S., Amoura, F.K.: A comparative study of electric fields beneath compact and non-compact transmission lines. Int. J. Electr. Eng. Educ. **44**(1), 76–83 (2007)
18. Marincu, A., Greconici, M., Musuroi, S.: The electromagnetic field around a high voltage 400 KV electrical overhead lines and the influence on the biological systems. Facta Univ. Ser. Electron. Energet. **18**(1), 105–111 (2005). https://doi.org/10.2298/FUEE0501105M
19. Abdel-Salam, M., Abdallah, H., El-Mohandes, M., El-Kishky, H.: Calculation of magnetic fields from electric power transmission lines. Elect. Power Syst. Res. **49**(2), 99–105 (1999)
20. Roshdy, R., Mazen, A.S., Abdel-Bary, M., Mohamed, S.: Laboratory validation of calculations of magnetic field mitigation underneath transmission lines using passive and active shield wires. Innov. Syst. Des. Eng. **2**(4), 218–232 (2011)
21. Yao, D., Li, B., Deng, J., Huang, D., Wu, X.: Power frequency magnetic fields of heavy current transmit electricity lines based on simulation current method. In: Proceedings of the 2008 IEEE World Automation Congress. Hawaii, HI, pp. 1–4 (2008)
22. Khawaja, A.H., Huang, Q.: Characteristic estimation of high voltage transmission line conductors with simultaneous magnetic field and current measurements. In: Conference Proceedings of 2016 IEEE International Instrumentation and Measurement Technology Conference Proceedings. Taipei, Taiwan, pp. 1–6 (2016)
23. Riba Ruiz, J.R., Espinosa, A.G.: Magnetic field generated by sagging conductors of overhead power lines. Comput. Appl. Eng. Educ. **19**(4), 787–794 (2011)
24. Memari, A.R., Janischewskyj, W.: Mitigation of magnetic field near power lines. IEEE Trans. Power Deliv. **11**(3), 1577–1586 (1996). https://doi.org/10.1109/61.517519

# Performance Evaluation of a Real Polycrystalline Photovoltaic Field Under Desert Conditions

Abdeldjalil Dahbi[1,2(✉)], Fatma Bouchelga[3], Abderrahmane Khelfaoui[1], Miloud Benmedjahed[1], Hocine Guentri[4,5], Ahmed Bouraiou[1], Tidjar Boudjemaa[1], Abdeldjalil Slimani[1], and Samir Mouhadjer[1]

[1] Unité de Recherche en Energies Renouvelables en Milieu Saharien (URERMS), Centre de Développement des Energies Renouvelables (CDER), 01000 Adrar, Algeria
Dahbi_j@yahoo.fr
[2] Laboratory of Sustainable Development and Computing, (L.D.D.I), University of Adrar, Adrar, Algeria
[3] Department of Automatics and Electromechanics, Faculty of Science and Technology, University of Ghardaia, Ghardaia, Algeria
[4] Department of Mechanic and Electromechanics, Institute of Science and Technologies, Abdelhafid Boussouf, University Centre, Mila, Algeria
[5] GE Laboratory, Saida University, Saida, Algeria

**Abstract.** This paper's goal is to examine the operation of the polycrystalline photovoltaic field at the Oued Nechou location in Ghardaia, Algeria. To comprehend the performances of this PV plant, panel, and array have been modelled and simulated under various input situations. In order to assess the PV plant and determine the solar energy potential in the examined region, the polycrystalline PV plant has also been simulated by using climate data. The results demonstrate how the performance of photovoltaic systems is significantly influenced by external conditions, mainly the temperature and irradiation. This causes changes in the performance ratio and output of electrical energy. Furthermore, they demonstrate that the area has a good solar potential which provides the production of an acceptable power despite the local environment, which promotes the development of new PV there.

**Keywords:** Solar potential · PV cell · Photovoltaic plant · Modeling · Simulink

## 1 Introduction

Electrical energy has become increasingly important as a result of technological advancements for serving both domestic and commercial needs. Yet, the evolution of daily life, population increase, and industrial expansion have resulted in a significant need for energy. Regrettably, this latter has brought to a variety of environmental problems as well as the depletion of fossil fuel resources. The fossil fuels use has also prompted academics to hunt for alternatives, including the use of renewable energies such as solar energy [1].

M. Hatti (Ed.): IC-AIRES 2023, LNNS 984, pp. 490–501, 2024.
https://doi.org/10.1007/978-3-031-60629-8_47

Algeria is located in the center of the solar belt. Consiquently, solar energy has a lot of promise, especially in desert areas and other remote areas. Solar energy, in particular, is a useful and effective way to generate electricity in these locations. Since photovoltaic power output is particularly susceptible to weather, it is crucial to assess the solar plant's performance in saharan climate zones to create a just balance between production and load demands [2, 3]. The goal of this study is to examine the PV plant of Oued Nechou station in Ghardaia that is subject to the local meteorological conditions in southern Algeria [4–6]. As a part of Algeria's energy transition strategy, several photovoltaic stations have been set up; the station in Oued Nechou Ghardaia that is the focus of this study is one of these stations [7, 8].

This paper is organized on seven sections: The first paragraph of Sect. 2 describes the 1.1 MW Oued Nechou PV plant in Ghardaia. Afterwards, in the same facility, the characteristics of the polycrystalline photovoltaic field were installed. The modeling of the solar cell and array comes next. In section six, the features of the PV module have been evaluated under various temperatures and irradiance levels in order to comprehend their behaviour and the impact of weather conditions on the produced energy. The fifth and final section focuses on results of the PV plants based on real data to determine their annual production. The results have been examined. Some recommendations have been provided to increase the output for subsequent work.

## 2   Ghardaia's 1.1 MW Oued Nechou PV Plant

The investigated photovoltaic plant is located 15 km north of the city of Ghardaia, next to the town of Oued Nechou. It has a nominal power of around 1,100 kWp and a surface area of 10 Hectares. It belongs to SONELGAZ. This station provides electricity to the 30 kV line. It includes the polycrystalline, amorphous, thin-film, and monocrystalline solar cell technologies [4, 5]. This work is focused on the polycrystalline field (Fig. 1).

**Fig. 1.** Oued Nechou's plant photovoltaic panels

## 3   Characteristics of the Poly Crystalline Photovoltaic Panels

In the sub-field, there are 1100 polycrystalline photovoltaic panels. A total of 55 panels are connected in parallel, while 20 panels are connected in serial. Each panel generates 235 W of STC power. 258.5 kWc of electrical energy are produced by the entire sub-field.

The characteristics of the polycrystalline photovoltaic panels installed in the Oued Nechou facility are shown in the image and table below [4, 5]: (Fig. 2 and Table 1).

**Fig. 2.** Oued Nechou's poly crystalline panels

**Table 1.** Real characteristics of silicium poly ocrystalline photovoltaic panels in studied plant

| Parameter | Value |
| --- | --- |
| Rated power | 235 W |
| Voltage in MPP | 29.04 V |
| Current in MPP | 8.1 A |
| Open circuit voltage | 36.94 V |
| Short-circuit voltage | 8.64 A |
| Yield | 14.43 |

To understand the behavior of this kind of cell technology, it is required to model it at first.

## 4   Photovoltaic Cell Modelling

With a current source and a P-N junction diode, the circuit of the solar panel may be made to behave as it does in real life. A parallel resistor $R_{sh}$ also serves as a representation of the leakage current. To reflect the internal resistance to current flow, a serial resistor $R_s$ is subsequently added to the circuit (Fig. 3) [9].

It is anticipated that the PV cell and module model parameters will alter in this scenario. The current $I_{Ph}$ of the PV cell can be computed using [4]:

$$I_{ph} = I + I_d + I_p \tag{1}$$

Hence, the current of output can be calculated by:

$$I = I_{ph} - I_d - I_p \tag{2}$$

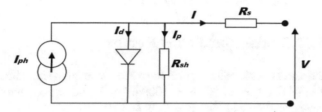

**Fig. 3.** Electrical equivalent PV cell circuit

Where $I_p$ and $I_d$ can be given:

$$I_p = \frac{V + R_s.I}{R_{sh}} \tag{3}$$

$$I_d = I_0 \left( EXP^{(\frac{q.V}{N_s k.n.T})} - 1 \right) \tag{4}$$

with: $V$: Voltage; $K = 1.38 * 10{-}23$: Constant of Boltzmann; $q = 1.602 * 10^{-19}$: electron charge; $T$: absolute temperature in $°K$; $I_0$: diode saturation current; $I_{ph}$: photo current; $I_d$: Diode current.

## 5   PV Array Modelling

PV cells are combined to form PV modules. PV modules can be arranged serially or parallelly to form PV arrays, which are then used in PV production systems to generate electricity. The equivalent circuit for a PV array is shown in Fig. 4 [10].

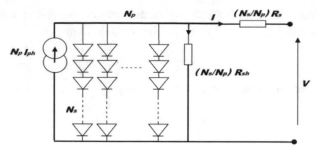

**Fig. 4.** Electrical equivalent circuits of the PV array

# 6 Studied PV Module Simulation Results

At first, the parameters of the real PV panel have been entered in simulation. The results are presented in (STC; T = 25 °C; E = 1000 W/m²), and various temperatures and irradiations values, as shown in Figs. 5, 6, 7, 8, and 10.

Firstly, the PV panel's settings are entered into the simulation parameters. Then, the simulation results are displayed in Figs. 5, 6, 7, 8, and 10 for Standard Test Conditions, after that, the simulation for a variable range of temperatures and irradiations.

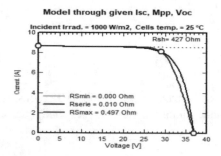

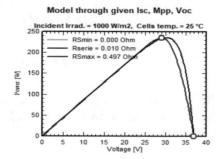

**Fig. 5.** Characteristics I(V) plot of the panel (STC)

**Fig. 6.** Characteristics P(V) plot of the panel (STC)

## 6.1 Result Analysis

Figures 5 and 6 depict the simulated characteristics of the investigated poly ocrystalline 235 W solar panel under STC. As it can be observed, 8.64 A is the short circuit current and 36.94 V is the open circuit voltage; they are exactly like the identical numbers that were shown in the actual characteristics. It is observed that the maximum power of 235 W corresponds to a value less than Voc. Furthermore, when the voltage level increases, the power value increases till the maximum value ($P_{max}$). After, it falls to zero. The curves' characteristics are slightly altered by the movement of $Rs$, therefore the temperature can have an indirect impact on them.

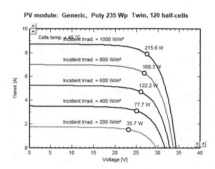

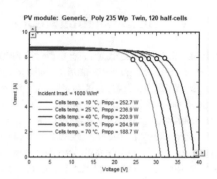

**Fig. 7.** Irradiance effect on the I(V) PV panel characteristics

**Fig. 8.** Temperature effect on I(V) the PV panel characteristics

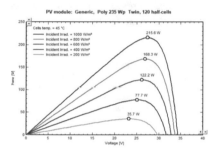

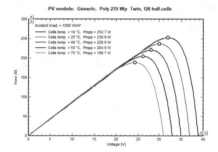

**Fig. 9.** Irradiance effect on P(V) PV panel characteristics

**Fig. 10.** Temperature effect on P(V) PV panel characteristics

It is obvious from Fig. 7, which presents the current-voltage characteristic under various irradiations, that the irradiation has a direct impact on the current value (proportional). However, it has a significant impact on the voltage value.

The value of the voltage is proportional to the value of the temperature, as it can be seen in Fig. 8. Contrarely, the current almost stays constant. As a result, the fluctuation in voltage and current implies the variation in the maximum power value.

The effect of irradiation variation on a power characteristic is shown in Fig. 9. It can be seen that the increasing in irradiation corresponds to an increasing in power level.

Figure 10 shows that at a constant temperature (Irradiation = 1000 W/m$^2$), it has no a big impact on the power level, unlike the irradiation parameter; power rises as temperature rises.

After running the simulation and analyzing the panel under various conditions, now the simulation of the ply crystal photovoltaic plant is realized.

# 7 PV Plant Simulation Results

This studied fixed polycrystalline subfield, has 1100 modules. 20 modules are coupled in series with 55 parallel chains. Moreover, a single inverter with an operational voltage range of 430 to 760 V is connected to the system. The orientation of the PV system should be optimized in order to maximize the energy from fixed panels. Otherwise, a sun-tracking system should be used. The photovoltaic modules' angular relationship to the sun directly influences how much electricity they produce. They should therefore be in a good position to maximize their performance. The PV orientation is very important where the active face of the PV panel is turned during each season or according to an ideal year angle. The inclination, which is measured in degrees, represents the panel's angle with respect to the horizontal plane. A solar panel should be oriented toward the equator in order to function optimally.

This determines whether the orientation faces south. It is important to consider the inclination and the less sunny times of the year in order to maximize the generation of energy. The parameters have been set as they are presented in Table 2.

**Table 2.** Poly ocrystalline PV plant parameters

| Simulation parameters | | System type | No 3D scene defined, no shadings | | |
|---|---|---|---|---|---|
| Collector Plane Orientation | | Tilt | 28° | Azimuth | 0° |
| Models used | | Transposition | Perez | Diffuse | Perez, Meteonorm |
| Horizon | | Free Horizon | | | |
| Near Shadings | | No Shadings | | | |
| User's needs : | | Unlimited load (grid) | | | |
| | | | | | |
| PV Array Characteristics | | | | | |
| PV module | Si-poly | Model | Poly 235 Wp Twin, 120 half-cells | | |
| Custom parameters definition | | Manufacturer | Generic | | |
| Number of PV modules | | In series | 20 modules | In parallel | 55 strings |
| Total number of PV modules | | Nb. modules | 1100 | Unit Nom. Power | 235 Wp |
| Array global power | | Nominal (STC) | 259 kWp | At operating cond. | 231 kWp (50°C) |
| Array operating characteristics (50°C) | | U mpp | 536 V | I mpp | 432 A |
| Total area | | Module area | 1790 m² | Cell area | 3128 m² |
| | | | | | |
| Inverter | | Model | PVS800-57-0250kW-A | | |
| Original PVsyst database | | Manufacturer | ABB | | |
| Characteristics | | Operating Voltage | 450-825 V | Unit Nom. Power | 250 kWac |
| Inverter pack | | Nb. of inverters | 1 units | Total Power | 250 kWac |
| | | | | Pnom ratio | 1.03 |

The PVSys software's recommended inclination, with a due south direction and annual optimization is 28° for the yearly fixed inclination, as illustrated in Fig. 12. The system's electrical components were then chosen after that. Namely, the technology of the inverters, photovoltaic panels, etc. The simulation's findings are shown below. The PV plant at Ghardaia's displays monthly variations in meteorological and electric variables throughout the year are shown in Table 3.

**Table 3.** Polycrystalline PV plant monthly input/output parameters

**Ouednechou47_poly fix**
**Balances and main results**

| | GlobHor kWh/m² | DiffHor kWh/m² | T_Amb °C | GlobInc kWh/m² | GlobEff kWh/m² | EArray MWh | E_Grid MWh | PR |
|---|---|---|---|---|---|---|---|---|
| January | 119.5 | 19.70 | 10.63 | 189.5 | 185.6 | 42.13 | 40.92 | 0.835 |
| February | 133.6 | 24.00 | 13.10 | 187.1 | 183.2 | 40.89 | 39.74 | 0.822 |
| March | 193.1 | 32.50 | 17.93 | 233.9 | 229.0 | 49.38 | 47.97 | 0.793 |
| April | 220.5 | 41.90 | 21.18 | 233.7 | 227.1 | 48.13 | 46.81 | 0.775 |
| May | 246.3 | 53.60 | 26.41 | 235.7 | 228.5 | 47.12 | 45.89 | 0.753 |
| June | 247.2 | 52.20 | 31.01 | 225.9 | 218.7 | 43.88 | 42.76 | 0.732 |
| July | 255.8 | 50.70 | 35.36 | 238.8 | 231.6 | 45.13 | 44.02 | 0.713 |
| August | 229.5 | 52.40 | 33.91 | 233.9 | 227.3 | 44.73 | 43.63 | 0.722 |
| September | 186.4 | 45.20 | 28.18 | 213.0 | 207.3 | 42.34 | 41.26 | 0.749 |
| October | 154.7 | 37.00 | 23.57 | 203.3 | 198.4 | 41.74 | 40.65 | 0.773 |
| November | 125.1 | 20.00 | 15.81 | 190.5 | 187.0 | 41.17 | 40.03 | 0.813 |
| December | 109.2 | 17.40 | 11.99 | 181.1 | 177.5 | 40.13 | 39.02 | 0.833 |
| Year | 2220.9 | 446.60 | 22.48 | 2566.4 | 2501.0 | 526.77 | 512.70 | 0.773 |

Legends:  GlobHor    Horizontal global irradiation              GlobEff    Effective Global, corr. for IAM and shadings
         DiffHor     Horizontal diffuse irradiation              EArray     Effective energy at the output of the array
         T_Amb      Ambient Temperature                        E_Grid     Energy injected into grid
         GlobInc     Global incident in coll. plane               PR         Performance Ratio

As seen in Table 3, in the summertime is when all types of irradiation increase because the sky is clear. The ambient temperature varies from a minimum of 10.63 °C in January to a maximum of 35.36 °C in July, with a monthly average of 22.48 °C. The energy injected into the grid is inversely proportional to the irradiation, although other factors also have an impact (such as fog, dust, temperature, irradiation, and shodow…). A minimum of 40.13 MWh was recorded in December, and a maximum of 49.38 MWh was observed in March [11]. The Figures below present the additional parameters:

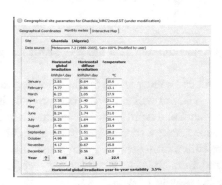

**Fig. 11.** Ghardaia's climatic Data

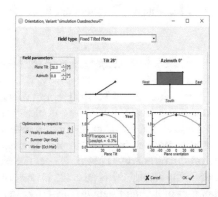

**Fig. 12.** Setting of panels orientation

As it can be observed in Fig. 11, the examined site at Ghardaia has a significant solar potential. The yearly average is 6.08 kWh/m²/day, ranging from 3.52 kW/m² in December to 8.25 kW/m² in July. PV panels can benefit of this potential by adopting the ideal orientation angle (28°), as shown in Figs. 12 and 13.

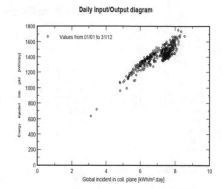

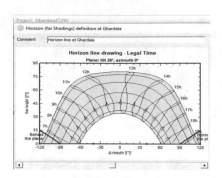

**Fig. 13.** Horizon line

**Fig. 14.** Produced Daily energy in function of irradiation

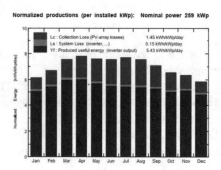

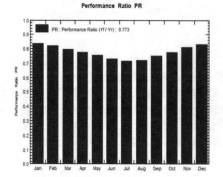

**Fig. 15.** Annual production rate histogram

**Fig. 16.** Monthly performance ratio

The electrical energy injected into the grid and the radiation are proportionate, as seen in Fig. 14.

From Fig. 15 and 16, the amount of the produced energy varies throughout the year. Because the weather in spring is closer to the STC, in which the PV can produce its rated power, it reaches its maximum levels in March and April [12]. The summer's high temperatures has a negative impact on the produced energy, cause a drop in the amount of the produced energy. Then, this later decreasses more from September to January than it does in the summer. Because the impact of irradiation on the generated energy is bigger than that of ambient temperature (Fig. 15). In the months of January and December, the performance ratio is at its highest point, about 0.83.5%. However, it decreases also in all subsequent months till 0.713% in July. This is because of the high temperatures during this time, which results in performance decreasing of photovoltaic panels (Fig. 16).

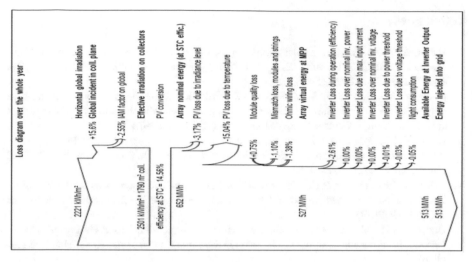

**Fig. 17.** Diagram of loss over the year

As it can be seen in Fig. 17, the produced energy is only about 78,7% of the nominal energy of the array at STC, because of several losses, which account for roughly 21,3% of the total energy produced, like the Ohmic wiring losses, electric connections, temperature effect, and component loss.

The photovoltaic system demonstrates its viability in a variety of climatic circumstances [13–19]. This encourages to install more PV plants, which helps to address environmental issues by lowering $CO_2$ emissions [15–18].

## 8 Conclusions

This article analyzes, studies, and simulates a real polycrystalline photovoltaic plant in Oued Nechou, in Algerian Saharan region. The purpose of this work is to determine the solar energy potential in this area under investigation; and to evaluate the effectiveness and performance of the polycrystalline photovoltaic (PV) plant that was installed in southern Algeria. The solar cell modeling, panel, and plant have been provided. The PV panel simulation was then examined and debated in STC.

In order to determine how climate variables affect PV panel performances and characteristics, the PV panel was then simulated for various irradiance and temperature values. The sub-field of the polycrystalline PV plant at Oued Nechou was then modelled using climate data from the same area. A good solar potential was discovered in the area under study, as evidenced by the values of irradiation (GlobHor: 255.8 kWh/m$^2$ in July,109.2 kWh/m$^2$ in December). However; it has been discovered that the variations in the in the temperature and irradiance, have a big impact on the energy output. The amount of energy added to the grid is related to irradiations, it fluctuates between a minimum of 40.13 MWh in December and a maximum of 49.38 MWh observed in March. However, it is impacted by additional factors (such as the outside temperature, radiation, shadow, fog, dust, etc.). These results promote the investments in solar systems in this area and provide a starting point for studies to address issues that lower yield and efficiency.

# References

1. Dahbi, A., et al.: Sizing of a solar parking system connected to the grid in Adrar. In: Hatti, M. (ed.) ICAIRES 2019, vol. 102, pp. 506–514. Springer, Cham (2020). https://doi.org/10.1007/978-3-030-37207-1_54
2. Boussaid, M., Belghachi, A., Agroui, K.: Contribution to the degradation modeling of a polycrystalline photovoltaic cell under the effect of stochastic thermal cycles of a desert environment. Int. J. Control Energy Electr. Eng. (CEEE) **6**, 66–72 (2018)
3. Boussaid, M., Dahbi, A., Lahcena, A., Elkaiem, L.M.: The interest of connecting mini solar stations to the public electricity grid in a desert environment. In: Proceedings IRSEC. IEEE (2019). ISBN: 978-1-7281-5152-6
4. Bahri, A.: Modélisation et simulation d'un générateur photovoltaïque sous Matlab/Simulink Etude pratique site Oued Nechou à Ghardaïa. El-Wahat pour les Recherches et les Etudes **10**(1), 1–19 (2017)
5. Oudjana, S.H., Mosbah, M., Mahammed, I.H.: Load forecasting based PV power in Oued Nechou. In: Proceedings of the 4th International Seminar on New and Renewable Energies (2016)
6. Tati, F., Talhaoui, H., Aissa, O., Dahbi, A.: Intelligent shading fault detection in a PV system with MPPT control using neural network technique. Int. J. Energy Environ. Eng. **13**(4), 1147–1161 (2022). https://doi.org/10.1007/s40095-022-00486-5
7. Aoun, N.: Performance analysis of a 20 MW grid-connected photovoltaic installation in Adrar, South of Algeria. In: Advanced Statistical Modeling, Forecasting, and Fault Detection in Renewable Energy Systems, pp. 1–12 (2022). https://doi.org/10.5772/intechopen.89511
8. Chabachi, S., Necaibia, A., Abdelkhalek, O., Bouraiou, A., Ziane, A., Hamouda, M.: Performance analysis of an experimental and simulated grid connected photovoltaic system in southwest Algeria. Int. J. Energy Environ. Eng. **13**(2), 831–851 (2022). https://doi.org/10.1007/s40095-022-00474-9
9. Benabdelkrim, B.: Modeling and parameter extraction of PV cell using single- and two-diode model. Int. J. Energetica (IJECA) **2**(2), 06–14 (2017). ISSN: 2543-3717. https://www.ijeca.info
10. Nguyen, X.H., Nguyen, M.P.: Mathematical modeling of photovoltaic cell/module/arrays with tags in Matlab/Simulink. Environ. Syst. Res. **4**, 24 (2015). https://doi.org/10.1186/s40068-015-0047-9
11. Ghenai, C., Bettayeb, M.: Design and optimization of grid-tied and off-grid solar PV systems for super-efficient electrical appliances. Energ. Effi. **13**(2), 291–305 (2019). https://doi.org/10.1007/s12053-019-09773-3
12. Said, M.R., El-Samahy, A.A., El Zoghby, H.M.: Cleaning frequency of the solar PV power plant for maximum energy harvesting and financial profit. Int. J. Power Electron. Drive Syst. (IJPEDS) **14**(1), 546–554 (2023). https://doi.org/10.11591/ijpeds.v14.i1.pp546-554
13. Abdelkader, B., Merabti, A., Yamina, B.: Using PSO algorithm for power flow management enhancement in PV-battery grid systems. Int. J. Power Electron. Drive Syst. (IJPEDS) **14**(1), 413–425 (2023). https://doi.org/10.11591/ijpeds.v14.i1.pp413-425
14. Mahmmoud, O.N., Mehdi, S.R., Gaeid, K.S., Al-Tameemi, A.L.S.: Solar cell split source inverter for induction motor with computer control. Int. J. Power Electron. Drive Syst. (IJPEDS) **14**(1), 174–184 (2023). https://doi.org/10.11591/ijpeds.v14.i1.pp174-184
15. Arango-Aramburo, S., Veysey, J., Martínez-Jaramillo, J.E., Díez-Echavarría, L., Calderón, S.L., Loboguerrero, A.M.: Assessing the impacts of nationally appropriate mitigation actions through energy system simulation: a Colombian case. Energy Effi. **13**(1), 17–32 (2019). https://doi.org/10.1007/s12053-019-09826-7

16. Palaniappan, K., Veerapeneni, S., Cuzner, R.M., Zhao, Y.: Viable residential DC microgrids combined with household smart AC and DC loads for underserved communities. Energy Effi. **13**(2), 273–289 (2018). https://doi.org/10.1007/s12053-018-9771-0

17. Al-Shetwi, A.Q., Sujod, M.Z.: Modeling and design of photovoltaic power plant connected to the MV side of Malaysian grid with TNB technical regulation compatibility. Electr. Eng. **100**(4), 2407–2419 (2018). https://doi.org/10.1007/s00202-018-0726-4

18. Ayang, A., et al.: Least square estimator and IEC-60891 procedure for parameters estimation of single-diode model of photovoltaic generator at standard test conditions (STC). Electr. Eng. **103**, 1253–1264 (2021)

19. Bylykbashi, B., Filkoski, R.V.: Modelling of a PV system: a case study Kosovo. Int. J. Power Electron. Drive Syst. (IJPEDS) **14**(1), 555–561 (2023). https://doi.org/10.11591/ijpeds.v14.i1.pp555-561

# Cooling Load Forecasting Through an Innovative Environmental Awareness Module

Amine Belkhir[(✉)]

Renewable Energy Department, University of Blida "Saad Dahleb" Blida, Ouled Yaïch, Algeria
A.Belkhir@outlook.com

**Abstract.** Accurately predicting cooling load is a significant stride towards enhancing energy efficiency. This prediction relies on a multitude of factors. This paper explains the idea behind Environmental Awareness, elucidating a comprehensive methodology for selecting the most fitting forecasting technique for the Weather Awareness Module (WAM). Exponential Smoothing and the Nonlinear Autoregressive Neural Network with External Input (NARX) are analyzed, utilizing time series data pertaining to specific humidity and dry-bulb temperatures within the Boughezoul region. This work underscores the significance of data analysis, and the findings showcase that the NARX approach is particularly suitable for short-term forecasting objectives.

**Keywords:** Cooling Load · Forecasting · Renewable Energy · Time Series · Neural Networks · Exponential Smoothing · Nonlinear autoregressive neural network

## 1 Introduction

During the mid-1980s, a surge of worldwide concern emerged regarding the depletion of the Earth's ozone layer, recognizing it significant impact and the irreversible damage to the planet [1]. The dedication of scientific community to raise awareness while finding the practical solutions to the situation has led to a series of measures to contain the predicament. Today the consequences of global warming are clearly observable in the form of extreme weather condition, such as floods, wildfires and droughts. Notable instances in this regard include the experiences of Algeria and Turkey within recent years [2, 3].

Some of the measures to limit the impact of global warming steering humanity toward a more sustainable trajectory, include the transition towards the use of renewable energies, embracing the utilization of environmentally congenial refrigerants in the context of vapor compression cooling mechanisms [4]. Sorption based air handling units hold great potential to be used for sustainable buildings applications since it can be powered by solar thermal energy and uses water vapor as refrigerant. Nevertheless, a prevailing constraint of these systems is their inability to provide precise assessments of the fluctuations in cooling load, a metric contingent upon a range of variables (namely, occupancy, infiltrations, and weather conditions), each of which poses distinct challenges requiring

M. Hatti (Ed.): IC-AIRES 2023, LNNS 984, pp. 502–510, 2024.
https://doi.org/10.1007/978-3-031-60629-8_48

resolution. While this paper omits discussions on occupancy and infiltrations, it does delve into the subject of artificial intelligence-aided weather condition forecasting as a focal point.

An ideal system should have what we call "environment awareness" which in our case can be divided to a set of three separate modules as illustrated in Fig. 1. Each module should be able to quantify and output the required parameters to forecast the cooling load, this is then fed to the system controller in order to deliver the required air flow and blowing temperature. It is also an important factor to determine the electrical load especially during summer time.

Forecasting the cooling load lacks deterministic precision, this motivates this work to develop the environment awareness module using a combination of conventional forecasting technics coupled with artificial intelligence to account for the randomness and chaotic nature of the data. In this paper we will discuss mainly the approach used in the Weather Awareness Module (WAM) with some outlines for other modules.

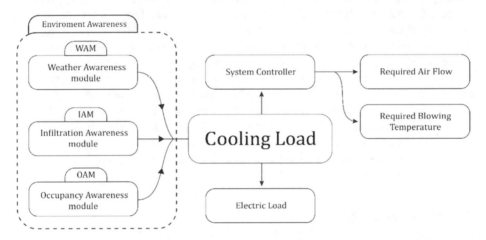

**Fig. 1.** Flowchart of the implementation of the environment awareness module

## 2   Related Research Work

This section encapsulates key contributions and initial efforts in the field of time series forecasting. C. C. Holt's seminal work in 1957 [5] marked a pivotal point with the introduction of the concept of seasonality in time series data. This phenomenon is a primary contributor to the emergence of white noise and randomness in cyclic data. Holt not only conceptualized this but also provided the method's formulation. Building upon Holt's foundation, P. R. Winters made strides in 1960 [6] by enhancing Holt's formulation. Winters presented a more lucid mathematical structure and a comprehensive flowchart of the method, supplemented by illustrative examples. Further advancements and refinements emerged over time, culminating in a compilation of improved variations and models in 2008 by R. Hyndman et al. [7]. Recent progress in the domain is evidenced by the work of Dudek, G et al. in 2021 [8]. Here, a hybrid model combining exponential smoothing and Long-Short Term Memory (LSTM) architecture is proposed. The outcomes demonstrated competitive performance against classical models as well as other machine learning methodologies.

Today, the internet serves as a vast data repository. With the rise of big data and its colossal daily influx, novel data-mining techniques harness this abundance to train potent models. In 1998, D. J. Hand [9] laid the groundwork for data mining fundamentals. In 2022, Y.A. Barrera-Animas [10] conducted a comparative analysis of machine learning algorithms for rainfall prediction, highlighting the prowess of Bidirectional-LSTM while finding Stacked-LSTM underperformed. MS Bakay et al. (2021) [11] explored multiple ML algorithms for forecasting greenhouse gas emissions in Turkey using electricity production time series. Lastly, A. Altan et al. (2019) [12] examined the impact of kernel parameters in support vector machine learning for financial time series prediction.

More into optimization algorithm which constitute the backbone of all machine learning, alongside Levenberg (1944) [13] algorithm. Baumann, P (2021) [14], proposed two new algorithms and compared them to existing algorithms. They found that similarity based algorithms perform better than non-similarity ones. L. Yang et al. (2020) [15] discussed the hyper-parameter optimization HPO in comparison with the Particle swarm optimization PSO, results showed that HPO is closing the gap in performance. Tarutani et al. (2016) [16] predicted air conditionner effeciency in a data center using a set of temperature sensors, they have achieved a reduction of 30% of the total power consuption of the data center. Always in the context of cooling applications K. Amasyali et al. (2021) [17]. They compared various machine learning models from classification and regression trees CART to ensemble bagging trees EBT and ANNs to predict cooling energy consumption based on occupant behavior, results show that ANNs with optimal layer number outperforms other models.

## 3   Experimental Setup

### 3.1   Modules Description

In the context of cooling load, the environment awareness can be composed into three different modules each have its own characteristics:

- Occupancy Awareness Module (OAM)

Occupancy emerges as the principal metric for assessing internal cooling load. This module's key role is to quantify occupant count, activity intensity, and clothing attire, complemented by an electronics component monitor. The strategic integration of machine learning algorithms within this module is of paramount significance. This amalgamation facilitates the discernment of distinct patterns and behaviors specific to each cooled area.

- Infiltration Awareness Module (IAM)

Infiltration governs the entry of external, unconditioned air into a space through apertures. Thus, it necessitates quantifying the impact level attributed to each opening, encompassing factors such as doors and windows. Additionally, this module should ascertain the thermal resistance of walls to complete its assessment.

- Weather Awareness Module (WAM)

The WAM is used essentially to forecast outdoor condition i.e. weather conditions in general (Dry bulb temperature, wind speed and direction, solar radiation, precipitation and specific humidity). It takes in location coordinates and require a set of historical time series.

## 3.2 Proposed Approach

In this paper we are going to explore the statistical forecasting of time series utilizing an exponential smoothing series to forecast cyclic behavior in climatic data, then using a nonlinear autoregressive neural network with external input (NARX) to account for seasonality of our prediction.

- Neural Network Architecture

The architecture features a feed forward ANN, with a single input layer of two time series data of equal length namely dry bulb temperature T(t) and specific humidity $\omega$(t). With a delay of 24 representing 24 h four hourly time series which feed into a 10 hidden layers using the Logsig() activation function. Which is then fed to the output layer of a purelin() activation function to finally obtain our final temperature prediction.

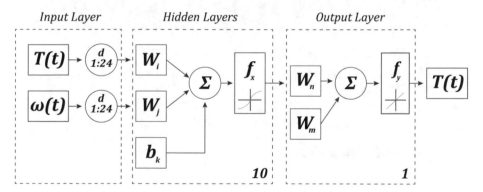

**Fig. 2.** Schematic of the proposed Artificial Neural Network

- Exponential Smoothing methodology

Using the Holt-Winters [5, 6] method, we can perform a forecasting of cyclic data. Such as.

Forecast equation $\hat{y}_{t+h} = l_t + hb_t + s_{t+h-m(k+1)}$.

Where $t$ is time, $h$ is the forecasting horizon, $m$ is cyclic frequency and $k$ is the integer part of $(h-1)/m$.

Level Component equation $l_t = \alpha(y_t - s_{t-m}) + (1-\alpha)(l_{t-1} + b_{t-1})$

Trend Component equation   $b_t = \beta(l_t - l_{t-1}) + (1 - \beta)\beta_{t-1}$

Cyclic Component equation   $s_t = \gamma(y_t - l_{t-1} - b_{t-1}) + (1 - \gamma)s_{t-m}$

where $\alpha, \beta$ and $\gamma$ are the smoothing parameters of the level, Trend and Cyclic respectively.

- Historical and Training Data

For reasons of the scarcity of measured data, openly available data that will be used are obtained through NASA open data programs [18]. The chosen site of study is Boughzoul, Medea, Algeria. (35°42'N; 2°50'E) As it is situated at the Algerian highlands and represent an anchor point between the desert and coastal region in addition to having high levels of humidity due to two large water surfaces.

The data exhibit a prominent cyclic pattern, as depicted in Fig. 3. However the trend cannot be shown and require a larger time series.

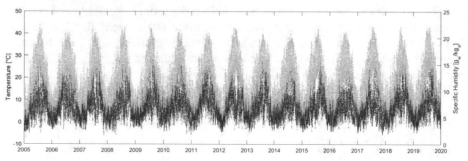

**Fig. 3.** Time Series data visualization of the dry bulb Temperature and Specific Humidity for the period from 2005–2020

## 4   Results and Discussion

### 4.1   Data Analysis

Fortunately, the time series are said to be cleaned from missing and faulty values. Otherwise a rigorous analysis should've been made. Consequently, we swiftly progressed towards conducting correlation and autocorrelation analysis. As depicted in Fig. 4, the autocorrelation pattern for both temperature and specific humidity appears analogous. A decline in correlation is evident, culminating in the weakest correlation at lag 12. This value is deemed undesirable and thus avoided when devising the neural network architecture. Notably, the most favorable autocorrelation is observed at or near a zero lag, accompanied by a distinct cyclic pattern. While temperature's autocorrelation exhibits a consistent amplitude, specific humidity displays a slightly damped pattern. This prompts the selection of a delay value of 24, as a zero lag proves ineffectual for forecasting applications.

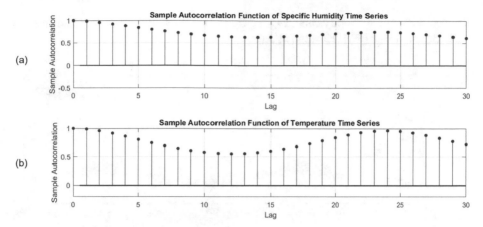

**Fig. 4.** Autocorrelation of (a) Temperature time series (b) Specific humidity time series

The correlation analysis between the two properties is revealing a logarithmic relationship, as depicted in Fig. 5. In light of this observation, the logsig() activation function is opted for.

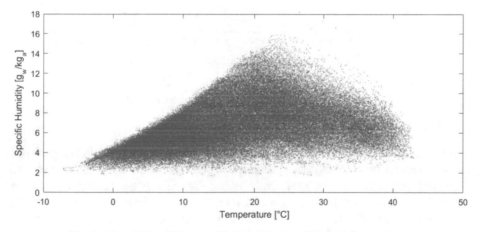

**Fig. 5.** Correlation of the specific humidity as a function of temperature

## 4.2   Exponential Smoothing

The implementation of the Holt-Winters model exhibited high computational demands and substantial random access memory requirements. These limitations confined our computations to a 16 GB capacity, enabling the analysis of solely a two-year span of hourly time series data. The outcomes are illustrated in Fig. 6. Initially, a notable overestimation is observed, which subsequently damped to capture the recurring behavior of the time series. Upon closer examination, it becomes evident that the model exaggerates the amplitude while adeptly capturing the frequency. In sum, the model yields a Root Mean Square Error (RMSE) value of 5.68.

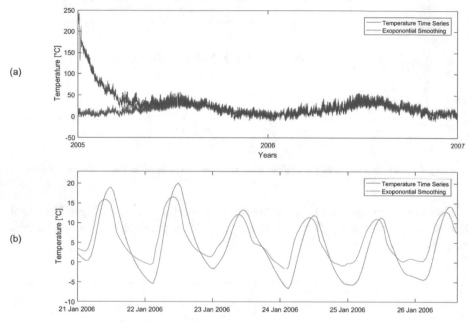

**Fig. 6.** Response of exponential smoothing model

## 4.3 Autoregressive Neural Network with External Input

In contrast to exponential smoothing, this Artificial Neural Network (ANN) necessitates a time series for both the target response and a predictor variable. In our context, we employed temperature as the target and specific humidity as the predictor, based on established thermodynamic and psychometric relationships. The dataset was partitioned into training data (70%), testing data (15%), and validation data (15%). The model underwent training utilizing the Levenberg–Marquardt Algorithm. Following 603 iterations, the model exhibits an average Mean Squared Error (MSE) score of 0.1337.

The neural network's performance is assessed by predicting a time step delay of 24, equivalent to one day, using a novel series distinct from the training data. It becomes apparent that there is an initial overshooting during the early hours. However, the model closely approximates the overall mean temperature for the day.

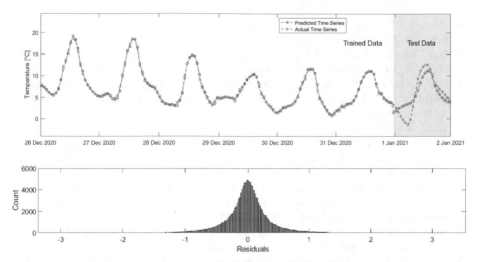

**Fig. 7.** Response and residuals for the NARX Neural Network

## 5 Conclusions

The analysis of time series data yields valuable insights crucial for designing an effective neural network. The statistical forecasting using exponential smoothing heavily relies on extensive datasets, demanding significant computational resources. Conversely, machine learning forecasting methods prove more resource-efficient due to their optimized algorithms. Implementing these algorithms to ES will help determine more accurate smoothing coefficients.

The overall performances results show that NARX method is well optimized for short-term forecasting which enough for our application in cooling load estimation. For higher horizon forecasting exponential smoothing hold better potential and need to be further investigated.

Other awareness modules should be able to store their own time series and utilize the NARX neural network, in order to develop a larger interconnected neural network.

## References

1. Molina, M.J., Rowland, F.S.: Stratospheric sink for chlorofluoromethanes: chlorine atom-catalysed destruction of ozone. Nature **249**(5460), 810–812 (1974)
2. Wodon, Q., Burger, N., Grant, A., Joseph, G., Liverani, A., Tkacheva, O.: Climate change, extreme weather events, and migration: Review of the literature for five Arab countries. People on the Move in a Changing Climate: The Regional Impact of Environmental Change on Migration, (pp. 111–134) (2014)
3. Abbasnia, M., Toros, H.: Trend analysis of weather extremes across the coastal and non-coastal areas (case study: Turkey). J. Earth Syst. Sci. **129**, 1–13 (2020)
4. Häfele, W.: Energy in a Finite World: A Global Systems Analysis (Volume 2) (Vol. 2). Ballinger (1981)
5. Holt, C.C.: Forecasting seasonals and trends by exponentially weighted moving averages. Int. J. Forecast. **20**(1), 5–10 (2004)

6. Winters, P.R.: Forecasting sales by exponentially weighted moving averages. Manage. Sci. **6**(3), (pp. 324–342) (1960)
7. Hyndman, R., Koehler, A.B., Ord, J.K., Snyder, R.D.: Forecasting with exponential smoothing: the state space approach. Springer Science & Business Media (2008)
8. Dudek, G., Pełka, P., Smyl, S.: A hybrid residual dilated LSTM and exponential smoothing model for midterm electric load forecasting. IEEE Trans. Neural Netw. Learn. Syst. **33**(7), 2879–2891 (2021)
9. Hand, D.J.: Data mining: statistics and more?. American Stat. **52**(2), 112–1180 (1998)
10. Barrera-Animas, A.Y., Oyedele, L.O., Bilal, M., Akinosho, T.D., Delgado, J.M.D., Akanbi, L.A.: Rainfall prediction: a comparative analysis of modern machine learning algorithms for time-series forecasting. Mach. Learn. Appl. **7**, 100204 (2022)
11. Bakay, M.S., Ağbulut, Ü.: Electricity production based forecasting of greenhouse gas emissions in Turkey with deep learning, support vector machine and artificial neural network algorithms. J. Clean. Prod. **285**, 125324 (2021)
12. Altan, A., Karasu, S.: The effect of kernel values in support vector machine to forecasting performance of financial time series. J. Cogn. Syst. **4**(1), 17–21 (2019)
13. Levenberg, K.: A method for the solution of certain non-linear problems in least squares. Quart. Appl. Math. **2**(2), 164–168 (1944)
14. Baumann, P., Hochbaum, D.S., Yang, Y.T.: A comparative study of the leading machine learning techniques and two new optimization algorithms. Europ. J. Oper. Res. **272**(3), (pp. 1041–1057) (2019)
15. Yang, L., Shami, A.: On hyperparameter optimization of machine learning algorithms: Theory and practice. Neurocomputing **415**, 295–316 (2020)
16. Tarutani, Y.: Reducing power consumption in data center by predicting temperature distribution and air conditioner efficiency with machine learning. In: 2016 IEEE International Conference on Cloud Engineering (IC2E), (pp. 226–227). IEEE (2016, April)
17. Amasyali, K., El-Gohary, N.: Machine learning for occupant-behavior-sensitive cooling energy consumption prediction in office buildings. Renew. Sustain. Energy Rev. **142**, 110714 (2021)
18. NASA Open data portal, data.nasa.gov

# Author Index

M. Hatti (Ed.): IC-AIRES 2023, LNNS 984, pp. 511–512, 2024.
https://doi.org/10.1007/978-3-031-60629-8

Printed in the United States
by Baker & Taylor Publisher Services